Composite Materials:
Engineering and Science

Composite Materials: Engineering and Science

F. L. Matthews

and

R. D. Rawlings

Imperial College of Science, Technology and Medicine, London, UK

CRC Press
Boca Raton   Boston   New York   Washington, DC

WOODHEAD PUBLISHING LIMITED
Cambridge England

Published by Woodhead Publishing Limited,
80 High Street, Sawston, Cambridge CB22 3HJ, UK
www.woodheadpublishing.com

Woodhead Publishing India Private Limited, G-2, Vardaan House, 7/28 Ansari Road,
Daryaganj, New Delhi – 110002, India
www.woodheadpublishingindia.com

Published in North America by CRC Press LLC, 6000 Broken Sound Parkway, NW
Suite 300, Boca Raton FL 33487, USA

First published 1994, Chapman & Hall
Reprinted 1999, 2002, 2003, 2006, 2007, 2008, Woodhead Publishing Limited and CRC Press LLC
© 1999, F. L. Matthews and R. D. Rawlings
The authors have asserted their moral rights.

British Library Cataloguing in Publication Data
A catalogue record for this book is available from the British Library.

Library of Congress Cataloging in Publication Data
A catalog record for this book is available from the Library of Congress.

Woodhead Publishing ISBN 978-1-85573-473-9
CRC Press ISBN 978-0-8493-0251-0
CRC Press order number: WP0621

Printed by Lightning Source

Contents

Preface

In order to meet the demand for training in understanding the behaviour of composite materials, the Centre for Composite Materials at Imperial College has, for a number of years, organized a course of lectures entitled 'Introduction to Fibre-Reinforced Composites'. Most of the notes associated with that course have been edited, adapted and expanded here by the present authors, who wish to acknowledge the following colleagues who also contributed to the course notes: Dr P. Cawley, Dr J. M. Hodgkinson, Professor A. J. Kinlock, Professor P. L. Pratt, Professor J. G. Williams (all of Imperial College), Dr P. T. Curtis, Dr D. Dorey (of DERA, Farnborough). Without their original material this text would not have been written.

The purpose of the present book is to give an up-to-date appreciation of the underlying science and the engineering performance of composite materials. The most widely used composites are those with polymer matrices and with fibrous reinforcement and the text inevitably concentrates on these, although two chapters are devoted to metal and ceramic matrix systems. A significant portion of the book deals with methods for calculating stiffness and strength, the text being supplemented where appropriate by worked examples and representative data. The final chapters deal with aspects of mechanical behaviour such as toughness, fatigue and impact resistance and the properties of joints. The important topic of non-destructive evaluation is also dealt with in detail.

R. D. Rawlings
F. L. Matthews
Imperial College, London

Overview

1.1 INTRODUCTION

In the continuing quest for improved performance, which may be specified by various criteria including less weight, more strength and lower cost, currently-used materials frequently reach the limit of their usefulness. Thus materials scientists, engineers and scientists are always striving to produce either improved traditional materials or completely new materials. *Composites* are an example of the latter category.

Not that composites are really new. A composite is a material having two or more distinct constituents or phases and thus we can classify bricks made from mud reinforced with straw, which were used in ancient civilizations, as a composite. A versatile and familiar building material which is also a composite is concrete; concrete is a mixture of stones, known as aggregate, held together by cement. In addition to synthetic composites there are naturally occurring composites of which the best known examples are bone, mollusc shells and wood; bone and wood are discussed later in the chapter.

Within the last forty years there has been a rapid increase in the production of synthetic composites, those incorporating fine fibres in various plastics (polymers) dominating the market. Predictions suggest that the demand for composites will continue to increase steadily with metal and ceramic based composites making a more significant contribution (Figure 1.1).

The spur to this rapid expansion over the last few decades was the development in the UK of carbon fibres and in the USA of boron fibres in the early 1960s. These new fibres, which have high elastic constants, gave a significant increase in the stiffness of composites over the well-established glass fibre containing materials, and hence made possible a wide range of applications for composites. One of the key factors was the very high strength-to-weight and stiffness-to-weight ratios possessed by these new composites (section 1.6). Early employment in aviation has led, in more recent times, to wide use in other areas such as the leisure and sports

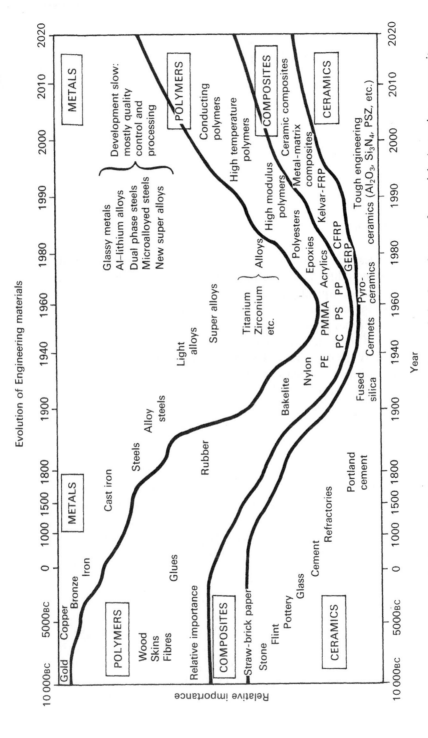

Figure 1.1 Schematic diagram showing the relative importance of the four classes of materials (ceramics, composites, polymers and metals) in mechanical and civil engineering as a function of time. The time scale is nonlinear. (Source: Ashby, 1987.)

industry. Similarly, the development of high strength silicon carbide, SiC, and alumina, Al_2O_3, fibres which maintain their properties at elevated temperatures initiated much of the current interest in composites based on metal and ceramics.

Composite materials can be studied from a number of different viewpoints each of which requires a different kind of expertise. Thus, the development of a composite material to resist a corrosive environment is primarily within the field of materials science and chemistry. In contrast the design of a load bearing structure requires the expertise of an engineer. It is therefore essential for the full exploitation of currently available composites, and for the future development of composites, that experts in one field are able to understand the problems of those in another. The study of composites is truly interdisciplinary and this will become increasingly clear as the reader proceeds through this book.

1.2 DEFINITIONS AND CLASSIFICATION

We have already stated that a composite is a mixture of two or more distinct constituents or phases. However this definition is not sufficient and three other criteria have to be satisfied before a material can be said to be a composite. First, both constituents have to be present in reasonable proportions, say greater than 5%. Secondly, it is only when the constituent phases have different properties, and hence the composite properties are noticeably different from the properties of the constituents, that we have come to recognize these materials as composites. For example plastics, although they generally contain small quantities of lubricants, ultra-violet absorbers, and other constituents for commercial reasons such as economy and ease of processing, do not satisfy either of these criteria and consequently are not classified as composites. Lastly, a man-made composite is usually produced by intimately mixing and combining the constituents by various means. Thus an alloy which has a two-phase microstructure which is produced during solidification from a homogeneous melt, or by a subsequent heat treatment whilst a solid, is not normally classified as a composite (Figure 1.2(a)). However if ceramic particles are somehow mixed with a metal to produce a material consisting of the metal containing a dispersion of the ceramic particles, then this is a true composite material (Figure 1.2(b)).

We know that composites have two (or more) chemically distinct phases on a microscopic scale, separated by a distinct interface, and it is important to be able to specify these constituents. The constituent that is continuous and is often but not always, present in the greater quantity in the composite is termed the *matrix*. The normal view is that it is the properties of the matrix that are improved on incorporating another constituent to produce a composite. A composite may have a ceramic, metallic or polymeric matrix.

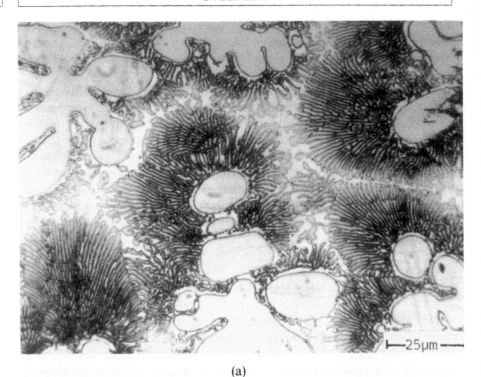

(a)

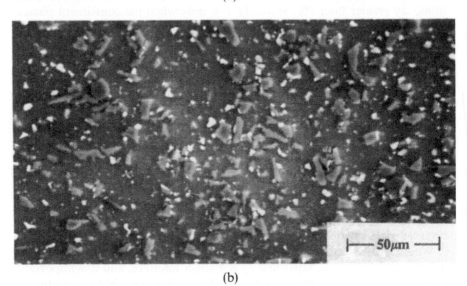

(b)

Figure 1.2 (a) Micrograph of a cast Co–Cr–Mo–Si alloy with a multiphase micro-structure. (Source: Halstead and Rawlings, 1986.) (b) Scanning electron micrograph of an aluminium alloy (2014) reinforced with angular particles of silicon carbide. The white particles are a second phase in the aluminium alloy matrix. (Courtesy D.J.B. Greenwood.)

The mechanical properties of these three classes of material differ considerably as demonstrated in Table 1.1. As a generalization polymers have low strengths and Young's moduli, ceramics are strong, stiff and brittle, and metals have intermediate strengths and moduli together with good ductilities, i.e., they are not brittle.

The second constituent is referred to as the *reinforcing phase*, or *reinforcement*, as it enhances or reinforces the mechanical properties of the matrix. In most cases the reinforcement is harder, stronger and stiffer than the matrix, although there are some exceptions; for example, ductile metal reinforcement in a ceramic matrix and rubberlike reinforcement in a brittle polymer matrix. At least one of the dimensions of the reinforcement is small, say less than 500 μm and sometimes only of the order of a micron. The geometry of the reinforcing phase is one of the major parameters in determining the effectiveness of the reinforcement; in other words, the mechanical properties of composites are a function of the shape and dimensions of the reinforcement. We usually describe the reinforcement as being either *fibrous* or *particulate*. Figure 1.3 represents a commonly employed classification scheme for composite materials which utilizes this designation for the reinforcement (Figure 1.3, block A).

Particulate reinforcements have dimensions that are approximately equal in all directions. The shape of the reinforcing particles may be spherical, cubic, platelet or any regular or irregular geometry. The composite illustrated in the micrograph of Figure 1.2(b) has angular particles of about 10 μm as the reinforcement. The arrangement of the particulate reinforcement may be *random* or with a *preferred orientation*, and this characteristic is also used as a part of the classification scheme (block B). In the majority of particulate reinforced composites the orientation of the particles is considered, for practical purposes, to be random (Figure 1.4(a)).

A fibrous reinforcement is characterized by its length being much greater than its cross-sectional dimension. However, the ratio of length to the cross-sectional dimension, known as the aspect ratio, can vary considerably. In single-layer composites long fibres with high aspect ratios give what are called *continuous* fibre reinforced composites, whereas *discontinuous* fibre composites are fabricated using short fibres of low aspect ratio (block C). The orientation of the discontinuous fibres may be random or preferred (Figures 1.4 (b) and (c)). The frequently encountered preferred orientation in the case of a continuous fibre composite (Figure 1.4(d)) is termed *unidirectional* and the corresponding random situation can be approximated to by *bidirectional* woven reinforcement (Figure 1.3, block D).

Multilayered composites are another category of fibre reinforced composites. These are classified as either laminates or hybrids (block E). *Laminates* are sheet constructions which are made by stacking layers (also called plies or laminae and usually unidirectional) in a specified sequence. A typical laminate may have between 4 to 40 layers and the fibre orientation changes

Table 1.1 Comparison of room temperature properties of ceramics, metals and polymers

	Density (Mg/m³)	Young's modulus (GPa)	Strength[a] (MPa)	Ductility (%)	Toughness K_{IC} (MPa m$^{1/2}$)	Specific modulus [(GPa)/(Mg/m³)]	Specific strength [(MPa)/(Mg/m³)]
CERAMICS							
Alumina Al$_2$O$_3$	3.87	382	332	0	4.9	99	86
Magnesia MgO	3.60	207	230	0	1.2	58	64
Silicon Nitride Si$_3$N$_4$		166	210	0	4.0		
Zirconia ZrO$_2$	5.92	170	900	0	8.6	29	152
β-Sialon	3.25	300	945	0	7.7	92	291
Glass-ceramic Silceram	2.90	121	174	0	2.1	42	60
METALS							
Aluminium	2.70	69	77	47		26	29
Aluminium-3%Zn-0.7%Zr	2.83	72	325	18		25	115
Brass Cu-30%Zn	8.50	100	550	70		12	65
Nickel-20%Cr-15%Co	8.18	204	1200	26		25	147
Steel mild	7.86	210	460	35		27	59
Titanium-2.5% Sn	4.56	112	792	20		24	174
POLYMERS							
Epoxy	1.12	4	50	4	1.5	4	36
Melamine formaldehyde	1.50	9	70			6	47
Nylon 6.6	1.14	2	70	60		18	61
Polyetheretherketone	1.30	4	70			3	54
Polymethylmethacrylate	1.19	3	50	3	1.5	3	42
Polystyrene	1.05	3	50	2	1.0	3	48
Polyvinylchloride rigid	1.70	3	60	15	4.0	2	35

[a]Strength values are obtained from the test appropriate for the material, e.g., flexural and tensile for ceramics and metals respectively.

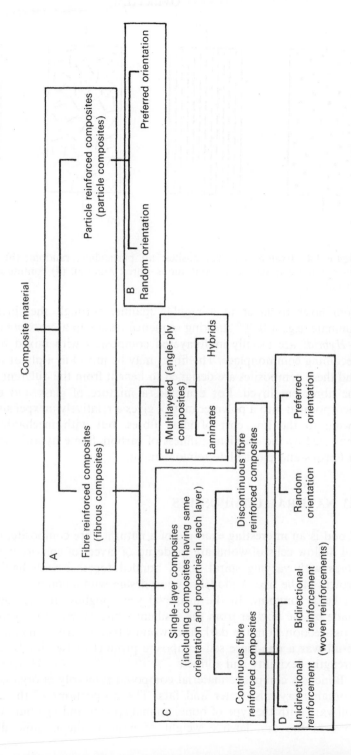

Figure 1.3 Classification of composite materials.

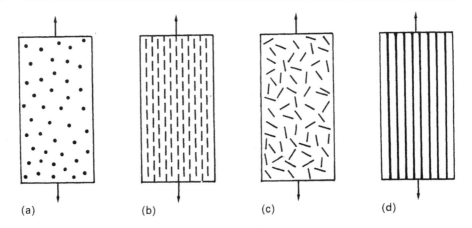

Figure 1.4 Examples of composites: (a) particulate, random; (b) discontinuous fibres, unidirectional; (c) discontinuous fibres, random; (d) continuous fibres, unidirectional.

from layer to layer in a regular manner through the thickness of the laminate, e.g., a 0/90° stacking sequence results in a *cross ply* composite.

Hybrids are usually multilayered composites with mixed fibres and are becoming commonplace. The fibres may be mixed in a ply or layer by layer and these composites are designed to benefit from the different properties of the fibres employed. For example, a mixture of glass and carbon fibres incorporated into a polymer matrix gives a relatively inexpensive composite, owing to the low cost of glass fibres, but with mechanical properties enhanced by the excellent stiffness of carbon. Some hybrids have a mixture of fibrous and particulate reinforcement.

1.3 NATURAL COMPOSITES

Wood is an interesting example of a natural fibre composite; the longitudinal hollow cells of wood are made up of layers of spirally wound *cellulose fibres* with varying spiral angle, bonded together with lignin during the growth of the tree. A model of good design with a composite material is the medieval longbow. In the medieval yew longbow the tensile surface was made by the bowyer from the resilient white sapwood with the bulk of the cross-section from the dark heartwood which is strong in compression. The result was a one-piece self-composite product at two levels, of remarkable strength, flexibility and stiffness.

Bone is a composite material composed primarily of organic fibres, small inorganic crystals, water and fats. The proportions of these components will vary with the type of bone, animal species and age but typically about 35% of the dry, fat-free weight of bone is the organic fibre, *collagen.*

Collagen is a fibrous protein and located around the outside of the collagen fibres are small rod-like crystals of *hydroxyapatite* with dimensions of the order of $5 \times 5 \times 50$ nm. Therefore at this microscopic level we have a hydroxyapatite-reinforced collagen composite and this may be considered to be the basic 'building block' for bone. The different types of bone are simply composites on a macroscopic scale resulting from different arrangements of the collagen fibres; in woven bone the fibres are arranged more or less randomly whereas in lamellar bone the fibres are orientated locally in layers to form lamellae.

A long bone, such as the femur, has an outer shell of high density, low fat content bone which is termed *compact* or *cortical* bone. Cortical bone has good mechanical properties with flexural strength in the range 46–156 MPa and Young's modulus of the order of 20 GPa. Contained within the shell of cortical bone is a softer, spongy bone (*cancellous* or *trabecular* bone) which consists of a three-dimensional network of beams and sheets known as

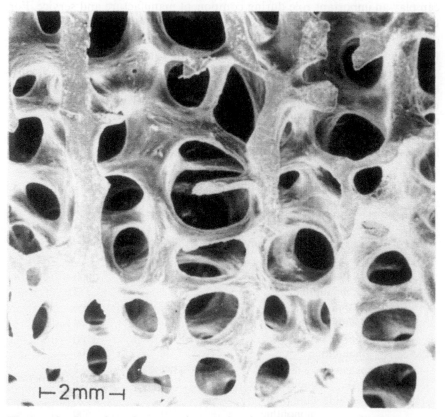

⊢2mm⊣

Figure 1.5 Scanning electron micrograph of cancellous bone. (Courtesy I. M. Thomas.)

trabeculae (Figure 1.5). The compressive strength and the Young's modulus of cancellous bone are typically in the ranges 5 to 20 MPa and 0.02 to 1.7 GPa respectively.

There are many more examples of composite materials in nature but the coverage given here on wood and bone is sufficient to illustrate this class of material and to indicate the high standards that synthetic composites have to achieve to be of comparable quality.

1.4 MORE ABOUT THE MATRIX AND REINFORCEMENT

Although there are many examples of composites employed mainly for their physical properties, such as the superconducting composites consisting of continuous Nb_3Sn fibres in a bronze matrix, most composites are designed to exploit an improvement in mechanical properties. Even for composites produced essentially for their physical properties, the mechanical properties can play an important role during component manufacture and service. For these reasons, in this section on the matrix and reinforcement, we will concentrate on mechanical behaviour. The properties of some fibres and their monolithic counterparts are compared in Table 1.2.

It has already been mentioned that the reinforcement is often a strong, stiff material and this is borne out by the data presented in Table 1.2. The reader may well ask why one should not make use of the high strength and high elastic constants of these reinforcements by producing mono- lithic components solely from the reinforcement. In fact this is done and, for example, components manufactured from the ceramics silicon carbide

Table 1.2 Comparison of the mechanical properties of fibres and their monolithic counterparts

	Young's modulus (GPa)	Strength[a] (MPa)
Alumina: fibre (Saffil RF)	300	2000
monolithic	382	332
Carbon: fibre (IM)	290	3100
monolithic	10	20
Glass: fibre (E)	76	1700
monolithic	76	100
Polyethylene: fibre (S 1000)	172	2964
monolithic (HD)	0.4	26
Silicon carbide: fibre (MF)	406	3920
monolithic	410	500

[a]tensile and flexural strengths for fibre and monolithic respectively.

and alumina are readily available. The problem with these materials is their brittleness; although able to sustain a high stress, when they fail they do so in a catastrophic manner without warning. Moreover the stress at which nominally identical specimens, or components, fail can vary markedly. The brittleness and the variation in strength is due to failure being initiated at flaws which normally occur at the surface. There is always a distribution of flaw sizes and, in an homogeneously stressed material, failure commences at the largest flaw. The larger the failure-initiating flaw the lower the failure stress (Figure 1.6). The smaller the volume of material, the smaller on average the size of the maximum flaw and the greater the strength. Therefore the fracture stress of a small ceramic particle or fibre is greater than that of a large volume of the same material.

We will see in the next chapter that the mechanical properties of fibres of a given type and dimensions are also determined by the structure. For example the properties of the organic fibre, aramid, are dependent on the orientation of the molecular structure whereas for carbon

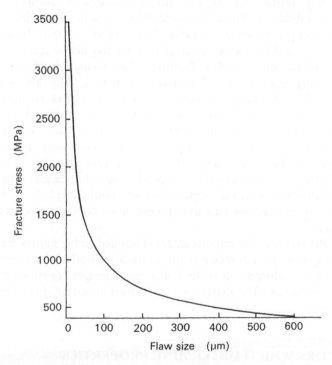

Figure 1.6 Effect of flaw size, a, on fracture stress, σ, of a brittle material with $K_{IC} = 10 \, \text{MPa} \, \text{m}^{1/2}$. The fracture stress is calculated from the equation $K_{IC} = Y\sigma a^{1/2}$ with the parameter Y, which depends on geometry and loading, arbitrarily put equal to unity.

fibres the perfection and alignment of the graphite crystals are para-mount.

There are of course many properties other than strength that we have to take into account when selecting a reinforcement. In the case of fibres the flexibility is important as it determines whether the fibres may be easily woven or not and influences the choice of method for composite manufac-ture. The flexibility of a fibre depends mainly on the Young's modulus, E_f, and diameter, D, of the fibre:

$$\text{flexibility} \propto \frac{1}{E_f D^4} \tag{1.1}$$

Thus large diameter fibres with a high Young's modulus are not flexible.

Clearly, single fibres, because of their small cross-sectional dimensions, are not directly usable in structural applications. On the other hand the possibility exists of trying to exploit the high strength of single fibres by physically entwining the fibres in a similar manner to the manufac-ture of rope from hemp fibres. Unfortunately such a fabrication method would greatly restrict the shape and dimensions of components and furthermore would introduce a considerable amount of surface damage leading to a degradation in strength. These problems may be overcome by embedding the fibres in a material to hold the fibres apart, to protect the surface of the fibres and to facilitate the production of components. The embedding material is, of course, the matrix. The amount of re-inforcement that can be incorporated in a given matrix is limited by a number of factors. For example with particulate reinforced metals the reinforcement content is usually kept to less than 40 vol. % (0.4 volume fraction) because of processing difficulties and increasing brittleness at higher contents. On the other hand, the processing methods for fibre reinforced polymers are capable of producing composites with a high proportion of fibres, and the upper limit of about 70 vol. % (0.7 volume fraction) is set by the need to avoid fibre–fibre contact which results in fibre damage.

Finally, the fact that the reinforcement is bonded to the matrix means that any loads applied to a composite are carried by both constituents. As in most cases the reinforcement is the stiffer and stronger constituent; it is the principal load bearer. The matrix is said to have transferred the load to the reinforcement.

1.5 FACTORS WHICH DETERMINE PROPERTIES

The fabrication and properties of composites are strongly influenced by the proportions and properties of the matrix and the reinforcement. The proportions can be expressed either via the weight fraction (w), which is relevant to fabrication, or via the volume fraction (v), which is commonly used in property calculations (for example, Chapter 8).

The definitions of w and v are related simply to the ratios of weight (W) or volume (V) as shown below.

Volume fractions: $v_f = V_f/V_c$ and $v_m = V_m/V_c$.   (1.2a)

Weight fractions: $w_f = W_f/W_c$ and $w_m = W_m/W_c$,   (1.2b)

where the subscripts m, f and c refer to the matrix, fibre (or in the more general case, reinforcement) and composite respectively.

We note that

$$v_f + v_m = 1, \text{ and}$$

$$w_f + w_m = 1.$$

We can relate weight to volume fractions by introducing the density, ρ, of the composite and its constituents. Now

$$W_c = W_f + W_m$$

which, as $W = \rho V$, becomes

$$\rho_c V_c = \rho_f V_f + \rho_m V_m,$$

or

$$\rho_c = \rho_f(V_f/V_c) + \rho_m(V_m/V_c)$$

$$= \rho_f v_f + \rho_m v_m.$$ (1.3)

It may also be shown that

$$\frac{1}{\rho_c} = \frac{w_f}{\rho_f} + \frac{w_m}{\rho_m}$$ (1.4)

Also we have

$$w_f = \frac{W_f}{W_c} = \frac{(\rho_f V_f)}{(\rho_c V_c)} = \frac{\rho_f}{\rho_c} v_f,$$

and similarly

$$w_m = \frac{W_m}{W_c} = \frac{(\rho_m V_m)}{(\rho_c V_c)} = \frac{\rho_m}{\rho_c} v_m.$$ (1.5)

We can see that we can convert from weight fraction to volume fraction, and vice versa, provided the densities of the reinforcement (ρ_f) and the matrix (ρ_m) are known.

Equation 1.3 shows that the density of the composite is given by the volume fraction adjusted sum of the densities of the constituents. This equation is not only applicable to density but, in certain circumstances, may apply to other properties of composites. A generalized form of the equation is

$$X_c = X_m v_m + X_f v_f, \qquad (1.2)$$

where X_c represents an appropriate property of the composite, and, as before, v is the volume fraction and the subscripts m and f refer to the matrix and reinforcement respectively. This equation is known as the *Law of Mixtures*.

Most properties of a composite are a complex function of a number of parameters as the constituents usually interact in a synergistic way so as to provide properties in the composite that are not fully accounted for by the law of mixtures. The chemical and strength characteristics of the *interface* between the fibres and the matrix is particularly important in determining the properties of the composite. The interfacial bond strength has to be sufficient for load to be transferred from the matrix to the fibres if the composite is to be stronger than the unreinforced matrix. On the other hand, as the reader will learn in Chapter 11, if we are also concerned with the toughness of the composite, the interface must not be so strong that it does not fail and allow toughening mechanisms such as debonding and fibre pull-out to take place.

Other parameters which may significantly affect the properties of a composite are the shape, size, orientation and distribution of the reinforcement and various features of the matrix such as the grain size for polycrystalline matrices. These, together with volume fraction, constitute what is called the *microstructure* of the composite.

However it should be noted that even for properties which are microstructure dependent, and which do not obey the law of mixtures, the volume fraction still plays a major role in determining properties. The volume fraction is generally regarded as the single most important parameter influencing the composite's properties. Also, it is an easily controllable manufacturing variable by which the properties of a composite may be altered to suit the application.

A problem encountered during manufacture is maintaining a uniform distribution of the reinforcement. Ideally, a composite should be homogeneous, or uniform, but this is difficult to achieve. *Homogeneity* is an important characteristic that determines the extent to which a representative volume of the material may differ in physical and mechanical properties from the average properties of the material. Non-uniformity of the system should be avoided as much as possible because it reduces those properties that are governed by the weakest part of the composite. For example, failure in a non-uniform material will initiate in an area of lowest strength, thus adversely affecting the overall strength of a component manufactured from that material.

The orientation of the reinforcement within the matrix affects the *isotropy* of the system. When the reinforcement is in the form of equiaxed particles, the composite behaves essentially as an isotropic material whose elastic properties are independent of direction. When the dimensions of the reinforcement are unequal, the composite can behave as if isotropic, provided the reinforcement is randomly oriented, as in a randomly oriented, short fibre-reinforced composite (Figure 1.4(c)). In other cases the manufacturing process may induce orientation of the reinforcement and hence loss of isotropy; the composite is then said to be *anisotropic* or to exhibit *anisotropy* (Figure 1.7).

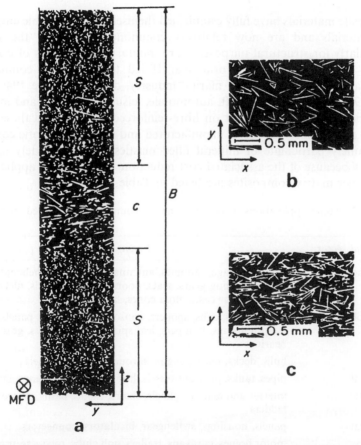

Figure 1.7 Micrographs of injection moulded discontinuous glass fibre reinforced PEEK showing orientation effects (a) across the thickness of the moulding, (b) and (c) across surface layers showing increase in fibre orientation with increased fibre content; (a) and (c) 18vol. % glass and (b) 11vol. % glass. (Source: Friedrich *et al.*, 1986.)

In components manufactured from continuous fibre reinforced composites, such as unidirectional or cross-ply laminates, anisotropy may be desirable as it can be arranged for the maximum service stress to be in the direction that has the highest strength. Indeed, a primary advantage of these composites is the ability to control the anisotropy of a component by design and fabrication.

It is clear from the preceding discussion that many factors may be important in determining the properties of composites; this should be borne in mind when studying theoretical models as all the factors are seldom accounted for in the development of the models.

1.6 THE BENEFITS OF COMPOSITES

Composite materials have fully established themselves as workable engineering materials and are now relatively commonplace around the world, particularly for structural purposes. Early military applications of polymer matrix composites during World War II led to large-scale commercial exploitation, especially in the marine industry, during the late 1940s and early 1950s. Today, the aircraft, automobile, leisure, electronic and medical industries are quite dependent on fibre-reinforced plastics, and these composites are routinely designed, manufactured and used. Less exotic composites, namely particulate or mineral filled plastics, are also widely used in industry because of the associated cost reduction. Some typical applications of polymer matrix composites are listed in Table 1.3.

Table 1.3 Some applications of polymer matrix composites. (Adapted from Hull, 1981)

Industrial sector	Examples
Aerospace	wings, fuselage, radomes, antennae, tail-planes, helicopter blades, landing gears, seats, floors, interior panels, fuel tanks, rocket motor cases, nose cones, launch tubes
Automobile	body panels, cabs, spoilers, consoles, instrument panels, lamp-housings, bumpers, leaf springs, drive shafts, gears, bearings
Boats	hulls, decks, masts, engine shrouds, interior panels
Chemical	pipes, tanks, pressure vessels, hoppers, valves, pumps, impellers
Domestic	interior and exterior panels, chairs, tables, baths, shower units, ladders
Electrical	panels, housings, switchgear, insulators, connectors
Leisure	motor homes, caravans, trailers, golf clubs, racquets, protective helmets, skis, archery bows, surfboards, fishing rods, canoes, pools, diving boards, playground equipment

The success of fibre composites with thermosetting (epoxies and poly-esters) or thermoplastic matrices, largely as replacements for metals, results from the much improved mechanical properties of the composites compared with the matrix materials. The good mechanical properties of the composites are a consequence of utilizing the special properties of glass, carbon and aramid fibres (Table 1.2).

It can be seen from a comparison of Tables 1.1 and 1.4 that on the basis of strength and stiffness alone, fibre-reinforced polymer composites do not have a clear advantage over conventional materials particularly when it is noted that their elongation to fracture is much lower than that for metallic alloys of comparable strength. The advantage that polymer matrix compos-ites have over metals is their low density, ρ. The benefit of a low density becomes apparent when the Young's modulus per unit mass, E/ρ (*specific modulus*), and tensile strength per unit mass, $\hat{\sigma}_T/\rho$ (*specific strength*), are considered. The higher specific modulus and specific strength of these composites means that the weight of certain components can be reduced. This is a factor of great importance in moving components, especially in all forms of transport where reduction in weight results in greater efficiency leading to energy and hence cost savings.

However the specific modulus and specific strength alone are only capable of specifying the performance under certain service conditions. Let us illustrate this point by first considering specific modulus. The specific modulus is an acceptable parameter for comparing materials for rod-like components under tensile loading as the lightest rod that will carry a given tensile force without exceeding a given deflection is that made from material with the greatest value of E/ρ. On the other hand under compression loading of a rod, elastic buckling may be the limiting factor and the lightest rod that will support a compressive force without buckling is constructed from material with the greatest value of $E^{1/2}/\rho$. Finally if we consider a panel loaded in bending, we find that the lightest panel which will support a given load with minimum deflection should be manufactured from material with the greatest value of $E^{1/3}\rho$.

It is concluded therefore that a single *performance indicator*, such as specific modulus, is insufficient for assessment of materials. It is preferable to compare materials, and classes of materials, by means of *materials property charts* (Figure 1.8) where one property (Young's modulus in this case) is plotted against another (the density) on logarithmic axes, as these can incorporate a number of performance indicators. On the chart of Figure 1.8 are plotted guide lines for material selection for the three service situations, and hence performance indicators, previously described. Consider the guide line drawn for $E^{1/3}/\rho$; this corresponds to a specific set of service conditions for a panel supported at both ends and subject to a central force. Any change in these conditions, e.g., change in the load, just moves the line up or down the chart whilst maintaining the same gradient. All materials which lie

Table 1.4 Properties of polymer matrix composites

	Density (Mg/m^3)	Young's modulus (GPa)	Tensile strength (MPa)	Ductility (%)	Flexural strength (MPa)	Specific modulus [(GPa)/(Mg/m^3)]	Specific strength [(MPa)/(Mg/m^3)]
Nylon66+40%carbon fibre	1.34	22	246	1.7	413	16	184
Epoxide+70%glass fibres							
unidirectional – longitudinal	1.90	42	750		1200	22	395
– transverse	1.90	12	50			6	26
Epoxy+60%Aramid	1.40	77	1800			55	1286
Polyether imide+52%Kevlar		54	253				
Polyester+glass CSM	1.50	7.7	95		170	5	63
Polyester+50%glass fibre							
unidirectional – longitudinal	1.93	38	750	1.8		20	389
– transverse	1.93	10	22	0.2		5	11

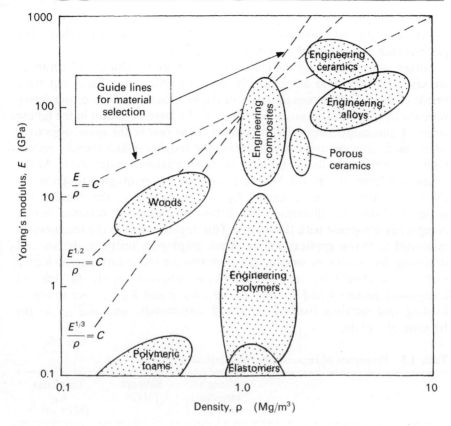

Figure 1.8 An example of a materials property chart which illustrates the good characteristics of polymer matrix composites when analysed in terms of the performance indicators E/ρ, $E^{1/2}/\rho$ and $E^{1/3}/\rho$ (Source: Ashby, 1989.)

on the line will perform equally well under the specified conditions for that line. The performance of materials above the line is better and that of materials below the line is worse. It can be seen from the relative position of the polymer matrix composites on the chart with respect to engineering polymers that the composites are preferable for all three applications. Comparison between the composites and the alloys indicate that most composites out-perform alloys under conditions corresponding to the $E^{1/2}/\rho$ and $E^{1/3}/\rho$ guide lines but only some composites are better than alloys for E/ρ service-limited situations.

Similar materials property charts with various combinations of properties used for the axes, e.g., strength against density and Young's modulus against strength, and which are applicable to a wide range of service conditions are available. These demonstrate the benefits to be obtained in using composites for many applications. It would be impossible for all these performance indicators to be presented in the figures and tables in this book. In most instances the important role of density is demonstrated by values for the specific strength and stiffness but it should always be remembered that other performance indicators have to be considered.

Earlier, when discussing the characteristics of monolithic (bulk) ceramics we saw that they have a reasonably high strength and stiffness but were brittle (Table 1.1). The consequences of the brittleness, or lack of *toughness*, were the likelihood of catastrophic failure and a large variation in the failure stress of nominally identical components. Thus one of the main objectives when reinforcing ceramics is to increase the toughness and thereby reduce their susceptibility to flaws and improve their reliability under stress. As the reader will learn in Chapter 11, one parameter for quantifying toughness is the *critical stress intensity factor*, K_{IC}; the greater K_{IC}, the tougher the material. Table 1.5 illustrates the better toughness of ceramic matrix composites compared with the matrix. This improvement in the toughness is beneficial in many applications as, when employing brittle materials and designing for minimum weight, the performance indicators K_{IC}/ρ, $K_{IC}^{2/3}/\rho$ and $K_{IC}^{1/2}/\rho$ should be maximized. These indicators apply to different component geometry and loading, namely, K_{IC}/ρ and $K_{IC}^{2/3}/\rho$ refer to tensile loading and bending loading of a rod respectively, and $K_{IC}^{1/2}/\rho$ to the bending of a plate.

Table 1.5 Properties of ceramic matrix composites

	Young's modulus (GPa)	Strength (MPa)	Toughness K_{IC} (MPa m$^{1/2}$)
Alumina (99%purity)	340	300	4.5
Alumina + 25%SiC whiskers	390	900	8.0
Borosilicate glass (Pyrex)		70	0.7
Pyrex + 40%Al$_2$O$_3$ CF^a		305	3.7
LAS^b glass-ceramic	86	160	1.1
LAS + 50%SiC CF	135	640	17.0
Mullite		244	2.8
Mullite + 20%SiC whiskers		452	4.4

^{a}CF is continuous fibres
^{b}LAS is lithiumaluminosilicate

Furthermore it should be noted that a ceramic matrix composite may also have improved strength and stiffness. An example of an application of a ceramic matrix composite which arises from the enhanced toughness and strength, together with the intrinsic hardness associated with the matrix, is industrial cutting tools; tools manufactured from alumina reinforced with SiC whiskers are said to improve the rate of cutting greatly and to have a longer life compared with conventional tool components.

Another factor that can play a role in determining the efficiency of tools working at high cutting rates is their ability to maintain their properties to high temperatures. In general the main three classes of structural materials may be ranked in terms of their high temperature capabilities, in descending order, as ceramics, metals and polymers. It follows that if ceramic matrix composites are to replace ceramics for applications at elevated temperatures, it is essential that the reinforcement should not adversely affect the high temperature properties. This means that the properties of the reinforcement should also be retained to high temperatures and that there should be no hostile chemical reaction between the matrix and the reinforcement. To date the composites with the best high temperature performance are *carbon–carbon* composites which, under vacuum or an inert atmosphere, can be used at temperatures well in excess of 2000 °C.

In a similar manner to polymers and ceramics, metals may be reinforced by either continuous fibres, discontinuous fibres or particulates. Silicon carbide and alumina are the most frequently employed reinforcements and a wide range of composites have been produced. However, most metal matrix composites are still under development and there are only a few examples of commercial components currently in service (Table 1.6).

Recent interest has concentrated on transport applications and consequently the light metals, particularly aluminium and its alloys, have received the most attention. The Young's modulus of aluminium and its alloys is relatively low for metals and hence there is considerable potential for improvement by reinforcement. However the reader will recall that the forte of metals is their good ductility and toughness and obviously it is desirable that these properties are not degraded in metal matrix composites.

Table 1.6 Some applications of ceramic and metal matrix composites

Industrial sector	Application	
	Ceramic matrix	Metal matrix
Aerospace	afterburners, brakes, heat shields, rocket nozzles	struts, antennae
Automobile	brakes	piston crowns
Manufacturing	thermal insulation, cutting tools, wire drawing dies	
Electrical		superconductors, contacts, filaments, electrodes
Medical	prostheses, fixation plates	

Table 1.7 Properties of metal matrix composites

	Young's modulus (GPa)	Yield stress (MPa)	Tensile strength (MPa)	Ductility (%)
Al–Cu–Mg (2618A)	74	416	430	2.5
Al–Cu–Mg + 20%Al_2O_3	90		383	0.8
Al–Zn–Mg	~70		273	11.5
Al–Zn–Mg + 25%Al_2O_3	80+		266	1.5
Titanium (wrought)	120	200	400	25
Titanium + 35%SiC	213		1723	<1
Ti–Al–V (wrought)	115	830	1000	8
Ti–Al–V + 35%SiC	190		1434	0.9

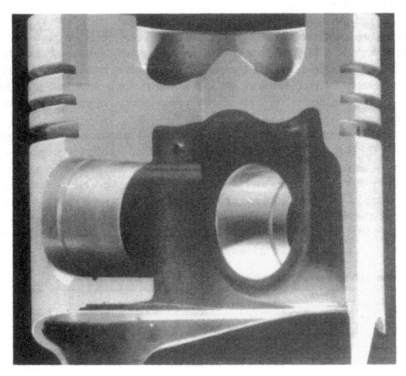

Figure 1.9 Metal matrix composite insert (shown dark) in the region of the upper piston ring groove of a diesel engine piston. (Courtesy of Art Metal/ Toyota.)

Unfortunately, as can be seen from Table 1.7, although there are significant improvements in Young's modulus for both aluminium alloy and titanium alloy matrix composites, this is achieved at the expense of ductility.

Because the reinforcements commonly used are the ceramics SiC and Al_2O_3, which have good high temperature mechanical properties, it is generally found that composites maintain their strength to higher temperatures than the matrix alloy alone. This attribute has been exploited in diesel engines; the first mass-produced structural component incorporating a metal matrix composite was the piston in a passenger diesel car where a fibre reinforced aluminium alloy was inserted in the region of the upper piston ring groove (Figure 1.9).

However metal matrix components, whose prime function depends on physical rather than mechanical characteristics, were in commercial use earlier. For example, composites with particles of tungsten or molybdenum in silver and copper matrices are used for heavy duty electrical contacts. These contacts require materials with properties such as high thermal and electrical conductivities, relatively high melting points, and low friction and wear resistance. The matrices are largely responsible for the high thermal and electrical conductivities whereas the reinforcement increases the hardness and hence the wear resistance. These materials are also used for electrodes and related applications in the welding industry.

1.7 SUMMARY

We have seen that a composite generally consists of two, but sometimes more, distinct constituents. One of these, the reinforcement, is said to enhance the properties of the other constitutent which is known as the matrix.

Natural composites, such as wood, and some synthetic composites have been used by humans for thousands of years but the major developments have taken place in recent times. The important feature of composites is that their properties are markedly different from those of the constituents and are determined by the properties of the constituents, the microstructure (which includes the volume fraction and morphology of the reinforcement) and the properties of the matrix–reinforcement interface. Young's modulus, strength and toughness are the commonly quoted mechanical properties but for a full assessment of the benefit of composites other parameters, such as the specific modulus or specific strength, should also be considered.

FURTHER READING

General

Agarwal, B. D. and Broutman, L. J. (1990) *Analysis and Performance of Fibre Composites*, 2nd edn., John Wiley and Sons.

Chawla, K. K. (1987) *Composite Materials*, Springer-Verlag.
Harris, B. (1986) *Engineering Composite Materials*, Institute of Metals.

Specific

Ashby, M. F. (1989) On the Engineering Properties of Materials. *Acta Metall.*, **37**, 1273.
Ashby, M. F. (1987) Technology of the 1990: advanced materials and predictive design. *Phil. Trans. R. Soc. Lond.*, **A322**, 393.
Friedrich, K., Walter, R., Voss, H. and Karger-Kocsis, J. (1986), *Composites*, **17**, 205.
Halstead, A. and Rawlings, R. D. (1986) *Inst. Phys. Conf. Ser. No. 75*, Ch. 10, p. 989.
Hull, D. (1981) *An Introduction to Composite Materials*. Cambridge University Press.

PROBLEMS

1.1 Compare the mechanical properties of the three main classes of structural materials, namely ceramics, metals and polymers. Discuss, in general terms, how the properties of these materials are affected by incorporating a reinforcement.

1.2 State, giving your reasoning, which composite materials you would select for the following applications:

(a) a tie-rod carrying a tensile load for a room temperature transport application,

(b) a tool to cut wood at high speeds,

(c) a component to operate within a vacuum furnace at 2100 °C,

(d) a component of diesel engine which is subject to wear.

SELF-ASSESSMENT QUESTIONS

Indicate whether statements 1 to 8 are true or false.

1. Usually the matrix has a lower Young's modulus than the reinforcement.

(A) true
(B) false

2. The most widely used composites are metal matrix composites.

(A) true
(B) false

3. The performance indicator $E^{1/2}/\rho$ is applicable when considering the possibility of buckling under the action of a compressive force.

(A) true
(B) false

4. A hybrid has a mixed metal and ceramic matrix reinforced with polymer fibres.

(A) true
(B) false

5. The main objective in reinforcing a metal is to lower the Young's modulus.

(A) true
(B) false

6. A laminate is an example of a particle reinforced composite.

(A) true
(B) false

7. The properties of a composite are essentially isotropic when the reinforcement is randomly orientated, equiaxed particles.

(A) true
(B) false

8. Materials property charts always have Young's modulus for one of the axes.

(A) true
(B) false

For each of the statements of questions 9 to 15, one or more of the completions given are correct. Mark the correct completions.

9. The matrix

(A) is always fibrous,
(B) transfers the load to the reinforcement,
(C) separates and protects the surface of the reinforcement,
(D) is usually stronger than the reinforcement,
(E) is never a ceramic.

10. Bone

(A) is a natural composite,
(B) consists of spirally wound cellulose fibres,
(C) contains hydroxyapatite,
(D) contains inorganic crystals,
(E) contains the organic fibre, collagen.

11. The specific modulus

(A) is given by $1/E$ where E is Young's modulus,
(B) is given by $E\rho$ where ρ is density,
(C) is given by E/ρ,
(D) is generally low for polymer matrix composites,
(E) is generally low for metallic materials.

12. Hybrids

(A) are composites with two matrix materials,
(B) are composites with mixed fibres,
(C) always have a metallic constituent,
(D) are also known as bidirectional woven composites,
(E) are usually multilayered composites.

13. The micrograph given below is of a

(A) multi-phase material,
(B) single-phase material,
(C) particulate reinforced composite,
(D) continuous fibre composite,
(E) cross-ply composite.

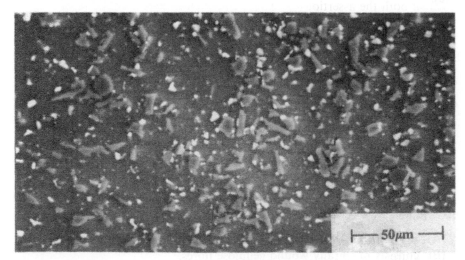

14. Metal matrix composites usually

(A) have a heavy metal for the matrix,
(B) have a poorer ductility than the matrix,
(C) retain their strength to higher temperatures than the matrix,
(D) have a lower Young's modulus than the matrix,
(E) are reinforced by polymer fibres.

15. Compared with a ceramic, a polymer normally has a

(A) greater strength,
(B) lower stiffness,
(C) lower density,
(D) better high temperature performance,
(E) lower hardness.

Each of the sentences in questions 16 to 21 consists of an assertion followed by a reason. Answer:

(A) if both assertion and reason are true statements and the reason is a correct explanation of the assertion,
(B) if both assertion and reason are true statements but the reason is not a correct explanation of the assertion,
(C) if the assertion is true but the reason is a false statement,
(D) if the assertion is false but the reason is a true statement,
(E) if both the assertion and reason are false statements.

16. The properties of a composite are influenced by the properties of the constituents *because* the properties of the composite are always given by the law of mixtures.

17. The properties of a continuous fibre composite are anisotropic *because* the arrangement of the fibres is often unidirectional.

18. Ceramic matrix composites have a lower toughness than monolithic ceramics *because* the reinforcement is usually fibres of low stiffness.

19. Polymer matrix composites have high values for the specific modulus *because* polymers are high strength materials.

20. Metal matrix composites usually retain their strength to higher temperatures than the matrix alloy *because* the reinforcement is normally a ceramic which has good mechanical properties at elevated temperatures.

21. It is preferable to use a ceramic matrix composite for a rod-like component subject to bending rather than the unreinforced matrix *because* the composite is likely to have the higher value for the performance indicator K_{IC}/ρ.

ANSWERS

Self-assessment

1. A; 2. B; 3. A; 4. B; 5. B; 6. B; 7. A; 8. B; 9. B, C; 10. A, C, D, E; 11. C, E; 12. B, E; 13. A, C; 14. B, C; 15. B, C, E; 16. C; 17. A; 18. E; 19. C; 20. A; 21. B.

Reinforcements and the reinforcement–matrix interface

2

2.1 INTRODUCTION

We have seen in the previous chapter that the reinforcement in a composite may be fibrous or particulate. A wide range of both these forms of reinforcement is available for use in the production of composite materials but most of the major developments in recent times have been in the area of fibrous reinforcement. The underlying philosophy in the design of fibre composite materials is to find or to make a fibre material of high elastic modulus and strength, and preferably low density, and then to arrange the fibres in a suitable manner to give useful engineering properties to the final product. There are naturally occurring fibres which are used to produce composites and these will be discussed briefly but more attention will be paid to the production, structure and properties of selected synthetic fibres. For completeness, silicon carbide particles will be used to exemplify the large number of particulate reinforcements available. The final sections of the chapter are devoted to the matrix–reinforcement interface and cover wettability, bonding and the measurement of bond strength.

2.2 NATURAL FIBRES

The reader will be familiar with many of the natural fibres such as cotton, silk, wool, jute, hemp and sisal as these are widely used for textiles, twine and rope throughout the world. These fibres are of course animal or plant products; the latter are essentially micro-composites consisting of cellulose

Table 2.1 Relative proportions of the major constituents and properties of some of the common natural fibres. (Adapted from Hodgson, 1989)

	Density (Mg/m^3)	Cellulose (%)	Hemicellulose (%)	Lignin (%)
Wood	1.5	40	40	20
Jute	1.3	72	14	14
Hemp		71	22	7
Sisal	0.7	74	–	26

	Young's modulus (GPa)	Tensile strength (MPa)	Specific modulus $[(GPa)/[Mg/m^3)]$	Specific strength $[(MPa)/(Mg/m^3)]$
Wood		500		333
Jute	55.5	442	43	340
Hemp		460		
Sisal	17	530	24	757

fibres in an amorphous matrix of lignin and hemicellulose (Table 2.1) and often have a high length to diameter ratio, called the *aspect ratio*, of greater than 1000. The strength and stiffness of these fibres are low compared with the synthetic fibres currently available (compare the data presented in Tables 2.1 and 2.2).

Although the plant-based organic fibres are occasionally used in synthetic composites, the natural fibre which has been exploited most widely is *asbestos*. Asbestos is the name given to a group of minerals which exist in fibrous form and are often found separated from the surrounding rock. The best source of asbestos is one of the serpentine minerals known as *chrysotile*, which is a hydrated magnesium silicate, $Mg_3Si_2O_5(OH)_4$. This occurs as silky fibres of up to several centimetres in length, which have good flexibility, stiffness and strength (Table 2.3).

The other forms of asbestos are not so commercially important. Of these *crocidolite*, or blue asbestos, which belongs to the amphibole minerals, is the best known. Its fibres are usually short and although they have good stiffness and tensile strength, as can be seen from the data in Table 2.3, their flexibility is poor.

Asbestos is used to reinforce some synthetic resins but mainly cement and plasters such as gypsum. In this context it is worth mentioning that asbestos is resistant to the very corrosive cement matrix environment during both setting and service. Asbestos has excellent thermal resistance and retains its properties to intermediate temperatures of about 600 °C and 400 °C for chrysotile and crocidolite respectively. However asbestos does have a major disadvantage of being a hazard to health if ingested. The concentration in air is limited by health standards in many countries; in the UK the limit is

Table 2.2 Properties of synthetic fibres; asbestos (chrysotile) has been included for comparison

	Density ρ (Mg/m^3)	Young's modulus, E_f (GPa)	Tensile strength, (MPa) $\hat{\sigma}_{Tf}$	Failure strain $(\%)$	E_f/ρ	$E_f^{1/2}/\rho$	$\hat{\sigma}_{Tf}/\rho$
Asbestos	2.56	160	3100	1.9	63	4.94	1213
E-glass	2.54	70	2200	3.1	27.6	3.29	866
Aramid (Kevlar 49)	1.45	130	2900	2.5	89.7	7.86	2000
SiC (Nicalon)	2.60	250	2200	0.9	96.2	6.08	846
Alumina (FP)	3.90	380	1400	0.4	97.4	4.99	359
Boron	2.65	420	3500	0.8	158.5	7.73	1321
Polyethylene (S1000)	0.97	172	2964	1.7	177.3	13.5	3056
Carbon (HM)	1.86	380	2700	0.7	204.3	10.5	1452

2 fibres per millilitre except for crocidolite where the limit is even more restrictive, namely 0.2 fibres per millilitre. Nevertheless the good mechanical, physical and chemical properties of asbestos and in particular chrysotile, set the standard by which synthetic fibres discussed in the following section should be judged.

Table 2.3 Properties of asbestos fibres

Property	Chrysotile	Crocidolite
Young's modulus (GPa)	160	190
Tensile Strength (MPa)	3100	3500
Density (Mg/m^3)	2.56	3.43
Specific modulus [(GPa)/(Mg/m^3)]	62.5	55.4
Specific Strength [(MPa)/(Mg/m^3)]	1211	1020
Maximum service temperature (°C)	600	400

2.3 INTRODUCTION TO SYNTHETIC FIBRES

Many different fibres are manufactured for the reinforcement in composites and some typical properties are given in Table 2.2. The values for stiffness and strength given in the table should be viewed with some caution. The manufacture of the fibres involves a number of processing steps and variability of properties from one fibre to another is large even when made by the same process. Between fibres of the same material made by different processes, the resulting microstructure and properties can differ even more markedly. Furthermore, the high tensile strength of freshly made fibres is normally reduced by surface damage during subsequent handling and storage. Finally any variation in size leads to a range of strength values; as explained in section 1.4 the bigger the diameter and the longer the length of the fibre the larger is the flaw that is likely to be found and thus the lower the strength (Figure 2.1).

As far as toughness is concerned most fibres are brittle and show only elastic extension before fracture (Figure 2.2). Thus the quoted failure strains correspond to $(\hat{\sigma}_{Tf}/E_f) \times 100\%$ where $\hat{\sigma}_{Tf}$ and E_f are tensile strength and Young's modulus respectively. The exception is the aramid fibres, which neck down before fracture with a significant amount of local drawing and reduction in area in the neck.

The reader will recall from the previous chapter that the specific modulus is one of the important performance indicators and the fibres are listed in Table 2.2 in order of increasing specific modulus. Although the fibres chosen for this table are representative of their class, the order of the fibres is not definitive, for example if carbon (HS) instead of carbon (HM) were chosen to represent carbon fibres then carbon would lie

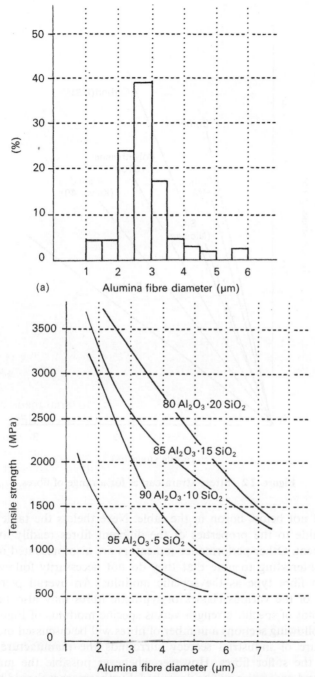

Figure 2.1 (a) Typical variation in the diameter of alumina fibres (average diameter is 2.8 μm) and (b) tensile strength as a function of fibre diameter. (Courtesy Itoh & Co., Ltd.)

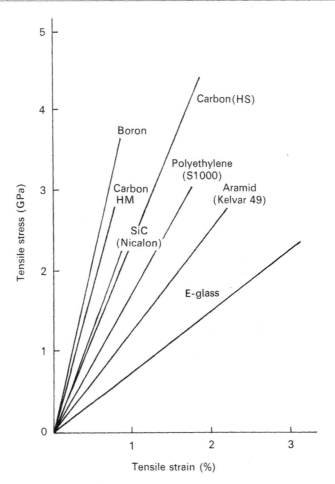

Figure 2.2 Stress–strain curves for a range of fibres.

above and not below boron in the table. Nevertheless the table is a good general guide to the properties of the different fibres readily available as reinforcement. Other performance indicators are also presented in the table and it is interesting to note that they do not necessarily follow the same trend with fibre type as the specific modulus. An overall picture of the properties of the various classes of synthetic fibres may also be obtained from the plot of specific strength versus specific modulus of Figure 2.3.

In the following sections a number of fibres will be discussed more fully. A fair measure of industrial secrecy surrounds the manufacture of fibres, especially the stiffer fibres. However whenever possible the manufacture, structure and properties will be described. Furthermore it should be borne in mind that there are many fibre manufacturers and it is beyond the scope of this book to cover all the subtle differences in the fibres of a given type. Instead it is

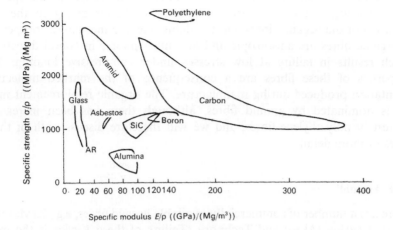

Figure 2.3 A plot of specific strength versus specific modulus for easy comparison of the properties of synthetic fibres. Asbestos has been included for comparison. Data for SiC whiskers and carbon produced from rayon precursor have been omitted as their properties are not typical for their class and distort the areas on the graph.

the intention to present sufficient details so that the reader may appreciate the variety of available fibres of a given type and to instil an understanding of the main advantages and disadvantages of the different types of fibre.

2.4. SYNTHETIC ORGANIC FIBRES

The data for aramid and polyethylene in Table 2.4 demonstrate that organic fibres of good strength and stiffness may be produced. Their tensile strengths

Table 2.4 Properties of commercially available organic fibres

	Density ρ (Mg/m^3)	Young's Modulus, (GPa) E_f	Tensile Strength, (MPa) $\hat{\sigma}_{Tf}$	E_f/ρ	$\hat{\sigma}_{Tf}/\rho$
ARAMID					
Kevlar 29	1.44	59	2640	41.0	1833
Kevlar 49	1.45	130	2900	89.7	2000
Kevlar 149	1.47	146	2410	99.3	1639
Technora HM50	1.39	74	2990	53.2	2151
Twaron	1.44	80	2800	55.6	1944
Twaron HM	1.45	115	3150	79.3	2172
POLYETHYLENE					
Spectra 900	0.97	117	2585	121	2665
Spectra 1000	0.97	172	2964	177	3056

and Young's moduli are not as high as those for asbestos but the specific properties ($\hat{\sigma}_{Tf}/\rho$ and E_f/ρ) are comparable or better owing to the low density ρ of the organic fibres. On the debit side, the mechanical properties of organic fibres are anisotropic and are much poorer in lateral directions which results in failure at low stresses under compressive loading. The properties of these fibres are a consequence of the marked molecular orientation produced during manufacture. The organic reinforcement market is dominated by aramid fibres, although there has been increasing interest in polyethylene fibres, and we will therefore discuss both of these fibres in more detail.

2.4.1 Aramid

There are a number of commercially available *aramid* fibres, e.g., Kevlar (Du Pont), Twarlon (Akzo) and Technora (Teijin); of these *Kevlar* is the most well known. The aramids can be viewed as nylon with extra benzene rings in the polymer chain to increase stiffness (Figures 2.4(a) and 2.4(b)); alternative nomenclature that the reader might encounter is aromatic polyamide or poly(phenylene terephthalamide).

The stiff aramid molecule is aligned to a certain extent even in solution or in the melt. The extended chains are arranged approximately parallel to

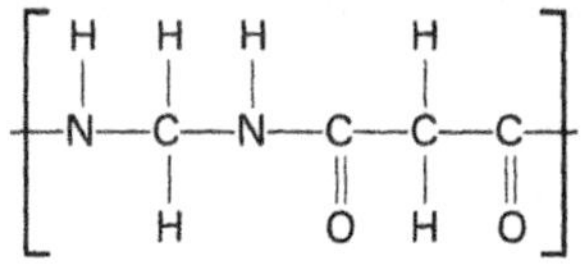

(a)

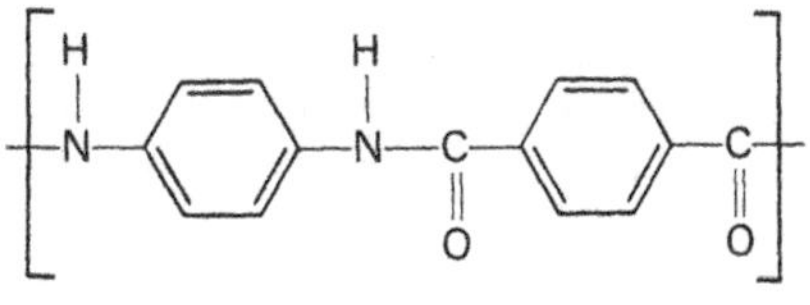

(b)

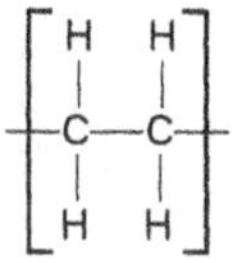

(c)

Figure 2.4   The molecular structure of: (a) nylon; (b) aramid; (c) polyethylene.

their long axis although the chain centres are randomly distributed; this structure is known as a *nematic liquid crystal* and is schematically illustrated in Figure 2.5(a). The alignment of the molecules is enhanced during fibre production by spinning and extrusion, and by any subsequent mechanical treatment (Figure 2.5(b)). The production route is as follows. A solution of the aramid in a suitable solvent such as sulphuric acid is held at a low temperature and then extruded at an elevated temperature into a coagulation bath to remove the residual solvent. This results in fibres of relatively low strength and Young's modulus of about 850 MPa and 5 GPa respectively. Finally the fibre is stretched and cold drawn to align the structure further and to increase the mechanical properties to those quoted in Table 2.4.

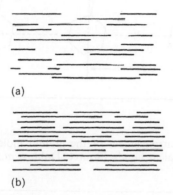

(a)

(b)

Figure 2.5 Schematic illustration of aramid: (a) in solution: nematic liquid crystal; (b) after alignment.

The molecules are arranged in planar sheets (Figure 2.6) with strong covalent bonds in the molecular chains parallel to the fibre axis but weak hydrogen interchain bonding in the transverse direction. The planar sheets are in turn pleated and stacked to give the three-dimensional orientated structure with radial symmetry of Kevlar 49 shown schematically in Figure 2.7. The low modulus Kevlar (Kevlar 29) is simply a structural variation; the structure of Kevlar 29 is less perfect and less orientated but can be converted to the Kevlar 49 structure by a suitable heat treatment.

Aramid fibres have good high temperature properties for a polymeric material. They have a glass transition temperature of about 360 °C, burn with difficulty and do not melt like nylon. A loss in performance due to carbonization occurs around 425 °C but they can be used at elevated temperatures for sustained periods and even at 300 °C for a limited time. Aramid fibres do have a tendency to degrade in sunlight. Their dimensional stability is good as the coefficient of thermal expansion is low (approximately $-4 \times 10^{-6} \, \text{K}^{-1}$). Other properties which are important in certain applications are low electrical and thermal conductivity and high thermal capacity.

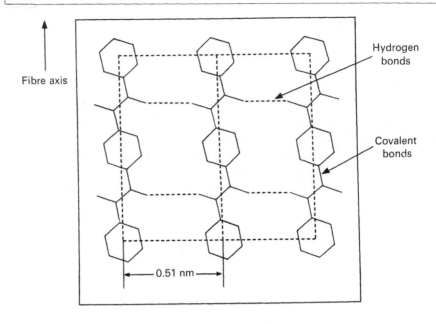

Figure 2.6 The planar arrangement of the aramid molecules showing the weak hydrogen and strong covalent bonds. (Source: Dobb *et al.*, 1980.)

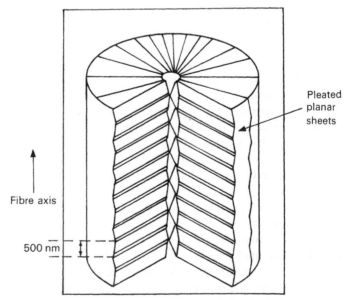

Figure 2.7 The three-dimensional orientated structure of an aramid fibre (Kevlar 49) showing the planes illustrated in Figure 2.6 arranged in a radially pleated morphology. (Source: Dobb *et al.*, 1980.)

2.4.2 Polyethylene

In the late 1980s a number of *polyethylene* fibres became commercially available including Spectra (Allied Signal) and Dyneema (Dutch State Mines, DSM). As can be seen from Figure 2.4(c), the polyethylene molecule with its carbon–carbon (C–C) backbone is simpler than that of aramid. The theoretical elastic modulus of the covalent C–C bond in the fully extended polyethylene molecule is 220 GPa along the direction of the chain. Although this is not fully achieved by the orientated structure of the fibres, the value of 172 GPa is impressive. Polyethylene has the lowest density of any readily available fibre and hence its specific properties are good and superior to those of Kevlar. However polyethylene has a low melting point of 135 °C and readily creeps at elevated temperature, so use is restricted to temperatures below 100 °C. The fibres can be surface treated with gas-plasma to replace H atoms with polar groups which gives improved bonding in composites and hence increased strength and stiffness but reduced toughness.

2.5 SYNTHETIC INORGANIC FIBRES

Inorganic fibres are well established, particularly glass and carbon fibres, but development is still taking place leading to enhanced performance.

2.5.1 Glass

The structure of *glass* will be discussed fully in Chapter 4. For present purposes it is sufficient to know that glass is a noncrystalline material with a *short-range network structure*. As such it has no distinctive microstructure and the mechanical properties, which are determined mainly by composition and surface finish, are isotropic. There are many groups of glasses, for example silica, oxynitride, phosphate and halide glasses, but from the point of view of composite technology only the *silica glasses* are currently of importance. However even within this group of glasses the composition, and hence properties, vary considerably; the compositions of some typical glasses which are used in the manufacture of fibres are given in Table 2.5.

The E in *E-glass* is an abbreviation for electrical. This glass is based on the eutectic in the ternary system $CaO–Al_2O_3–SiO_2$ with some B_2O_3 substituting for SiO_2 and some MgO for CaO. The B_2O_3 lowers the liquidus temperature substantially giving a longer working temperature range and consequently making fibre manufacture by drawing easier. Molten glass is exuded under gravity from a melting tank through an orifice and rapidly pulled to draw it down to a 10 μm diameter fibre. Normally 204 orifices are used on the same melting-tank giving fibres which are gathered

Table 2.5 Typical composition and properties of glass fibres

	E-glass (%)	S-glass (%)	AR-glass (%)
SiO_2	54	65	64
TiO_2			3
ZrO_2			13
Al_2O_3	14	25	1
B_2O_3	9		
MgO	5	10	
CaO	18		5
Na_2O			14
Modulus (GPa)	70	80	75
Strength (MPa)	2200	2600	1700
Density (Mg/m^3)	2.54	2.49	2.70

into a *strand*, sized with starch-oil emulsion to minimize surface damage and wound on to a drum at speeds up to 50 m/s. These strands can be chopped into 2.5–5 cm lengths to make *chopped strand mat* (CSM), wound parallel to give *roving* or ribbon, or twisted to form *yarn* for weaving into glass-fibre cloth or woven roving (WR). Freshly drawn and carefully handled fibres have tensile strengths of approximately $E_f/20$ but a typical value may be nearer to $E_f/50$, where E_f is Young's modulus.

A size to minimize surface damage has already been mentioned, but other coatings may also be applied. Active coatings which enhance the wetting and bonding between fibres and matrix are frequently used. These coatings, known as *coupling agents*, are complex organosilanes, the exact chemistry of which is adjusted to suit the type of matrix, i.e., the coupling agent for a polyester matrix is different from that for, say, an epoxy matrix.

S-glass, known as R-glass in Europe, is based on the SiO_2–Al_2O_3–MgO system; this fibre has higher stiffness and strength (hence the designation S) than E-glass (Table 2.5). It also retains its improved properties to higher temperatures. However it is more difficult to draw into fibres due to its limited working range and is therefore more expensive. S-glass fibres are still used for some specialist applications but they have been largely superseded by fibres such as carbon and aramid with superior mechanical properties.

Although the performance of the widely used general purpose E-glass is satisfactory in near neutral aqueous solutions, it is liable to degradation in environments which are highly acidic or alkaline. For this reason a number of more resistant glasses have been developed, namely *C-glass* (chemical glass), *E-CR-glass* (electrical-corrosion resistant glass) and *AR-glass* (alkali resistant glass). The latter was developed as a replacement for asbestos in cement and contains ZrO_2 to impart corrosion resistance. The alkaline cement environment leaches silica from the surface of the fibres leaving a ZrO_2-rich surface which inhibits the corrosion process. However the ZrO_2 layer, which is fairly porous, does continue to grow slowly with time with the result that a loss of about 50% in strength may be experienced after ten years in cement. The ZrO_2 has an adverse effect on the ease of glass forming and so the CaO plus Na_2O content is increased to counterbalance this effect.

2.5.2 Alumina

Alumina (Al_2O_3) can adopt several crystal structures, some of which are metastable; the different alumina phases are designated alpha- (α), delta- (δ), gamma- (γ) and eta- (η) alumina. The familiar bulk alumina ceramic, used for example in spark plugs, is α-*alumina* which has a hexagonal close packed crystal structure (Figure 2.8). α-alumina is the predominant phase in some fibres, but the other forms, particularly δ-*alumina*, are also found as the major phase in certain fibres. Alumina fibres are polycrystalline and are not 100% Al_2O_3 but contain other oxides often in appreciable quantities. The most common oxide addition is SiO_2 and fibres with as much as 20% SiO_2 are manufactured. Aluminas with more than a few percentage of SiO_2 are called *debased* aluminas. Alumina fibres are used with a refractory binder in the form of mats, boards or tiles for thermal insulation for high temperature industrial furnaces. More recently they have been used in metal matrix composites, especially with aluminium alloys.

Alumina has a high melting point in excess of 2000 °C and a relatively low viscosity when molten and is therefore not amenable to fibre production by melt spinning techniques. The production methods employed can be

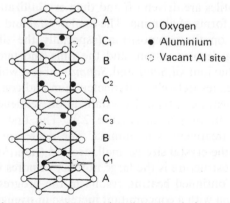

Figure 2.8 Crystal structure of α-alumina (hexagonal close packed structure). The aluminium sites (layers C_1, C_2 and C_3) which are only two-thirds full, are sited between the hexagonal layers (A, B) of oxygen atoms.

classified as slurry forming plus firing, and solution or sol-gel processing plus heat treatment. The first stage in the former process is to form a *slurry*, which is simply an aqueous suspension of fine alumina particles, together with additions to stabilize the suspension (deflocculents) and polymers to modify the viscosity. The slurry is then extruded into 'green' fibres and dried. Finally the 'green' fibres are fired during which additives are driven off and *sintering* occurs to produce solid fibres with a controlled amount of porosity and grain size. The most established fibre produced by this route is Al_2O_3-FP manufactured by Du Pont. This is a 99% α-alumina fibre of diameter 10–20 μm with a grain size of about 0.5 μm. The mechanical properties at room temperature are good, but the fibres are susceptible to grain growth and to creep at elevated temperatures.

There are some modifications that can be made to the α-alumina fibres produced by this method. We shall learn in the chapter on ceramic matrix composites (Chapter 4) that alumina may be transformation toughened by the addition of small particles of zirconia containing a small percentage of the stabilizing oxide yttria. This particulate composite material with α-alumina as the matrix is termed *zirconia toughened alumina* (ZTA), and fibres of ZTA with improved room temperature and elevated temperature strength have been produced.

In *solution* or *sol–gel* spinning a viscous, highly concentrated solution of an aluminium compound is used for spinning. The details of the solution and the spinning technique depend on whether continuous or discontinuous fibres are to be produced, e.g., continuous fibres produced by extrusion through 100–200 μm holes require a solution of high viscosity (10–100 Ns m^2) whereas the production of discontinuous fibres by gas-attenuation spinning generally employs solutions of lower viscosities of less than about 2 Ns m^2.

After spinning the fibres are dried and heat treated. During heat treatment the remaining volatiles are driven off and then crystallization takes place to the various crystal forms of alumina. The final crystal and microstructure of the fibre depends on the composition (especially the silica content), the atmosphere during decomposition, and the time and temperature of heat treatment. The behaviour of *Saffil* and *Safimax* fibres, which are short and long (~ 0.5 m) fibres respectively of diameter 3 μm and containing 4% SiO_2, will be used to illustrate a typical crystallization sequence and the corresponding changes in properties (Figure 2.9). The first crystalline phase observed on heat treatment is η-alumina. The degree of crystallization is only 50–55% and the crystal size is small, around 6 nm. Another noticeable feature of the microstructure is the large number of pores which gives a low effective density. Continued heating results in the progression through the γ-phase to δ-alumina with a concomitant increase in strength and toughness. An increasing proportion of α-alumina is the consequence of further heat treatment. This is accompanied by an increase in density and stiffness but a

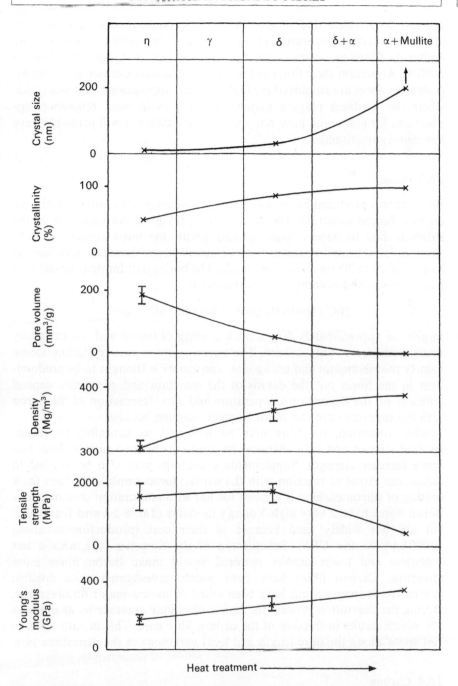

Figure 2.9 Effect of heat treatment on the structure and properties of alumina fibres (Saffil) produced by solution spinning. (Source: Dinwoodie and Horsfall, 1987.)

decrease in strength due to growth of the crystals to over 200 nm. The maximum service temperature of an η-alumina fibre is 900 °C, whereas the δ-alumina fibres are more stable and may be used at temperatures up to 1600 °C. At present these fibres are mainly used as reinforcements for metals. δ-alumina fibres are employed for high strength applications, and α-alumina where the hardness plays a major role such as in wear resistance; applications for η-alumina have not yet been identified but will probably have low density requirements.

2.5.3 Boron

Boron fibre is produced by *chemical vapour deposition* from boron trichloride on to a heated substrate. The temperatures involved are high and so the substrate has to have a high melting point; the most commonly used substrate is a heated tungsten wire of about 10 μm diameter. Continuous lengths of up to 3000 m have been made. The boron trichloride is mixed with hydrogen gas and decomposes according to:

$$2BCl_3 \text{ (gas)} + H_2 \text{ (gas)} = 2B \text{ (solid)} + 6HCl \text{ (gas)},$$

to give an approximately 50 μm thick coating of boron with an extremely small grain size of only 2–3 nm. The boron exists in two crystalline forms, namely rhombohedral and tetragonal. The former is thought to be predominant in the fibres but the details of the structure and properties depend critically on the deposition temperature and rate. Interaction of the boron with the tungsten core can occur to yield tungsten borides.

After formation, the fibre may be subjected to annealing to reduce residual stresses and to a chemical treatment to remove surface flaws and hence increase strength. Supplementary coatings may also be applied to reduce the extent of reaction with the matrix; for example, a 0.25 μm thick coating of silicon carbide is applied for use with aluminium alloy matrices. Boron fibres have a very high Young's modulus (Table 2.2 and Figure 2.2) but are not widely used because of their cost (production is about 50 000 kg/y in the USA). Substitution of the tungsten core with a less expensive and lower density material would make boron fibres more attractive. Carbon fibres have been widely investigated as a possible alternative to tungsten but have been found to have a major disadvantage. During the deposition process the boron fibre may elongate by as much as 5% which results in fracture of the carbon fibre core. This in turn leads to 'hot spots' along the fibre length and local variations in the deposition rate.

2.5.4 Carbon

Carbon fibres are produced by many companies and the world production capacity exceeds 12 000 tons (Table 2.6). In spite of this large production

Table 2.6 World-wide production capacity of carbon fibre as of December 1991. (Source: Lin, 1992)

Fibre	Region	Manufacturer	Trade name	Production capacity (ton)	Sub-total
PAN	Asia	Toray Co[b]	Torayca	2250	
		Toho Rayon[c]	Besfight	2020	
		Mitsubishi Rayon[d]	Pyrofil	500	
		New Asahi Carbon	Hicarbon	450	
		Taiwan Plastics	Tairyfil	230	
		Korea Steel	Kosca	150	
	Mid East	Afikim Carbon	Acif	100	
	America	Hercules	Magnamite	1750	
		BASF[c]	Ceilon	1485	
		Amoco	Thornel	1000	
		Akzo	Fortafil	360	
		Courtaulds-Grafil[d]	Grafil	360	
		BPAC[d]	Hitex	40	
		Zoltek	Panex	110	
		Textron	Avcarb	100	
	Europe	Courtaulds-Grafil	Grafil	350	
		Akzo[c]	Tenax	350	
		Sofica[b]	Torayca	300	
		Sigri	Sigrafil	25	
		R.K. Carbon	–	230	12160
Pitch	Japan	Kureha Chemical	Kureha	900	
		Osaka Gas	Donacarbo	300[a]	
		Mitsubishi Chemical	Dialead	50[a]	
		Nippon Oil	Granoc	50	
		New Nippon Steel	—	12[a]	
		Tonen	Forca	12	
		Petoca	Carbonic	12	
	America	Amoco	Thornel	230	1566

[a]Coal tar pitch, others are petroleum pitch
[b]Toray group, total production of 2550 tons, 21% of PAN fibre
[c]Toho Rayon Group, total production of 3855 tons, 32% of PAN fibre
[d]Mitsubishi Rayon provides raw material and technical assistance

capacity carbon fibres are still relatively expensive; for example carbon fibres cost more than the equivalent strength synthetic organic fibre. Nevertheless the usage for carbon fibres continues to increase as shown by the annual consumption of PAN based fibres increasing by almost an order of magnitude in the decade ending 1991 (Figure 2.10).

The structure and properties of these fibres vary considerably and new fibres are always under test. For example two recent introductions are *hollow* fibres and *coiled* fibres (Figure 2.11). The former are designed to impart

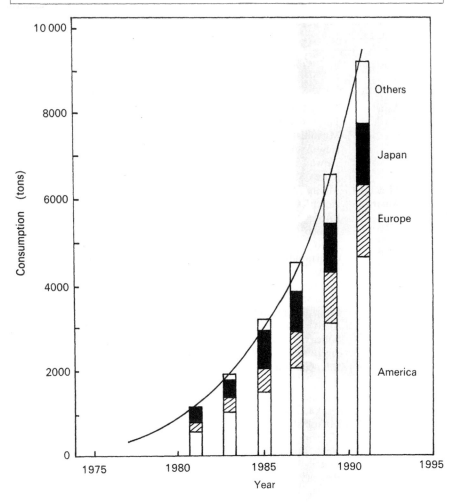

Figure 2.10 Annual consumption of carbon fibres produced from polyacrylonitrile PAN. (Source: Lin, 1992.)

better impact toughness to carbon reinforced polymers whereas the latter are capable of extending many times their original length without loss of elasticity. Clearly all the carbon fibres currently under investigation, or even just those commercially available, cannot be covered here and we only attempt to describe some of the common processing routes and structure–property relationships. However before ending this survey of carbon fibres we must briefly study the crystal structure of carbon.

Carbon has two well known crystalline forms (diamond and graphite) but it also exists in quasicrystalline and glassy states. As far as fibre technology is concerned *graphite* is the most important structural form of carbon. The graphitic structure consists of hexagonal layers, in which the bonding is

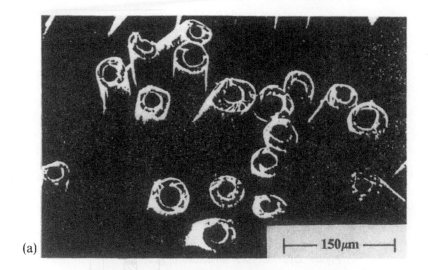

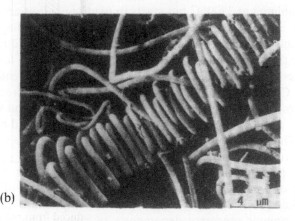

Figure 2.11 Some unusual carbon fibres: (a) hollow fibres produced from pitch; (b) coiled fibres produced from the vapour phase. (Source: Lin, 1992.)

covalent and strong ($\sim 525\,kJ/mol$); these layers, which are called the *basal* planes, are stacked in an ABAB---- sequence as shown in Fig. 2.12 with inter-layer bonds being weak ($< 10\,kJ/mol$) van der Waals bonds. It should be no surprise to the reader that a consequence of the marked difference in the intra-layer and inter-layer bonding is that the properties of graphite are very anisotropic. The theoretical elastic modulus of graphite is approximately $1000\,GPa$ in the basal plane and only $35\,GPa$ in the c-direction perpendicular to these planes. As we will see, alignment of the basal plane parallel to the fibre axis give stiff fibres which, because of the relatively low density of around $2\,Mg/m^3$, have extremely high values for specific stiffness (Figure 2.3).

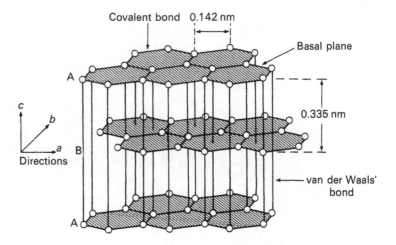

Figure 2.12   Crystal structure of graphite.

Graphite sublimes at 3700 °C but starts to oxidize in air at around 500 °C; carbon fibres can however be used at temperatures exceeding 2500 °C if protected from oxygen. Carbon is a good electrical conductor which, depending on the circumstances, can be advantageous or not; for example it permits the testing of carbon fibre composites by eddy current techniques (Chapter 15). Carbon fibres are produced from a variety of precursors. The mechanical properties vary greatly with the precursor used and the processing conditions employed as these determine the perfection and alignment of the crystals. There are three main precursors, although as we will see, they are not all of equal importance.

Controlled heating of *cellulose* or *rayon* fibre converts the fibre to graphite. The decomposition is complex but essentially loss of weight and shrinkage takes place during *pyrolysis* in the temperature range 200 to 400 °C as the organic precursor decomposes to carbon. This is followed by *carbonization*, i.e., gradual ordering of the structure, and finally *graphitization* at higher temperatures. The yield is low, only 15–30%, and although the fibre is crystalline the alignment is poor. Thus at this stage the fibres have low strength and stiffness, the latter being of the order of the modulus in the *c*-direction of the graphitic structure. However hot stretching by up to 50% at temperatures in the range 2700 to 3000 °C increases the modulus and strength by developing a preferred orientation in the fibres and reducing porosity (Table 2.7). The preferred orientation is such that there is a tendency for the basal planes to be aligned parallel to the axis of the fibre, although the alignment is far from perfect. This manufacturing route is declining because of the decreasing availability of the main source of suitable rayon, the low yield and the problems involved in stretching at such high temperatures.

Table 2.7 Properties of some commercially available carbon fibres

	Density (Mg/m^3)	Young's modulus (GPa)	Tensile strength (MPa)	Failure strain $(\%)$
RAYON				
Standard	1.60	40	500	1.25
HM	1.82	517	–	–
PAN				
HS	1.80	230	4500	2.0
IM	1.76	290	3100	1.1
HM	1.86	380	2700	0.7
PITCH				
P-25W	1.90	160	1400	0.9
P-75S	2.00	520	2100	0.4
P-120S	2.18	827	2200	0.3

Compared with rayon or cellulose fibres, those produced from *polyacrylonitrile* (Figure 2.13(a)), PAN, have a high degree of orientation in filament form. The PAN precursor also gives a higher yield of up to 50% on conversion to carbon. As for all commercial fibres the details of the production are confidential, but it is known that a three-stage process is involved in the conversion and the resulting structure is sensitive to the rates of heating.

The first stage involves stretching and oxidation. The fibres are initially stretched 500–1300% to improve molecular alignment and then heated in air, while still under tension, to 200–280 °C. This results in intramolecular rearrangement to give a ladder-like polymer; chemical cross-links between the oriented chains in three dimensions are formed so that chains can no longer bend or distort on heating (Figure 2.13(b)). This is followed by oxidation whereby some two-thirds of the oxygen sites are filled slowly with the release of H_2O (Figure 2.13(c)); this is sometimes called oxyPAN. Heating oxyPAN in nitrogen or argon at 900–1200 °C produces low modulus, high strength carbon fibres, which have fine well-oriented crystals parallel to the fibre axis. However they have a fair degree of porosity and the density is only 1.74 Mg/m^3. The idealized formation of the carbon ring structure is shown in Figure 2.13(d); it can be seen that there is an operational hazard at this stage as toxic volatiles such as HCN are given off.

Heating in argon up to 2800 °C causes *graphitization* and produces high modulus fibres of increased density up to 2.00 Mg/m^3 (the density of graphite is 2.26 Mg/m^3). A three-dimensional representation of the arrangement of the basal planes in PAN-based carbon fibre is given in Figure 2.14. The value of the modulus increases with increasing temperature of graphitization as shown in Figure 2.15. In fact the properties can be varied over a wide range by suitable heat treatment conditions but it is difficult to

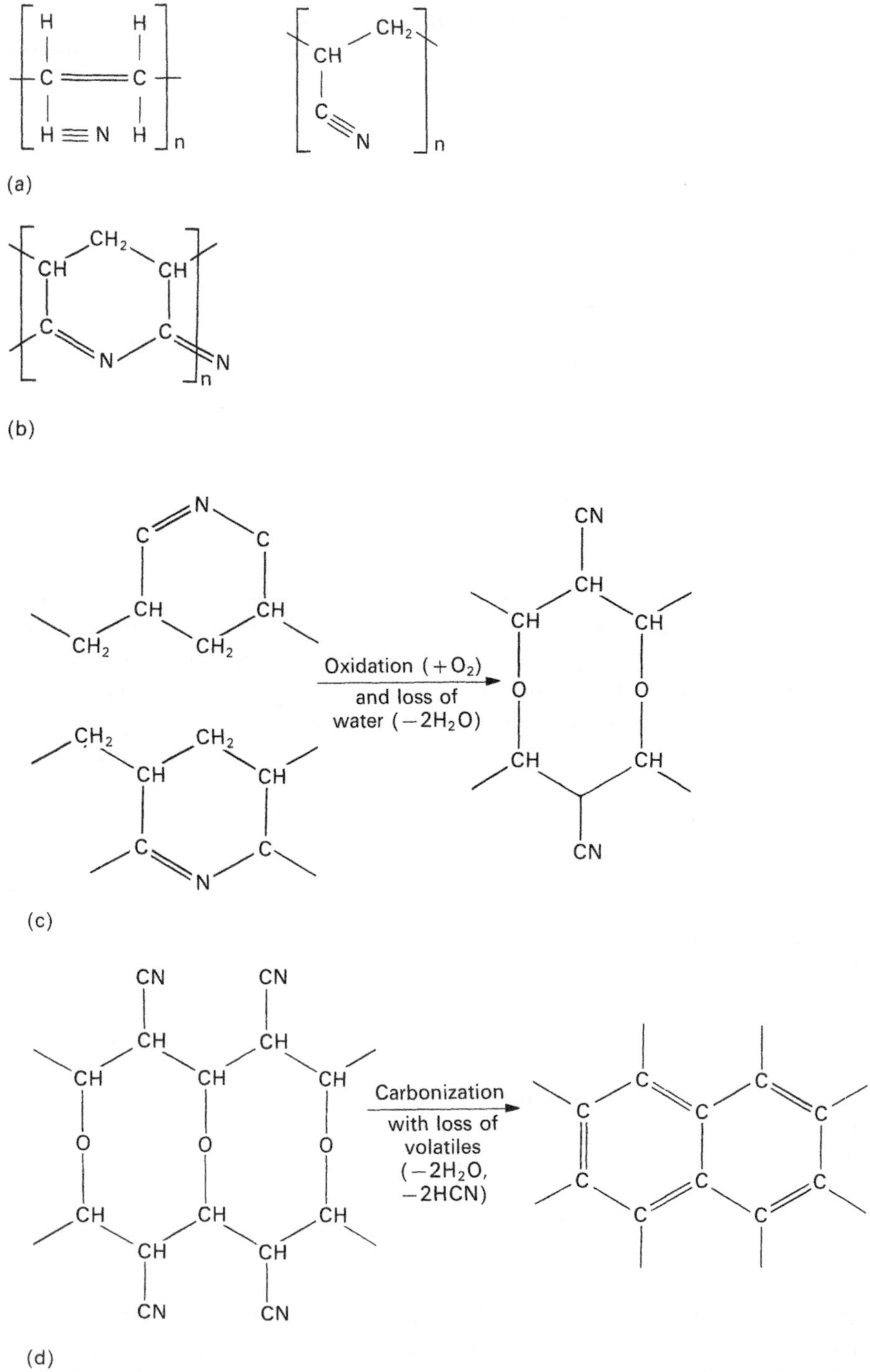

(a)

(b)

(c)

(d)

Figure 2.13 The changes in structure during the production of carbon fibres from PAN precursor. (a) polyacrylonitrile; (b) ladder-like polymer; (c) oxidation of ladder-like polymer to oxyPAN; (d) carbonization of oxyPAN to carbon ring structure.

Fibre
axis

Figure 2.14 Schematic representation of the three-dimensional structure of the orientation of the basal planes of the graphite in a carbon fibre. (Source: Bennett and Johnson, 1978.)

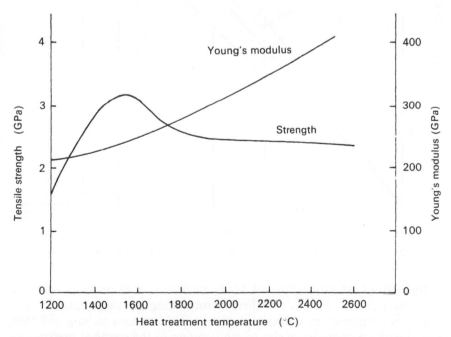

Figure 2.15 Effect of heat treatment temperature on the Young's modulus and strength of carbon fibres produced from a PAN precursor. (Source: Moreton *et al.*, 1967.)

combine high strength with high modulus. Refinements in fibre process technology over the past twenty years have led to considerable improvements in strength and in strain to fracture for PAN-based fibres. These can now be supplied in three basic forms, high modulus (HM), intermediate modulus (IM) and high strength (HS) as shown in Table 2.7. The most recent developments of the high strength fibres have led to what are known as high strain fibres, which have strain values approaching 2% before fracture. Figure 2.16 shows the stress–strain curves for a range of carbon fibres. The strain is elastic up to failure and a lot of stored energy, given by the area under the stress–strain curve, is released when the fibres break in a brittle manner.

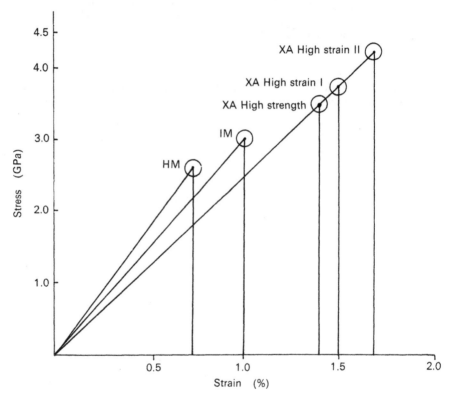

Figure 2.16 Stress–strain curves for carbon fibres produced from a PAN precursor. (Source: Hysol Grafil Ltd.)

These fibres are surface-treated during manufacture to promote adhesion with the polymer matrix, whether thermosetting or thermoplastic. The surface is roughened by chemical oxidation which causes etching and then coated with an appropriate size to aid bonding to the specified matrix.

Petroleum and *coal-tar pitches* contain a complex mixture of high molecular weight aliphatic and aromatic hydrocarbons; they have therefore a high carbon content and thus are suitable as precursors. The pitch is heated to above 350 °C to polymerize it to molecular weights of about 1000. The structure of the molecules is such that an ordered liquid–crystal or *mesophase* is formed. The polymer is extruded through the holes in a hot-walled metal cylinder whilst the cylinder rotates. This process, which is known as *melt-spinning*, orientates the hot mesophase pitch. After oxidation to induce cross-linking, and hence prevent the fibres remelting and sticking together, the fibres are carbonized at temperatures up to 2000 °C giving low modulus carbon fibres The carbon yield is high at about 80%. The degree of graphitization is controlled by heat treatment at temperatures up to 2900 °C. A high degree of perfection of the graphite may be obtained, together with good axial alignment of the basal planes, and not surprisingly this leads to high density, high modulus fibres (see for example P-120S in Table 2.7).

An unusual characteristic of carbon fibres is their very low, or even slightly negative, coefficients of longitudinal expansion. As for other properties, the coefficient of thermal expansion depends on the fabrication route and hence degree of graphitization and crystal orientation. Ultra-high modulus carbon fibres have negative coefficients of expansion of approximately $-1.4 \times 10^{-6} \mathrm{K}^{-1}$ and are employed in the production of polymer matrix composites with near net zero thermal expansion.

Finally it should be mentioned that although we have concentrated on the effect of graphitization and orientation on properties there are other structural features, such as voids, microcracks, surface flaws and impurities, which have to be controlled in order to obtain fibres of a consistent high quality.

2.5.5 Silicon based fibres

The last fibres we will discuss are those based on silicon, the most recent fibres in this class being *methylpolydisilylazane*, MPDZ, and *hydridopolysilylazane*, HPZ. These are amorphous fibres, consisting of Si, C, N and O, and are still in the experimental stage hence little is known about their effectiveness as reinforcement. Better established are *silicon carbide*, SiC, fibres which are attracting much interest as reinforcement for ceramics. The bonding in SiC is covalent and we would therefore expect the strong bonding to give brittle fibres with a high modulus; this is indeed the case as can be seen from the data of Table 2.2. However the density of SiC fibres is high and therefore their specific properties, with the exception of those for whiskers (section 2.6), are not so impressive and are inferior to those of carbon and polyethylene (Figure 2.3 and Table 2.8).

There are two methods for the production of SiC, one involves deposition on a substrate and the other decomposition of a precursor. First let us consider the former, which is a chemical vapour deposition process similar to that previously described for the production of boron fibres.

Table 2.8 Properties of silicon carbide fibres

	Density ρ (Mg/m^3)	Young's Modulus, (GPa) E_f	Tensile Strength, (MPa) $\hat{\sigma}_{Tf}$	E_f/ρ	$\hat{\sigma}_{Tf}/\rho$
Nicalon	2.60	250	2200	96.2	846
CVD mono-filament[b]	3.05	406	3920	133.1	1285
CVD mono-filament[c]	3.00	400	3450	133.3	1150
Tyranno	2.40	280	2000	116.7	833
Whisker	3.20	700	10000[a]	218.8	3125

[a] With a wide range around this value from around 2000 to 20000MPa.
[b] With tungsten core.
[c] With carbon core.

The SiC is made from a carbon-containing silane, a typical reaction being

$$CH_3SiCl_3 \text{ (gas)} \rightarrow SiC \text{ (solid)} + 3HCl \text{ (gas)}.$$

The substrate or core is usually tungsten, although carbon is also used without the major problems encountered with boron. The deposit consists of fine crystals of β-SiC orientated preferrentially with the {111} planes parallel to the fibre axis (Figure 2.17). With a tungsten core a reaction occurs at the SiC-core interface after prolonged use at temperatures above 1000 °C to give W_2C and W_5Si_3 which degrades the properties. These fibres, which are sometimes called *monofilaments*, are thick (100–150 μm diameter) and, depending on the size of

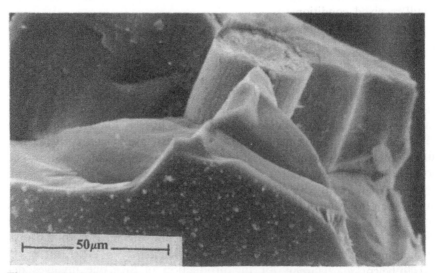

Figure 2.17 Scanning electron micrograph of fractured fibre of chemically vapour deposited SiC showing the fine grained structure and the tungsten core. The fine particles on the surface are due to some SiC formation in gas phase. (Courtesy G. Hocking and P. S. Sidky.)

the core and whether carbon or tungsten, can have a density as high as 3.35 Mg/m^3. Because of their large diameter and high Young's modulus, monofilaments are not flexible (equation 1.1) and as a consequence cannot be easily woven.

The fibres are sensitive to surface flaws and surface damage is reduced by the application of a thin coating of pyrolitic carbon to increase abrasion resistance.

The second method of production via precursors has some similarities with the methods used for carbon fibres. The first stage in making 'Nicalon' fibres is to thermally decompose polydimethylsilane in a complex process involving a high pressure stage ($\sim$10 MPa) in an autoclave followed by a lower temperature vacuum distillation treatment to give the polymer poly-carbosilane with a relatively low molecular weight of about 1500. The polymer is melt spun and oxidized in air at 200 °C to induce cross-linking. The temperature is then taken up slowly to 1300 °C to form SiC. The SiC is in the form of small ($\sim$2 nm) crystals of β-SiC. The fibre is not pure SiC as some oxygen remains from the low temperature heat treatment and also excess silicon and carbon are present; it is likely that some of the oxygen may be combined with silicon to give silica, SiO_2. Growth of the SiC crystals and changes in the proportions of the phases present can occur with prolonged heating at elevated temperatures. A typical composition (wt.%) for 'Nicalon' is 59%Si–31%C–10%O.

Another fibre, known by the trade name 'Tyranno', is produced by a similar route involving polytitanocarbosilane, thus the fibres contain up to 5wt.% titanium. It is thought that the titanium hinders crystallization and in the as-received condition the fibres are amorphous. However, the amorphous state is metastable and crystallization starts on heating to temperatures above 1000 °C. Fibres such as 'Nicalon' and 'Tyranno' with diameters of 12 μm and 8 μm respectively are much finer than their monofilament counterparts.

2.6 PARTICULATE AND WHISKER REINFORCEMENTS

The range of materials available in particulate form is far more extensive than as fibres because production is so much simpler – almost any material may be ground into a powder from the bulk. Not that all powders are produced in this way; other methods include precipitation from solution, gas atomization and sol-gel processing. Because of the large number of powders available a comprehensive survey will not be attempted. Instead one reinforcement, namely SiC, will be taken to illustrate the parameters that need to be considered when selecting either whisker or particulate reinforcement.

SiC is available in the form of *whiskers*, i.e., small single crystals. The morphology and dimensions of whiskers are extremely variable even when produced by the same process but typically they are a few tens of microns in length and less than one micron in diameter (Figure 2.18).

They are commonly grown from supersaturated gas phases at high temperatures and can be made purer than 'Nicalon' or 'Tyranno' fibres. Being single crystal and purer they are structurally more stable at elevated temperatures. SiC whiskers are also commercially produced by pyrolysis of rice husks. The husks contain cellulose, 15–20wt.%SiO_2 and other organic and inorganic matter. The volatile constituents are driven off by a coking treatment consisting of heating in an oxygen-free atmosphere at 700 °C. A higher temperature (1500–1600 °C) heat treatment in an inert or reducing atmosphere leads to the formation of SiC according to

$$3C + SiO_2 = SiC + 2CO.$$

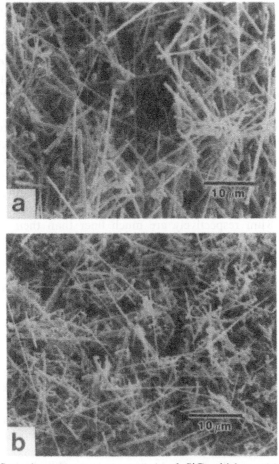

Figure 2.18 Scanning electron micrograph of SiC whiskers: (a) Silar-SC-9; (b) SCW-1. (Source: Homeny and Vaughn, 1987.)

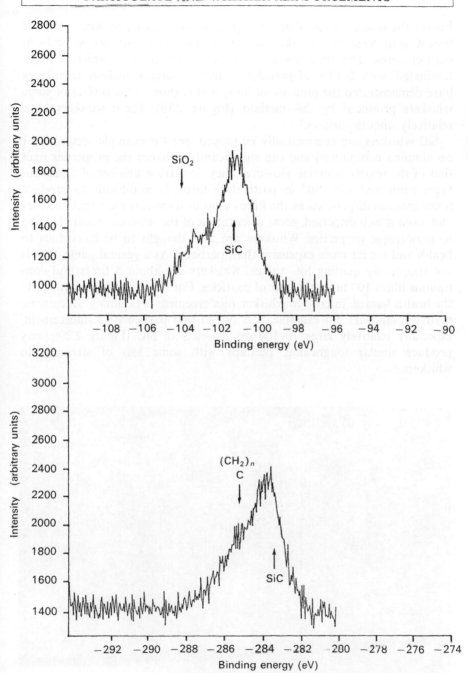

Figure 2.19 Electron spectroscopy for chemical analysis (ESCA) of SiC Silar-SC-9 whiskers showing some SiO₂ and C at the surface. (Source: Homeny and Vaughn, 1987.)

Finally the residue is heated at 800 °C to remove free carbon. About 10% of the SiC is in the form of whiskers and the remaining as particles, generally in platelet form. The products may be separated to give whiskers 'contaminated' with 5–10% of particles. Sensitive surface analysis techniques have demonstrated the presence of SiO_2 and carbon at the surface of some whiskers produced by this method (Figure 2.19). These whiskers have relatively smooth surfaces.

SiC whiskers are commercially employed (see for example section 4.4.1 on alumina composites) and can significantly enhance the properties over that of the matrix material. However they do have a number of disadvantages compared with SiC in particulate form. It is difficult to produce homogeneous dispersions as the fibres tend to form entwined agglomerates and, even if well dispersed, some orientation of the whiskers occurs leading to anisotropic properties. Whiskers are also thought to be hazardous to health and are far more expensive than particles. As a general guide, that is not specifically quoting SiC values, whiskers are about 5 times and continuous fibres 100 times the cost of particles. For these reasons, particularly the health hazard, interest in whisker reinforcement is declining. In ceramic matrix composites, for example, it is hoped that particulate reinforcement, especially relatively large 10–100 μm platelets of SiC (Figure 2.20), may produce similar toughening, perhaps with some loss of strength, to whiskers.

50μm

Figure 2.20 Scanning electron micrograph of SiC platelets. (Courtesy A. Selcuk.)

2.7 REINFORCEMENT–MATRIX INTERFACE

In our discussion of reinforcements it has been mentioned that sometimes a surface treatment is carried out to achieve the required bonding to the matrix. The reasons why the reinforcement–matrix bonding is so important have been stated in Chapter 1 but, because of their significance, the main points will be repeated here.

The load acting on the matrix has to be transferred to the reinforcement via the interface. Thus reinforcements must be strongly bonded to the matrix if their high strength and stiffness are to be imparted to the composite. The fracture behaviour is also dependent on the strength of the interface. A weak interface results in a low stiffness and strength but high resistance to fracture, whereas a strong interface produces high stiffness and strength but often a low resistance to fracture, i.e., brittle behaviour. Other properties of a composite, such as resistance to creep, fatigue and environmental degradation, are also affected by the characteristics of the interface. In these cases the relationship between properties and interface characteristics are generally complex.

The interface is important whether the reinforcement is in the form of continuous fibres, short fibres or whiskers or particles, although the exact role of the interface may differ with the type of reinforcement. For example in Chapter 10 we will see that the stress distribution along the interface is critical in determining the reinforcement efficiency of short fibre composites and the details of failure in the vicinity of the ends of broken continuous fibres.

In some cases a distinct phase, produced by a reaction between the matrix and the reinforcement, exists at the reinforcement–matrix interface. In other instances the interface can be viewed as a planar region of only a few atoms in thickness across in which there is a change in properties from those of the matrix to those of the reinforcement. Thus at the interface there is usually a discontinuity in chemical nature, crystal and molecular structure, mechanical properties and other properties. The characteristics of the interface are determined by the discontinuity in properties and are therefore specific to each reinforcement–matrix combination. The surface roughness of the reinforcement has also to be considered.

2.7.1 Wettability

Interfacial bonding is due to adhesion between the reinforcement and the matrix and to mechanical keying (see section 2.7.2). Clearly, for adhesion to occur during the manufacture of a composite the reinforcement and the matrix must be brought into intimate contact. At some stage in composite manufacture the matrix is often in a condition where it is capable of flowing and its behaviour approximates to that of a liquid. A key concept in this context is *wettability*. Wettability defines the extent to which a liquid will spread over a solid surface. Good wettability means that the liquid (matrix) will flow over the reinforcement covering every 'bump' and 'dip' of the rough surface of the reinforcement and displacing all air.

Wetting will only occur if the viscosity of the matrix is not too high and if wetting results in a decrease in the free energy of the system. Let us consider the latter in more detail by studying a thin film of liquid (matrix) spreading over the solid (reinforcement) surface. All surfaces have an associated energy and the free energy per unit area of the solid–gas, liquid–gas and solid–liquid interfaces are γ_{SG}, γ_{LG} and γ_{SL} respectively. For an increment of area dA covered by the spreading film, extra energy is required for the new areas of solid–liquid and liquid–gas interfaces. This extra energy is $(\gamma_{SL}\,dA + \gamma_{LG}\,dA)$, whereas $\gamma_{SG}\,dA$ is recovered as the solid surface is covered. For the spreading of the liquid to be spontaneous it has to be energetically favourable and therefore we must have

$$\gamma_{SL}\,dA + \gamma_{LG}\,dA < \gamma_{SG}\,dA.$$

Dividing by dA gives

$$\gamma_{SL} + \gamma_{LG} < \gamma_{SG}.$$

From this relationship the *spreading coefficient* SC is defined as

$$SC = \gamma_{SG} - (\gamma_{SL} + \gamma_{LG}), \qquad (2.1)$$

which has to be positive for wetting. It follows from equation 2.1 that if γ_{SG} is similar to or less than γ_{LG} then wetting will not occur. Thus an epoxy resin with γ_{LG} of about $40\,\text{mJ/m}^2$ might be expected to readily wet alumina ($\gamma_{SG} \sim 1100\,\text{mJ/m}^2$) but not polyethylene which has a low value of γ_{SG} of about $30\,\text{mJ/m}^2$.

The above analysis demonstrates the significance of the relative values for the surface energies involved in the wetting process but does not yield a parameter that specifies intermediate conditions of wetting. Figure 2.21 illustrates an example of a drop of liquid which has been allowed to reach

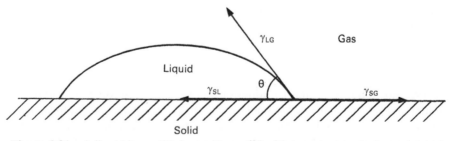

Figure 2.21   A liquid in equilibrium with a solid with a contact angle θ. γ_{SL}, γ_{LG} and γ_{SG} are the surface energies (surface tensions) of the solid–liquid, liquid–gas and solid–gas interfaces respectively.

equilibrium and has partially wet the solid. The free energy of an interface is measured in J/m^2 and can be shown to be equal to the surface tension, which has units of force per unit length (N/m). Therefore, as we are at equilibrium, the forces can be resolved horizontally to give.

$$\gamma_{SG} = \gamma_{SL} + \gamma_{LG} \cos \theta$$

where θ is called the *contact angle*. Rearranging gives

$$\cos \theta = \frac{(\gamma_{SG} - \gamma_{SL})}{\gamma_{LG}}. \tag{2.2}$$

θ may be used as a measure of the degree of the wettability. For a contact angle of 180°, the drop is spherical with only point contact with the solid and no wetting takes place. At the other extreme, $\theta = 0$ gives perfect wetting. For intermediate values of θ, i.e., $0° < \theta < 180°$ the degree of wetting increases as θ decreases. Often it is considered that the liquid does not wet the solid if $\theta > 90°$.

2.7.2 Interfacial bonding

Once the matrix has wet the reinforcement, and is therefore in intimate contact with the reinforcement, bonding will occur. A number of different types of bond may be formed. Furthermore, for a given system more than one bonding mechanism may be operative at the same time, e.g., mechanical and electrostatic bonding, and the bonding mechanism may change during the various production stages or during service, e.g., electrostatic bonding changing to reaction bonding. The type of bonding varies from system to system and depends on fine details such as the presence of surface contaminants or of added surface active agents (coupling agents). Let us now discuss the main features of the different bonding mechanisms.

(a) Mechanical bonding

A mechanical *interlocking* or *keying* of two surfaces, as shown in Figure 2.22(a), can lead to a reasonable bond. Clearly the interlocking is greater, and hence the mechanical bonding more effective, the rougher the interface. Also any contraction of the matrix onto the reinforcement is favourable to bonding.

The mechanical bond is most effective when the force is applied parallel to the interface, in other words, the shear strength may be considerable. On the other hand, when the interface is being pulled apart by tensile forces the strength is likely to be low unless there is a high density of features (designated A in Figure 2.22(a)) with re-entrant angles. In most cases a purely mechanical bond is not encountered and mechanical bonding operates in conjunction with another bonding mechanism.

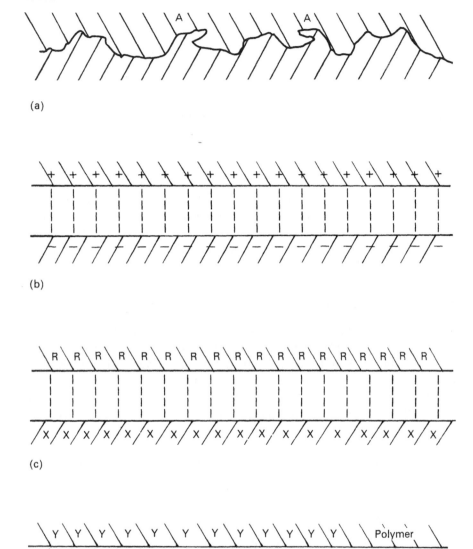

Figure 2.22 Schematic diagrams of the interfacial bonding mechanisms: (a) mechanical bonding (features designated A have a re-entrant angle); (b) electrostatic bonding; (c) chemical bonding (R and X represent compatible chemical groups); (d) chemical bonding as applied to a silane coupling agent; (e) reaction bonding involving polymers; (f) interfacial layer formed by interdiffusion.

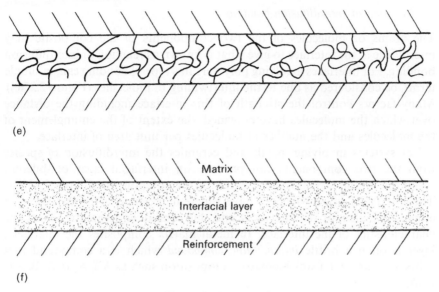

(e)

(f)

Figure 2.22 *(Contd.)*

(b) Electrostatic bonding

Bonding occurs between the matrix and the reinforcement when one surface is *positively charged* and the other *negatively charged* (Figure 2.22(b)). This leads to an electrostatic attraction between the components of the composite which will depend on the difference in charge on their surfaces. Electrostatic interactions are short range and are only effective over small distances of the order of atomic dimensions: the reader will therefore appreciate that it is essential that the matrix and reinforcement are in intimate contact and that surface contamination and entrapped gases will decrease the effectiveness of this bonding mechanism.

(c) Chemical bonding

In the context of composite science, chemical bonding is the bond formed between chemical groups on the reinforcement surface (marked X in Figure 2.22(c)) and compatible groups (marked R) in the matrix. Not surprisingly the strength of the chemical bond depends on the number of bonds per unit area and the type of bond.

It is thought that chemical bonding may account for the success of some coupling agents. For example *silanes* are commonly employed for coupling the oxide groups on a glass surface to the molecules of a polymer matrix. At one end (A) of the silane molecule a hydrogen bond forms between the oxide (silanol) groups on the glass and the partially hydrolyzed silane, whereas at the other end (B) it reacts with a compatible group in the polymer (Figure 2.22(d)).

(d) Reaction or interdiffusion bonding

The atoms or molecules of the two components of the composite may interdiffuse at the interface to give what is known as reaction or interdiffusion bonding. For interfaces involving polymers this type of bonding can, in simple terms, be considered as due to the intertwining of molecules (Figure 2.22(e)). Many factors control the strength of this interface including the distance over which the molecules have entwined, the extent of the entanglement of the molecules and the number of molecules per unit area of interface.

For systems involving metals and ceramics the interdiffusion of species from the two components can produce an interfacial layer of different composition and structure from either of the components (Figure 2.22 (f)). The interfacial layer will also have different mechanical properties from either the matrix or reinforcement and this consequently greatly affects the characteristics of the interface. In metal matrix composites the interfacial layer is often a brittle *intermetallic compound* which is a compound that exists at or around a stoichiometric composition such as AB, A_2B, A_3B, e.g., $CuAl_2$.

One of the main reasons why interfacial layers are formed is that ceramic matrix and metal matrix composite production invariably involve high temperatures. Diffusion is rapid at high temperatures as the rate of diffusion, or the *diffusion coefficient*, D_d, increases exponentially with temperature according to the Arrhenius-type equation

$$D_d = D_o \exp(-Q_d/RT), \qquad (2.3)$$

where Q_d is the activation energy for diffusion, D_o is a constant, R is the gas constant and T is the temperature. Temperature has a marked effect on the diffusion coefficient; substituting a typical value for Q_d of 250 kJ/mol into equation 2.3 it can be shown that D_d is 2×10^{34} greater at 1000 °C than at room temperature! The extent of the interdiffusion x, and hence the thickness of the reaction layer, depends on time, t, as well as temperature and is given by the approximate relationship

$$x = (D_d t)^{1/2}. \qquad (2.4)$$

An interface can change during service. In particular interfacial layers can form during service at elevated temperature. In addition, previously formed layers may continue to grow and complex multilayer interfaces may develop.

2.7.3 Methods for measuring bond strength

In view of the major role played by the interface in determining some of the most important features of the mechanical behaviour of composites, the reader will not be surprised to learn that numerous test methods have been employed to measure interfacial strength. Unfortunately it turns out that there is no simple, reliable method for determining this quantity. Besides the inherent problems associated with the different techniques, the results may be difficult to interpret because the failure induced during the test may not be *adhesive*, i.e., the two components may not have separated at the interface. The failure may take place close to the interface but in the reinforcement or in the matrix; this is a *cohesive* failure. The situation is further complicated if an interfacial layer of material C has formed between components A and B. In this situation there are two interfaces at which it is possible to have adhesive failure, namely at A–C and C–B, and three materials (A, B and C) which could fail cohesively.

It can be argued that, from the practical point of view, it does not matter whether failure during testing is adhesive or cohesive as long as the strength of the 'weak link' at the interface is being measured. This argument may be valid provided it is possible to ensure that the testing conditions reproduce the service conditions, but even then inadequate information on the type and strength of interfaces will undoubtedly limit our knowledge and the development of composites.

(a) Single fibre tests

The most direct test involves pulling a partially embedded single fibre out of a block of matrix material (Figure 2.23(a)). This test, although simple in principle, is difficult to carry out especially for thin brittle fibres. However if successful, from the resulting tensile stress versus strain plot the shear strength of the interface and the energies of debonding and pull-out may be obtained. These energies are discussed in more detail in Chapter 11 which considers toughening mechanisms.

The interfacial shear strength τ_I may also be evaluated using a specimen consisting of a block of matrix material with a single, completely embedded short fibre which is accurately aligned longitudinally in the centre of the specimen (Figure 2.23(b)). On testing in compression, shear stresses are set up at the ends of the fibres as a consequence of the difference in elastic properties of the fibre and matrix (see Chapter 10). The shear stresses eventually lead to debonding at the fibre ends and τ_I may be evaluated from the compressive stress σ_C at which debonding occurs as

$$\tau_I \sim 2.5\sigma_C.$$

The problem with this test is in determining the onset of debonding and hence σ_C. This is usually done by visual observation which means the test is only really suitable for transparent matrices.

Debonding induced by tensile stresses may be achieved if a curved neck specimen with a single continuous fibre is tested in compression (Figure 2.23(c)). With this test geometry, transverse tensile stresses, perpendicular

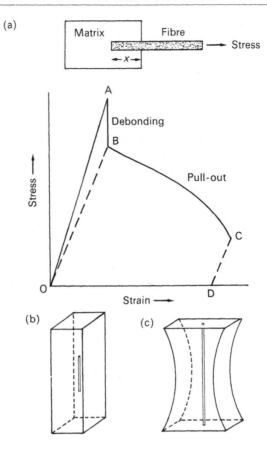

Figure 2.23 Single fibre tests: (a) fibre pull out test; (b) compression test for interfacial shear strength; (c) compression test for interfacial tensile strength.

to the interface, arise because of the differences in the Poisson's ratio of the fibre and matrix. At a compressive stress of σ_C the tensile strength σ_I of the interface is reached and tensile debonding occurs. σ_I is proportional to σ_C

$$\sigma_I = C\sigma_C$$

where C is a constant which depends on Poisson's ratio and Young's modulus of the fibre and matrix. As for the previous tests it is difficult to determine the onset of debonding and hence σ_C. Also both compression tests require accurate alignment of the fibre.

An alternative method is to carry out a tensile test on a dumbell-shaped specimen containing a single fibre. Application of the load causes the fibre to fragment; the limiting final fragment length is the critical length (Chapter 10), from which the interface shear strength may be calculated.

(b) *Bulk specimen tests*

We can acquire information on the interfacial strength using bulk composite specimens and modifications to conventional mechanical property tests such as torsion, tensile and three-point bend tests. For present purposes it is inappropriate to discuss all of these and the three-point bend and the Iosipescu tests have been selected as commonly used examples of this type of test.

Of these the *three-point bend* is the simplest and the most widely employed. The three-point bend geometry and the variation in the shear and tensile stresses in the loaded specimen are illustrated in Figure 2.24. At a given load P, the maximum tensile stress σ, which is parallel to the specimen length, lies on a line on the surface below the centre loading point (Figure 2.24(b)) and is given by

$$\sigma = \frac{3}{2} \frac{(PS)}{(BD^2)},$$ (2.5)

where B, D and S are defined in Figure 2.24(a); D and B are the height (thickness) and breadth of the specimen respectively, and S is the span between the outer loading points.

Using this testing arrangement with an aligned composite with the fibres perpendicular to the specimen length (Figure 2.24(d)) enables the tensile strength σ_1 of the interface to be evaluated by substitution of the failure load into equation 2.5.

We have to consider the shear stresses in a specimen if we wish to determine the interfacial shear strength. The shear stress has a maximum at the midplane of the specimen (Figure 2.24(c)) of

$$\tau = \frac{3}{4} \frac{P}{BD}.$$ (2.6)

Combining this equation with equation 2.5, gives

$$\frac{\tau}{\sigma} = \frac{D}{2S}$$ (2.7)

from which it can be seen that the ratio of the maximum shear stress to the maximum tensile stress depends on S and D, i.e., specimen geometry. A short span and a thick specimen increases the ratio and therefore enhances the likelihood of a shear failure. It follows that a short, thick specimen of an aligned fibre composite with the fibres parallel to the length of the specimen (Figure 2.24(c)) will fail by shear at the fibre–matrix interface provided $\tau_1/\hat{\sigma}_{TC}$, where $\hat{\sigma}_{TC}$ is the tensile strength of the composite, is less than $D/2S$. This test configuration is known as a *short-beam bend test* or the *interlaminar shear strength (ILSS) test*. Unfortunately both the transverse tensile strength and the shear strength obtained from three-point bend tests are

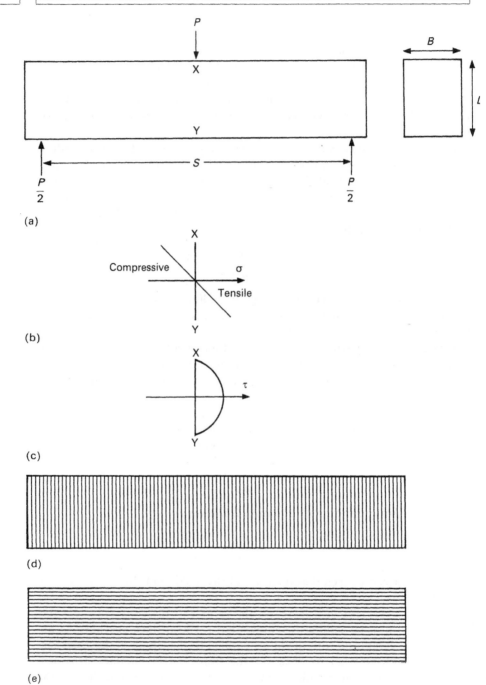

Figure 2.24 Three-point bend test: (a) specimen geometry; (b) tensile stresses; (c) shear stresses; (d) fibre alignment for determination of interfacial tensile strength; (e) fibre alignment for determination of interfacial shear strength.

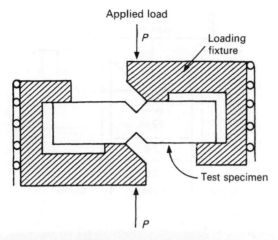

Figure 2.25 Diagram of Iosipescu testing fixture and specimen.

found to depend on the volume fraction of fibres and as a result cannot be considered as true values for the bond strengths.

A number of tests use specimens with a double-V-notch and the method attributed to *Iosipescu* falls into this category. The Iosipescu shear test (Figure 2.25) aims to achieve a state of pure shear in the mid-length of the specimen through the action of force couples which produce counteracting moments. The principle is that a constant shear force is induced through the mid-section of the specimen and that the moments exactly cancel at the mid-length of the specimen, thus producing a pure shear loading at that location. In practice however there is some non-uniformity of the shear stress distribution in the test section with probably small normal compressive stresses. Nevertheless this test geometry has been found to yield reliable shear strength data.

(c) Micro-indentation test

As the name suggests, this test employs a standard micro-indentation hardness tester. The advantages of the micro-indentation test are that the specimen is small and does not have to receive any special preparation except that one surface has to be polished to a finish suitable for microscopic examination. The indentor is loaded with a force P on to the centre of a fibre, whose axis is normal to the surface, and causes the fibre to slide along the fibre–matrix interface. The surface of the fibre is therefore depressed a distance u below the surrounding matrix surface (Figure 2.26). For a specimen of reasonable thickness the sliding distance is less than the specimen thickness and in these circumstances the interfacial shear strength τ_I is given by

$$\tau_I = P^2/4\pi u R^3 E_f,$$

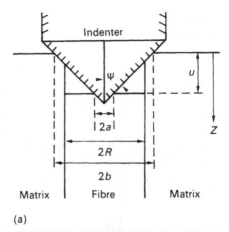

(a)

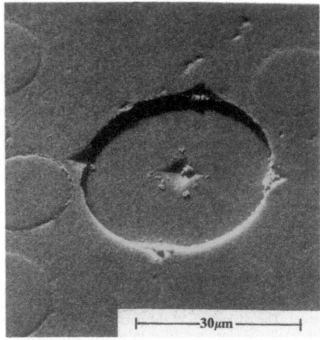

Figure 2.26 Micro-indentation test: (a) cross-sectional diagram of loaded indentor; (b) scanning electron micrograph of an indented SiC fibre in a lithium aluminosilicate glass-ceramic matrix. (Source: Marshall, 1984.)

where R and E_f are the radius and Young's modulus of the fibre, respectively. For a standard pyramid indentor the depression distance u may be calculated from

$$u = (b - a)\cot 74°,$$

where a is the half diagonal length of the indentation on the fibre and b is the half diagonal length of the indentation on the matrix surrounding the fibre. This method is particularly suitable for ceramic matrix composites and has been shown to be sensitive enough to detect changes in τ_I with heat treatment time and temperature for glass-ceramic matrix composites (Figure 2.27).

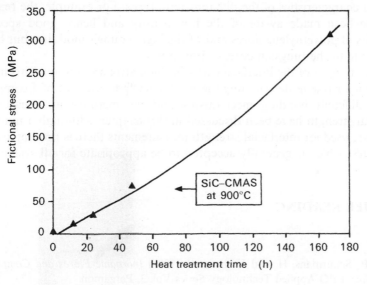

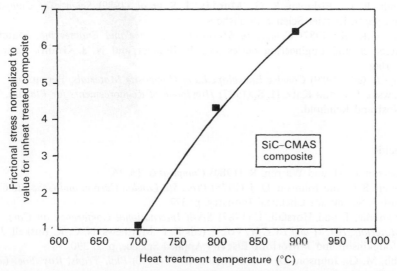

Figure 2.27 Micro-indentation data for the interfacial shear stress of SiC fibres in 'Silceram' glass-ceramic as a function of heat treatment: (a) time at 900 °C; (b) temperature for 12 h holding. (Source: Kim *et al.*, 1991.)

2.8 SUMMARY

In this chapter we have reviewed the principal types of fibre, such as carbon, glass and aramid, which are currently available for use as reinforcements. We pointed out that there are many fibre manufacturers and that it was not the intention to cover all the fibres of a given type but rather to demonstrate the main characteristics of the different fibre types. For example, the reader will have been made aware of the low density and hence good specific properties of polyethylene fibres and of the high Young's modulus, but high density due to the tungsten core, of boron fibres.

As the strength of the interface between the matrix and the reinforcement plays a major role in determining the mechanical performance of a composite, the different bonding mechanisms and the methods for measuring interfacial strength have been discussed in this chapter. Although a number of tests are used for interfacial strength measurements there is no simple and reliable test which is generally accepted to be appropriate for all systems.

FURTHER READING

General

Bracke, P., Schurmans, H. and Verhoest, J. (1984) *Inorganic Fibres and Composite Materials*, EPO Applied Technology Series Vol. 3, Pergamon.
Brandt, R. G., Fishman, S. G., Murday, J. S. *et al.* (1989) *Science of Composite Interfaces*, Elsevier Science Publishers.
Chawla, K. K. (1987) *Composite Materials – Science and Engineering*, Materials Research and Engineering Series (eds B. Ilschner and N. J. Grant), Springer-Verlag.
Kelly, A. (ed.) (1989) *Concise Encyclopedia of Composite Materials*, Pergamon.
Milewski, J. V. and Katz, H. S. (1987) *Handbook of Reinforcements for Plastics*, Van Nostrand Reinhold.

Specific

Anderson, C. H. and Warren, R. (1984) *Composites*, **15**, 16.
Bennett, S. C. and Johnson, D. J. (1978) *Proc. 5th London Carbon and Graphite Conf.*, Vol. 1, Society for Chemical Industry, p. 377.
Dinwoodie, J. and Horsfall, I. (1987) *Sixth International Conference on Composite Materials*, ICCM and ECCM, Vol. 2, (eds F. L. Matthews, N. C. R. Buskell, J. M. Hodgkinson and J. Morton), Elsevier Applied Science, p. 2, 390.
Dobb, M. G., Johnson, D. J. and Saville, B. P. (1980) *Phil. Trans. Roy. Soc. Lond.*, **A294**, 483.
Hodgson, A. A. (ed.) (1989) *Alternatives to Asbestos – the Pros and Cons*, published for the Society of Chemical Industry by Wiley.
Homeny, J. and Vaughn, W. L. (1987) *MRS Bulletin*, **12**, 66.

Kim, H. S., Yong, J. A., Rawlings, R. D. and Rogers, P. S. (1991) *Mat. Sci. and Technol.*, **7**, 155.
Lin, S. S. (1992) *SAMPE Journal*, **28**, 9.
Marshall, D. B. (1984) *J. Amer. Ceram. Soc.*, **67**, C259.
Moreton, R., Watt, W. and Johnson, W. (1967) *Nature (London)*, **213**, 690.

PROBLEMS

2.1 Tensile tests were carried out on alumina and SiC fibres of density 3.3 Mg/m^3 and 2.6 Mg/m^3 respectively. The deformation in all tests was elastic up to failure of the fibres and the mean tensile strengths and strains to failure were: alumina, 1500 MPa and 0.4%, and SiC, 2300 MPa and 1.0%. Calculate the specific modulus and the specific strength of the two types of fibre.
2.2 Three-point bend specimens were produced from an aligned continuous fibre composite of 4 mm thickness. The dimensions of the specimens were $4 \times 10 \times 120$ mm (specimen A) and $4 \times 10 \times 50$ mm (specimen B). The span between the lower loading points was 10 mm less than the specimen length. Account for the following experimental observations:

(a) Specimen A failed by tensile fracture from a line midway between the lower loading points at a force of 1164 N.
(b) Specimen B failed by shear at the midplane at a force of 2670 N.

SELF-ASSESSMENT QUESTIONS

Indicate whether statements 1 to 9 are true or false.

1. All natural fibres are organic and all synthetic fibres inorganic.

(A) true
(B) false

2. Polyethylene has the lowest density of any readily available synthetic fibre.

(A) true
(B) false

3. Glass fibres are crystalline and the crystal size is about 20 µm.

(A) true
(B) false

4. Alumina fibres always have the α-alumina structure.

(A) true
(B) false

5. Boron fibres are produced by vapour deposition from boron trichloride on to a heated substrate.

   (A) true
   (B) false

6. The elastic properties of graphite are anisotropic.

   (A) true
   (B) false

7. Polyacrylonitrile is a well-established precursor for SiC.

   (A) true
   (B) false

8. Natural fibres such as hemp, jute and cotton consist of cellulose fibres in an amorphous matrix of cellulose and hemicellulose.

   (A) true
   (B) false

9. All interfaces between the matrix and the reinforcement are stable and will never change during service even at high temperatures.

   (A) true
   (B) false

For each of the statements of questions 10 to 17, one or more of the completions given are correct. Mark the correct completions.

10. Aramid

   (A) is a form of collagen,
   (B) may be viewed as nylon with extra benzene rings,
   (C) fibres are stretched and drawn to align the structure,
   (D) fibres maintain their properties to temperatures in excess of 1000 °C,
   (E) fibres are produced by chemical vapour deposition from nylon.

11. Glass fibres

   (A) are made from silica glass,
   (B) have the highest specific modulus of any fibres,
   (C) are produced by vapour deposition,
   (D) have a short-range network structure,
   (E) may be chopped to make a chopped strand mat.

12. Alumina fibres

   (A) are produced from an organic precursor,
   (B) are produced via a slurry,
   (C) are produced by melt spinning,
   (D) cannot be produced from debased alumina,

(E) have extremely high values for the specific strength,

(F) may be predominately δ-phase.

13. Graphite

(A) is a non-crystalline form of carbon,

(B) has an hexagonal crystal structure,

(C) has covalent bonding in the basal planes,

(D) has weak bonds in the basal plane,

(E) starts to oxidize in air at around 500 °C.

14. Production of carbon fibres from polyacrylonitrile

(A) gives a high yield of up to 50%,

(B) involves initial stretching to 500–1300%,

(C) involves stretching by 50% at about 3000 °C,

(D) gives a Young's modulus which increases with increasing temperature of graphitization,

(E) can result in what are known as 'high strain fibres'.

15. SiC fibres

(A) are produced by melt spinning,

(B) are produced by chemical vapour deposition,

(C) are produced by decomposition of a precursor,

(D) sometimes have a tungsten core,

(E) usually contain a significant amount of Al_2O_3.

16. This scanning electron micrograph (Figure 2.28) is of a fibre

(A) that occurs naturally in wood,

(B) produced by chemical vapour deposition,

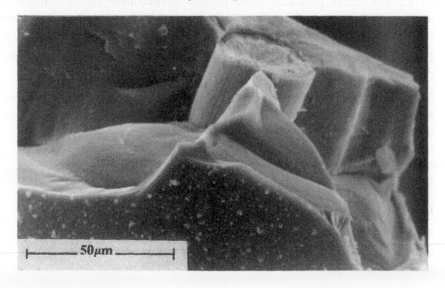

(C) produced from an organic precursor,
(D) produced on a core of another material,
(E) of asbestos,
(F) of polyethylene.

17. Wetting

(A) is energetically favourable when $\gamma_{SL} + \gamma_{LG} < \gamma_{SG}$,
(B) is energetically favourable when $\gamma_{SL} + \gamma_{LG} > \gamma_{SG}$,
(C) does not occur if the contact angle $\theta = 0°$,
(D) does not occur if the contact angle $\theta = 180°$,
(E) is the spreading of a liquid over a solid,
(F) is the flow of a liquid under gravity through a porous medium.

Each of the sentences in questions 18 to 22 consists of an assertion followed by a reason. Answer:

(A) if both assertion and reason are true statements and the reason is a correct explanation of the assertion,
(B) if both assertion and reason are true statements but the reason is not a true explanation of the assertion,
(C) if the assertion is true but the reason is a false statement,
(D) if the assertion is false but the reason is a true statement,
(E) if both the assertion and reason and are false statements.

18. Asbestos is a naturally occurring fibre with reasonable mechanical properties but its use is restricted *because* it can be a hazard to health.

19. Boron fibres are produced on a tungsten core and rarely on a carbon core *because* boron has a very low Young's modulus.

20. Carbon fibres produced from pitches have a low Young's modulus irrespective of the graphitization temperature *because* the fibres are non-crystalline.

21. SiC fibres are stiff *because* the fibres may be crystalline or amorphous.

22. An interface between the matrix and reinforcement which is bonded solely mechanically is weak in tension *because* strong covalent bonding usually results in strong, brittle materials.

ANSWERS

Problems

2.1 Specific modulus $\{(GPa)/Mg/m^3)\}$: alumina 113.6, silicon carbide 88.5. Specific strength $\{(MPa)/(Mg/m^3)\}$: alumina 454.5, silicon carbide 884.6.

2.2 (a) $\tau/\sigma = 0.018$; tensile strength of composite $= 1200$ MPa. (b) $\tau/\sigma = 0.05$; interlaminar shear failure at $\tau = 50$ MPa.

Self-assessment

1. B; 2. A; 3. B; 4. B; 5. A; 6. A; 7. B; 8. A; 9. B; 10. B, C; 11. A, D, E; 12. B, F; 13. B, C, E; 14. A, B, D, E; 15. B, C, D; 16. B, D; 17. A, D, E; 18. A; 19. C; 20. E; 21. B; 22. B.

<table>
<tr><td>**3**</td><td># Composites with
metallic matrices</td></tr>
</table>

3	# Composites with metallic matrices

3.1 INTRODUCTION

The production of fibres of boron and of silicon carbide in the 1960s and early 1970s enabled the reinforcement of light metals and particularly aluminium alloys to be considered seriously. Considerable research into boron fibre reinforced alloys was carried out in the USA in the early 1970s, leading to aerospace applications in the Space Shuttle and in military aircraft. Problems of chemical reaction between boron fibres and the matrix at temperatures above 600 °C restricted fabrication techniques to diffusion bonding of plasma-sprayed thin sheets. More recently the development of coatings to prevent fibre degradation and of more inert fibres like silicon carbide and alumina have enabled liquid metal process-ing routes to be developed (see section 3.2.2). Nevertheless, most metal matrix composites (MMCs) are still in the development stage, or the early stages of commercial production, and are not so widely established as polymer matrix composites.

The reader will learn in this chapter that metal matrix composites have many advantages over monolithic metals including a higher specific modulus, higher specific strength, better properties at elevated tempera-tures, lower coefficients of thermal expansion and better wear resistance. Because of these attributes MMCs are under consideration for a wide range of applications, as exemplified by the current and potential aero-space components listed in Table 3.1. However, on the debit side, their toughness is inferior to monolithic metals and they are more expensive at present.

In comparison with most polymer matrix composites, MMCs have certain superior mechanical properties, namely higher transverse strength

Table 3.1 MMCs for aerospace applications. (Source: Wei, 1992)

Matrix	Fibre	Application
Cu base	C	combustion chamber
	SiC	nozzle (rocket, space shuttle)
	W	NASP[a] heat exchanger
Fe base	W	tubing
Ni base and intermetallics	Al_2O_3 W	blades, discs
Ti base and intermetallics	SiC	housings, tubing
	TiB_2	blades, discs
	TiC	shafts, honeycomb
Al base	SiC	housings (pumps, instrumentation), mechanical connectors, satellite, structures
	Al_2O_3	fuselage
	C	structural members
	SiC	wings, blades
Mg base	Al_2O_3	structural members
Directionally solidified eutectics		blades, cable, NASP[a] heat exchanger, superconductors
Cu base	Nb	
Ni base	Carbide	
Ti base	Silicide	

[a]Hypersonic American National Aerospace Plane.

and stiffness, greater shear and compressive strengths, and better high temperature capabilities. There are also advantages in some of the physical attributes of MMCs such as no significant moisture absorption properties, non-inflammability, high electrical and thermal conductivities, and resistance to most radiations.

3.2 METAL MATRIX COMPOSITE PROCESSING

A wide variety of fabrication methods have been employed for MMCs but fortunately most of these can be conveniently classified into one of the following categories:

(a) solid state,
(b) liquid state,
(d) deposition,
(c) *in situ*,

and each of these will be discussed in turn.

3.2.1 Solid state processing

Over the last decade or so the potential for improved composite performance has been demonstrated primarily using solid state fabrication techniques although more recently liquid state processing has made considerable
strides. At the commencement of solid state processing the solid matrix
material is in the form of either particles or foil. In these forms there is a
large surface area of high energy solid–gas interface. Solid state processing
involves bringing the particles or foil into close contact with the reinforcement whence, on the application of a suitable combination of temperature
and pressure, the free energy of the system is reduced by the matrix
consolidating to give lower energy solid–solid interfaces. The actual mechanism by which material transport takes place during consolidation can
differ but it invariably involves diffusion. Methods using foil are usually
called *diffusion bonding* whereas those using particles tend to be referred to
as *powder metallurgy*.

The first stage in diffusion bonding is to sandwich a fibre mat, which has
the fibres held in place by a polymer binder, between two sheets of foil to
form a ply (Figure 3.1(a), (b)). In some cases this is followed by consolidation
of the ply. The plies are cut and stacked in the required sequence (Figure
3.1(c)). The stack is then hot pressed in a die to form the component (Figure
3.1(d)). Some titanium alloys are able to undergo very large plastic strains in
tension at elevated temperatures; they are said to be *superplastic*. For these
alloys the temperature of the diffusion bonding may be chosen in order to
benefit from the superplastic flow of the matrix around the fibres.

Diffusion bonding is an expensive process, generally limited to simple
shapes such as tubes and plates. More complicated structures require further
fabrication from these basic shapes, increasing the overall cost of the
composite and decreasing its attractiveness to the potential end user. On the
other hand the temperatures for solid state processing are lower than those
for liquid state processing and consequently the extent of undesirable
interface reactions can be negligible. The best known examples of MMCs
produced by diffusion bonding are titanium, nickel, copper and particularly
aluminium reinforced with boron fibres.

Powder metallurgy is especially suited for the production of discontinuous fibre, whisker or particulate reinforced metals. The components are
simply mixed and then pressed, often at an elevated temperature, to
consolidate. The mixing stage is critical if a homogeneous material with
good properties is to be produced. This stage is usually carried out under an
inert atmosphere in order to reduce the dangers of explosion and ingestion.
Also clean conditions are required because the large surface area of the
particles can lead to contamination; it is common practice to degas to
remove as much of the gases absorbed on the surface as possible and then to
can prior to consolidation. A range of mechanical working methods may be

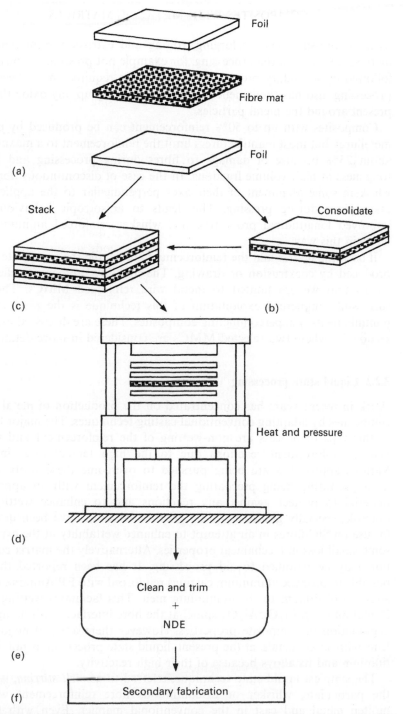

Figure 3.1 Diffusion bonding (a) starting components: fibre mat and sheets of foil; (b) form ply (sometimes the ply is consolidated); (c) plies are stacked; (d) hot press; (e) and (f) finishing.

used for pressing such as forging, rolling and extrusion. Sometimes two methods are used in the processing; for example hot pressing to consolidate followed by secondary processing to obtain the required shape. Secondary processing also has the beneficial effect of breaking up any oxide that was present around the metal particles.

Composites with up to 50% reinforcement can be produced by powder metallurgy but most manufacturers limit the reinforcement to a maximum of about 25% because of damage to fibres during processing and loss of toughness at high volume fractions. In the case of discontinuous fibres and whiskers some alignment of their axes perpendicular to the applied load takes place during pressing. This leads to anisotropic behaviour with improved longitudinal properties; SiC whisker reinforced aluminium behaves in this way.

If both the matrix and the reinforcement are ductile the composite can be produced by coextrusion or drawing. The ductility prerequisite effectively means that we are limited to metal wire reinforced MMCs. The most successful commercial exploitation of this technique is the production of multifilamentary superconducting composites. These are discussed further in section 3.5 where two selected MMCs are considered in more detail.

3.2.2 Liquid state processing

Work in recent years has concentrated on the production of metal matrix composites by adapting conventional casting techniques. The major barriers to this type of process are non-wetting of the reinforcement and adverse matrix–reinforcement reactions due to the high temperatures involved. Various approaches are being pursued to overcome these problems, the most promising being precoating the reinforcement with an appropriate material to protect against any reactions and to enhance wetting. For example, specially graded pyrolitic graphite coatings have been developed for use on SiC fibres in an attempt to enhance wettability at the expense of some small loss in mechanical properties. Alternatively the matrix composition may be modified to aid processing. It has been reported that it is possible to produce aluminium castings reinforced with FP Alumina fibre by addition of lithium to the aluminium melt. This facilitates wetting by the formation of an $Li_2O.5Al_2O_3$ spinel at the fibre interface, with no apparent degradation of composite properties. However there are still major problems with some metals; at the present, liquid state processing is not used for titanium and its alloys because of their high reactivity.

The simplest liquid state technique, referred to as *melt stirring*, is to mix the particulate, whisker or discontinuous fibre reinforcement with the molten metal and cast in the conventional manner. Even with stirring, uniform mixing is difficult to achieve because of differences in density between the molten matrix and reinforcement, although mixing is improved by allowing the melt to cool to a more viscous two-phase solid–liquid state for stirring. This modification of the melt stirring technique is known as *compocasting* or *rheocasting*. (To confirm that it is easier to mix particles uniformly into a fluid the more viscous the fluid try mixing sand (denser than water) or grass seeds (lighter than water) into water and then into more viscous custard!) There is a limit of approximately 20 vol. % since effective dispersion of the reinforcement becomes difficult above this level. The melt stirring process is limited to conventional casting alloys and the use of rheocasting is even more restricted to those alloys that have a wide solidification range over which the 'mushy' solid–liquid state exists.

In contrast to melt stirring and rheocasting, most liquid state processes involve the use of a pre-form which the liquid must infiltrate. If the fabrication is carried out at atmospheric pressure, then one is relying essentially on capillary action for infiltration. More efficient infiltration may be achieved by the use of a higher applied pressure than atmospheric. The applied pressure P, which has to overcome the forces due to the curvature of the meniscus at the infiltration front, is given by

$$P = \Sigma \gamma_{MG}/r_j,$$

where r_j are the principal radii of curvature at the molten metal front and γ_{MG} is the melt–atmosphere interfacial energy. The lower limit on r_j will be half the inter-fibre spacing. This equation is valid for axial infiltration of aligned preforms but for transverse infiltration and infiltration of, for example, less orderly arranged discontinuous fibre preforms the situation is more complex.

The pressure is usually obtained mechanically or through a gas but other methods have been studied including electromagnetic induction. Even when the process is carried out under pressure the pre-form cannot be too densely packed or infiltration is incomplete; consequently there is an upper limit of about 30 vol. % reinforcement. The pressure has to be chosen so that it is sufficient for infiltration but not so high that it damages the fibres or distorts the preform.

The pressure is applied mechanically in the technique known as *squeeze casting*, which is suitable for the production of small net shaped components. The squeeze casting technique may be understood by reference to the sequence of events (a) to (e) illustrated in Figure 3.2. First the pre-form is inserted into the die cavity (a) and a precise quantity of molten metal added (b). The pressure (70–100 MPa for aluminium alloys) is applied by means of a ram and forces the metal into the preform (c). In order to minimize the porosity the ram is not removed and the pressure released until solidification is finished (d). Finally the component is extracted from the die (e). Aluminium piston crowns locally reinforced with Saffil (a discontinuous alumina fibre – see Chapter 2) are currently made by

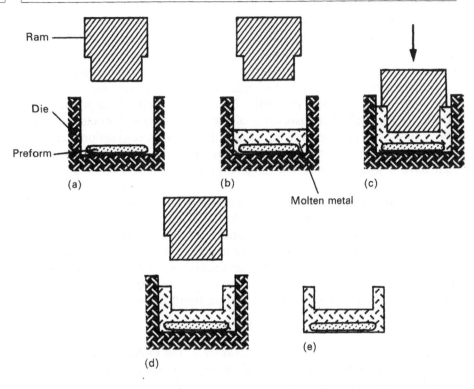

Figure 3.2 Squeeze casting: (a) insert preform into die cavity; (b) meter in a precise quantity of alloy; (c) close die and apply pressure; (d) remove ram; (e) extract component.

squeeze casting (Figure 1.9). A limitation on the component that can be produced by this process is imposed by the size of press that can be economically utilized.

The use of applied gas pressure to the melt effectively removes this problem of size. Another major advantage of using a gas is that it results in a more rapid process and therefore any fibre capable of withstanding contact with molten metal for relatively short periods of time may be used. This means that expensive fibre coatings may not be required and also a variety of different fibres, such as SiC, B, C and Al_2O_3, may be incorporated in a hybrid composite to realize a fully optimized design. Application of pressure by gas rather than by mechanical processes also reduces the extent of fibre breakage and misalignment.

Figure 3.3 is a diagram of liquid infiltration under a gas pressure. The perform is inserted into the die which is then closed (a). The chamber containing the molten metal is evacuated (b) and then a gas forced into the chamber under pressure. The pressure of the gas causes the molten metal to infiltrate the preform (c); the pressure is maintained for the short period of time required for solidification.

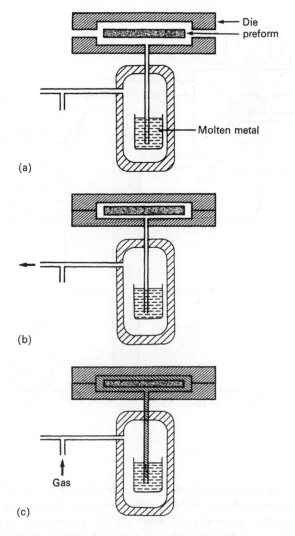

Figure 3.3 Liquid melt infiltration under gas pressure: (a) insert preform and close die; (b) evacuate air; (c) apply gas pressure and maintain during solidification.

3.2.3 Deposition

A deposition process which has considerable potential is *spray co-deposition* which is a modification of the Osprey deposition process. It involves atomizing a melt and introducing the reinforcement particles into the spray of fine metal droplets (Figure 3.4). The metal and the reinforcement particles are then co-deposited on to a substrate. The atomised metal exists as discrete droplets for short times, of the order of a few milliseconds, and the rapid solidification leads to a matrix with a fine microstructure and reduces

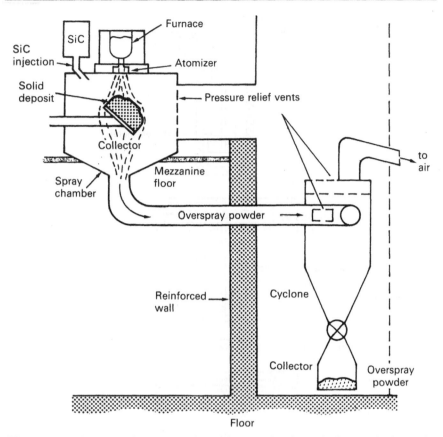

Figure 3.4 Diagram of spray co-deposition production of SiC particulate reinforced metal. (Source: Willis, 1988.)

the possibility of extensive chemical reaction. Control of atomization and of the particle feed enables an MMC with a uniform distribution of particles and of acceptable density (typically >95% theoretical density) to be produced at a reasonable rate. This technique has been mainly used for SiC particulate reinforcement of aluminium alloys. Some form of secondary processing is usually employed after co-deposition.

Many other deposition techniques have been tried for the fabrication of MMCs, namely chemical and physical vapour deposition (e.g., for tungsten), electroplating (mainly nickel matrices), sputtering and plasma spraying, but they have not been widely employed. In all cases the matrix is deposited throughout a fibre pre-form. An attractive feature of some of these techniques is that they operate at low temperature and, provided that the complete component is produced by deposition, reactions at the reinforcement–fibre interface are minimized. Sometimes however, deposition is simply used to produce a ply which is then consolidated by hot pressing as previously described for solid state processing. Another potential advantage of vapour deposition processes is that vapours are capable of penetrating high density preforms thus enabling the production of composites with a high volume fraction of reinforcement in excess of that typically attainable by liquid infiltration.

3.2.4 *In situ* processes

Unidirectional solidification of eutectic alloys can lead to a two-phase microstructure where one of the phases is in a lamellar or rod-like morphology and aligned approximately parallel to the direction of heat flow (Figure 3.5). The eutectic illustrated in Figures 3.5(a)–(c) is between two solid solutions of compositions α_e and β_e but for most *in situ* composites only the matrix is a solid solution and the reinforcement phase is an intermetallic compound or a carbide (see Figure 3.5 (d)). There is usually a preferred crystallographic growth direction and the interface between the phases is normally of low energy.

Unidirectional solidification is generally achieved by means of induction heating; the induction coil is moved up a bar of the eutectic alloy at a controlled rate. The parameters which are most important in determining the microstructure are the thermal gradient G at the liquid–solid interface and the growth rate R_G, i.e., the velocity at which the solid–liquid interface advances up the bar. The thermal gradient must be such that a planar solid–liquid interface is maintained and growth of the aligned structure is favoured rather than further nucleation of the phases. G is typically 10–100 K/cm. The inter-rod or inter-lamellar spacing λ_I can be shown to vary with growth rate according to

$$\lambda_I^2 R_G = C,$$

where C is a constant that depends on the particular eutectic system under consideration. λ_I is commonly in the range 0.1–10 μm.

The reader will recall the formal definition of a synthetic composite given in Chapter 1. Unidirectionally solidified eutectics are not true composite materials according to that definition, nevertheless their behaviour approximates to that of composites with aligned fibres and they have been widely investigated. For completeness they have been included in this chapter on MMCs.

3.3 INTERFACE REACTIONS

In the previous chapter it was pointed out that the matrix may interact with the reinforcement to give an interfacial layer. In MMCs interfacial layers may be formed during service but are more commonly formed during

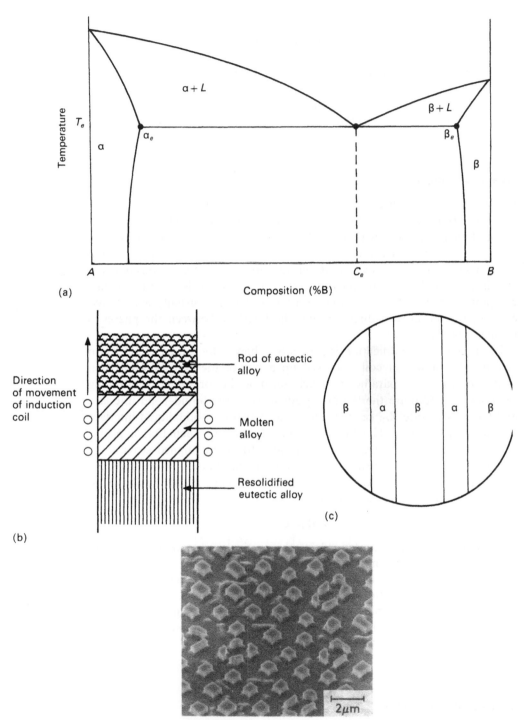

Figure 3.5 Production of *in situ* composite by unidirectional solidification: (a) eutectic phase diagram; (b) diagram showing induction heating used for unidirectional solidification; (c) alignment of the two-phase microstructure; (d) scanning electron micrograph of $\gamma'(Ni_3Al) - \alpha(Mo)$ composite. (Source: Funk and Blank, 1988.)

fabrication at high temperatures (Figure 3.6). According to equation 2.4 $(x = (D_d t)^{1/2})$ the extent of interdiffusion x increases with the square root of time t and we might therefore expect that interfacial layer thickness to follow the same time dependence. Plots of log(layer thickness) against log(time) are often linear in agreement with a power relationship, but the power is not always a half.

Figure 3.6 Transmission electron micrograph showing an interfacial layer of MgO in a magnesium alloy reinforced with eta-alumina (Safimax) fibres. The composite was produced by gas pressurised melt infiltration. (Courtesy S. Fox.)

The other important parameter in equation 2.4 is the diffusion coefficient D_d which increases exponentially with temperature. Not surprisingly it is found that high temperature liquid state processing is more likely to produce interfacial layers than lower temperature fabrication methods. Also D_d values for surface and grain boundary diffusion often give better agreement between predicted and observed layer thickness than the lower lattice diffusion values.

The high chemical reactivity of titanium and aluminium, the low density metals which are of major interest as matrices, can result in the formation of interfacial layers. However, whether or not interfacial layers are encountered depends on the composition of the matrix. For example there is negligible reaction between aluminium alloy matrices and most alumina fibres unless the matrix contains lithium or magnesium. In the case of Al–Li alloys, as previously mentioned in section 3.2.2., it has been suggested that the interface layer improves wetting.

Interfacial reactions are of concern as they can adversely affect the mechanical performance of an MMC. The effects of interfacial layer thickness on some mechanical properties of a titanium alloy reinforced with monofilament SiC, which had been coated with graphite and titanium diboride for use with titanium matrices, are presented in Figure 3.7. Both strength and toughness, as assessed by impact tests, are degraded by the presence of the interfacial layer. It is clear that, as a general rule, extensive interfacial reactions should be avoided if optimum mechanical performance is to be achieved with MMCs.

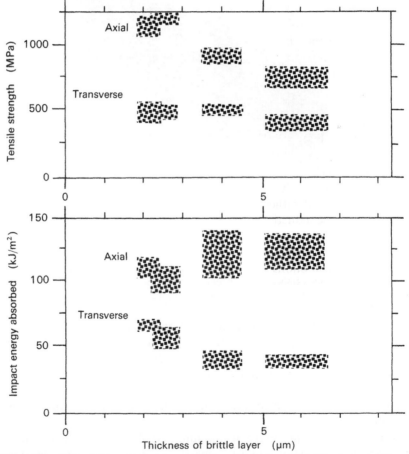

Figure 3.7 Effect of interfacial layer thickness on the mechanical properties of a Ti–6%Al–4%V alloy reinforced with 35% SiC monofilament fibres coated with C and TiB$_2$. (Source: Clyne and Flower, 1992.)

3.4 PROPERTIES OF MMCs

3.4.1 Physical properties

The advantages that MMCs have over polymer matrix composites and monolithic metals were mentioned in the introduction to this chapter. Selected physical properties will be discussed here to illustrate some of these advantages in more detail.

The *coefficient of thermal expansion* α of metals is large, consequently there are significant changes in dimensions with temperature which can lead to problems with metallic components with close tolerences. In contrast the coefficient of thermal expansion of ceramics is much lower and therefore it is not surprising that reinforcement with ceramic fibres or particles leads to a reduction in α. For example, α for SiC is about one-fifth and one-fourth of that for aluminium and magnesium respectively and, as shown in Figure 3.8, these metals when reinforced with SiC exhibit a smaller change in length with temperature than in the unreinforced state.

Similar reasoning applies to thermal and electrical conductivities: in both cases the conductivity of the reinforcement is less than that of the metallic matrix and hence composites have lower values for these parameters than the monolithic metal. Although there is a significant reduction in the conductivities on reinforcing metals, the values remain much greater than those for most polymers and polymer matrix composites (Table 3.2).

It is generally recognized that for a structural application a number of mechanical properties have to be considered when selecting the appropriate material. The same is also true of *functional* applications, i.e., applications that rely on physical rather than mechanical properties. For example the ratio of thermal conductivity to thermal expansion coefficient, which is called the *thermal deformation resistance,* is used as a parameter for evaluating materials in space applications demanding a high degree of dimensional stability under conditions of extreme temperature changes. Magnesium reinforced with carbon fibres has the unusual property of an extremely low thermal expansion over a wide range of temperatures – mainly because of the almost zero longitudinal expansion coefficient of the carbon fibres – and it has been possible to improve the thermal deformation resistance by over 6000% by reinforcing a magnesium alloy with carbon fibre.

3.4.2 Mechanical properties

(a) *Elastic properties*

Marked increases in the Young's modulus may be achieved by reinforcing metals. The improvement in the stiffness is particularly significant for those metals, such as aluminium and magnesium, that have a low Young's modulus. The data of Figure 3.9 illustrate this point; the stiffness of an

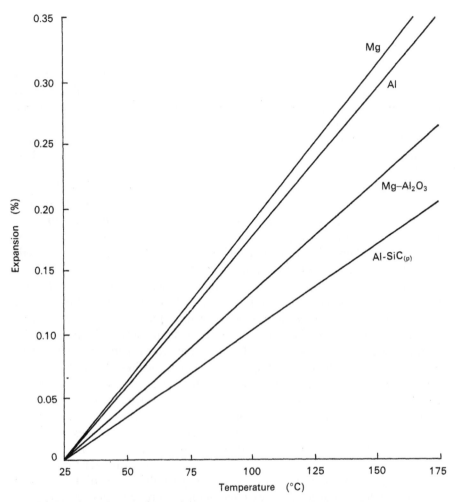

Figure 3.8   Comparison of the thermal expansion of metals and MMCs.

Table 3.2   Comparison of the room temperature thermal
conductivity of metals, polymers, MMCs and polymer matrix
composites

	Conductivity $(Wm^{-1} K^{-1})$
Aluminium	201
Al–15%SiC	140
Phenol formaldehyde	0.2
Phenolic–50% glass fibres	0.6
Epoxy	0.3
Epoxy–60%glass fibres	1.6
Epoxy–carbon fibres	5–100 plus[a]

[a]Depending on type and volume fraction of fibre.

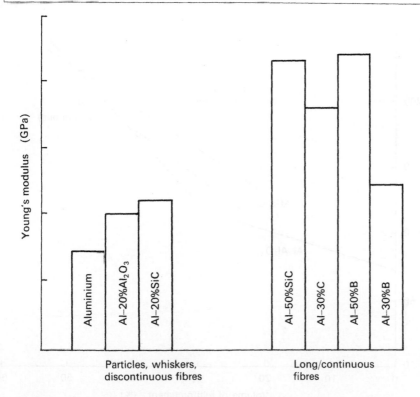

Figure 3.9 Effect of reinforcement on the Young's modulus of aluminium.

aluminium composite can be more than double that of aluminium. Not surprisingly composites becomes stiffer as the proportion of reinforcement increases (Figure 3.10). The moduli quoted so far for aligned continuous fibre composites have been *axial* or *longitudinal* moduli, i.e., measured parallel to the fibre axis. We will carry out a calculation later in the book (Chapter 8) to show that the modulus measured normal to the fibre direction (the *transverse modulus*) can be less than the longitudinal modulus. This has been observed for MMCs as demonstrated by the results shown in Figure 3.11.

(b) *Room temperature strength and ductility*

A high strength reinforcement and a strong reinforcement–matrix interface is needed to produce a high strength composite. A strong interface may not be obtained if there is an extensive interaction at the fibre–matrix interface during processing; this has been a major problem in many titanium based composites and has limited the strength (Figure 3.7) and fatigue resistance.

Problems, and some benefits, can arise from the large differences in the coefficients of thermal expansion between the matrix and the reinforcement. The mismatch in the expansion coefficients produces thermal stresses which

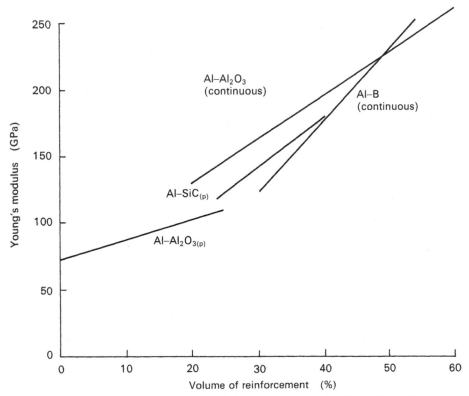

Figure 3.10 Effect of proportion of reinforcement on the Young's modulus of aluminium (a range of aluminium alloys were used for the matrix).

can be sufficient to deform the matrix plastically, and thus affect mechanical behaviour. Indeed the dislocation density resulting from the thermal stress induced deformation can reach $10^{14} \, m^{-2}$ and leads to strengthening in a similar manner to that obtained by cold working a monolithic alloy. However, if an MMC is thermally cycled it may not be possible to relieve all the stress by plastic deformation. In these circumstances micro-damage, such as cracks and/or voids at the reinforcement–matrix interface, has been observed and the mechanical properties are degraded.

As for stiffness, the strength is generally higher with increased volume fraction of reinforcement. However, because of processing difficulties, the strength is sometimes found to decrease at high levels of reinforcement as illustrated by the curve for Al–B in Figure 3.12. It can also be seen from the data of Figure 3.12 that, as would be expected, the longitudinal strength of a continuous fibre composite is better than that obtained with particulate reinforcement for a given volume of reinforcement.

When an aligned continuous fibre composite is tested in tension at an angle to the fibre axis a lower strength than the longitudinal strength is

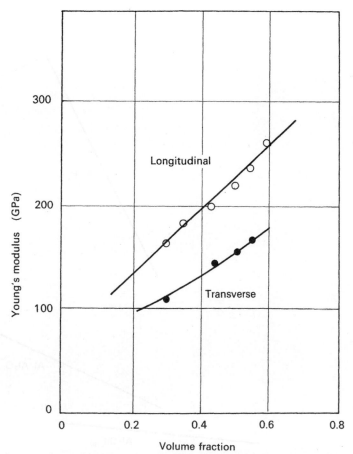

Figure 3.11 The difference in the longitudinal and transverse modulus for Al–Li reinforced with FP Al$_2$O$_3$. (Source: Champion *et al.*, 1978.)

observed. The lowest strength, called the *transverse strength*, is obtained when testing normal to the fibre axis (Figure 3.13). In most cases the transverse strength is largely dependent on the properties of the matrix and the fibre–matrix bonding rather than the properties of the reinforcement. As shown in Figures 3.13 and 3.14 the difference between the longitudinal and transverse strengths increases as the fibre content increases.

If we analyse the mechanical properties in terms of the specific strength and specific modulus, we find the performance of MMCs is superior to that of monolithic alloys as shown by the data for aluminium in Figure 3.15. This figure also shows the benefit of reinforcement by continuous fibres, provided the service stress is parallel to the fibre axis. The costs of fibres and composite manufacture are always large in the early stages of development and then tend to fall as output increases. It is therefore difficult to quote prices for these materials which are meaningful for any reasonable period of time. Nevertheless

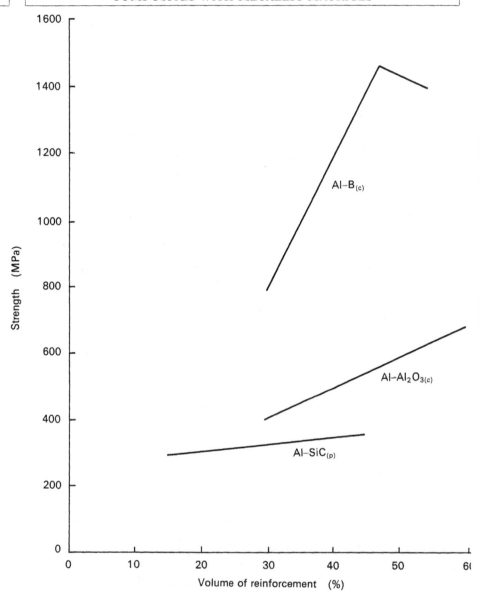

Figure 3.12 Effect of volume of reinforcement on the tensile strength of aluminium alloy matrix composites. 'c' and 'p' refer to continuous fibre and particulate composites respectively. The values for 'c' composites are longitudinal strengths.

it is apparent that the reinforcement and processing costs are likely to remain greater for continuous fibre reinforced composites than for whisker or particulate reinforced composites; this fact is also shown in Figure 3.15.

Metals are generally noted for their good ductility and toughness, but unfortunately both these properties are degraded by the incorporation of a

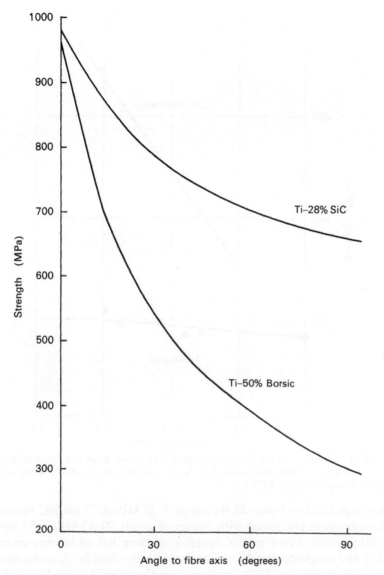

Figure 3.13 Graph showing the marked effect of angle between the tensile axis and the fibre axis on the strength of continuous fibre reinforced titanium alloy (Ti–6Al–4V). (Source: Metcalfe, 1974.)

reinforcement. Let us illustrate this by looking at the effect of reinforcement on aluminium and its alloys. The ductility of annealed aluminium exceeds 40% and although much lower values of around 10% are obtained from some alloys, especially if in the wrought condition, their ductilities remain superior to those of composites (Table 3.3).

The loss of toughness is also significant, the critical stress intensity factor

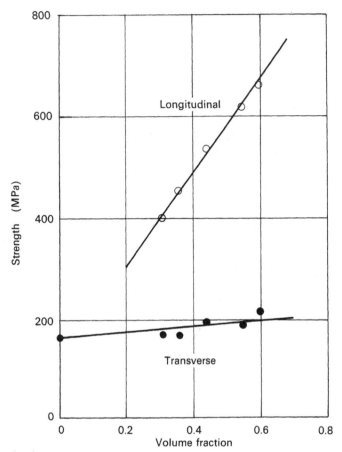

Figure 3.14   Graph showing the increasing difference between the longitudinal and transverse strengths with increasing FP Al_2O_3 content in an Al–Li alloy matrix. (Source: Champion *et al.*, 1978.)

K_{IC} being reduced to values in the range 5–25 MPa m$^{1/2}$ for SiC reinforced aluminium alloys compared with values of about 20–45 MPa m$^{1/2}$ for the monolithic alloys. Fibre–matrix interface reaction has an adverse effect on ductility and toughness but the reason for the low values for these parameters in the apparent absence of any significant reaction is not fully understood. The homogeneity of the reinforcement dispersion, the properties (particularly the surface properties) of the reinforcement, the cleanness of the matrix (for example the surface of powders can become contaminated), and the magnitude and inhomogeneity of the internal stress may all play some role.

(c) *Properties at elevated temperatures*

Provided there is no adverse reaction between the reinforcement and the matrix at elevated temperatures, and no micro-damage due to thermal cycling,

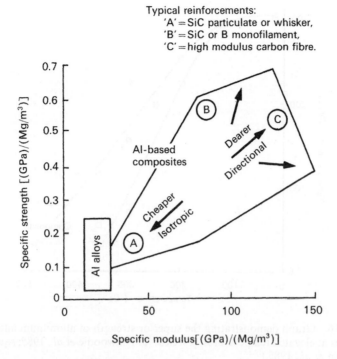

Figure 3.15 Range of specific strength and specific modulus values for aluminium alloy matrix composites compared with that for aluminium alloys. (Source: Feest, 1988.)

Table 3.3 Room temperature ductilities of composites of aluminium and its alloys containing 20 vol. % of either discontinuous alumina fibres or silicon carbide whiskers

Matrix	Reinforcement	Ductility (%)
Al	Alumina	4.0
Al–2.5%Mg	Alumina	3.3
Al–10%Mg	Alumina	1.3
Al–12%Si–1%Cu–1%Ni	Alumina	<1
6061 Al	Silicon carbide	7.0
7075 Al	Silicon carbide	4.2
2124 Al	Silicon carbide	4.0
5083 Al	Silicon carbide	0.6

the improvement in Young's modulus and strength observed at room temperature for composites is maintained, and even enhanced in some cases, at higher temperatures (Figures 3.16 and 3.17). The good high temperature performance of MMCs is not unexpected as the commonly employed ceramic reinforcements maintain their properties to higher temperatures than the matrix.

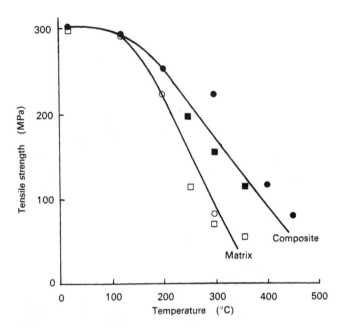

Figure 3.16 Graph demonstrating the superior strength of aluminium alloy matrix composites at elevated temperatures. (Circles – J. Dinwoodie *et al.*, 1985; squares – L. Ackermann *et al.*, 1985.)

The properties discussed so far have been obtained from short term tests. However in service a component may be under stress at an elevated temperature for long times and under these conditions we must consider the creep behaviour of the material. *Creep* is the increase in permanent strain with time which occurs with all materials under certain combinations of stress and temperature. The classical three-stage creep curve, which is applicable to metals, is shown in Figure 3.18(a); the linear portion of the creep curve (B to C in the figure) corresponds to *steady state* creep. The temperature T and stress σ dependence of the steady state creep rate $\dot{\varepsilon}$ for metals is often satisfactorily given by

$$\dot{\varepsilon} = \dot{\varepsilon}_o \, \sigma^n \exp(-Q_{cr}/RT), \qquad (3.1)$$

where Q_{cr} is the activation energy for creep, R is the universal gas constant, and $\dot{\varepsilon}_o$ and n are constants.

Let us now turn our attention to MMCs and see whether they exhibit the classical three-stage creep curve and if equation 3.1 is applicable. Creep curves of similar form are only found in continuous fibre MMCs if the matrix and reinforcement have similar melting points and consequently both creep at comparable rates. This effectively means the three-stage curves are only observed for metal fibre reinforced, and some *in situ*, composites.

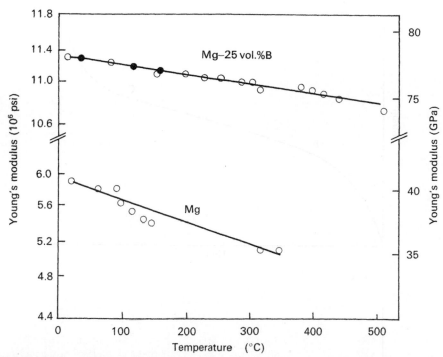

Figure 3.17 Graph of Young's modulus versus temperature for magnesium and magnesium reinforced with 25 vol. % B fibres. (Source: Huseby and Shyne, 1973.)

However even in these cases the steady state creep data are not well represented by equation 3.1; application of the equation gives values for both the activation energy Q_{cr} and the stress exponent n which apparently vary with stress and temperature.

The creep curves of continuous ceramic fibre reinforced metals are different (Figure 3.18(b)) because at the temperatures of interest for MMCs the creep rate of the fibres will be negligible in comparison with the creep rate of the matrix (the creep rates may differ by several orders of magnitude). It follows that the deformation of the fibres is essentially elastic and thus limited in extent. As a consequence, creep is hindered and instead of reaching steady state creep the creep rate of the composite falls progressively towards zero rate as illustrated in Figure 3.18(b). Since there is no steady creep regime, equation 3.1 is not applicable.

The complication of a creep rate that asymptotically approaches zero does not arise if the ceramic reinforcement is in particulate, discontinuous fibre or whisker form (or the composite is stressed transversely) as the matrix can then creep around the reinforcement. For example, for short fibres this leads to a finite steady state creep rate with the rate increasing with decreasing aspect ratio. Damage such as void formation at fibre ends contributes to the creep rate resulting in tertiary creep (region C–D in

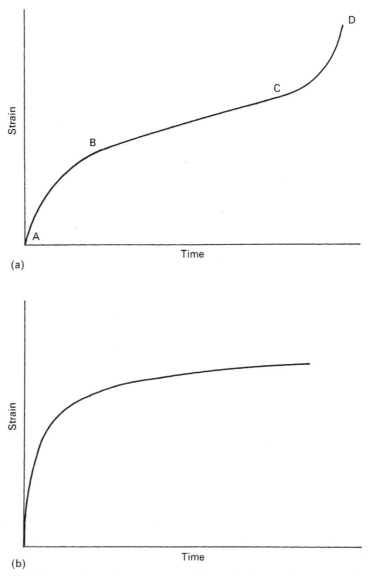

(a)

(b)

Figure 3.18 Creep curves: (a) classical three-state curve obtained from metals and some MMCs; (b) curve with decreasing creep rate obtained from MMCs reinforced with continuous ceramic fibres.

Figure 3.18(a)). A three-stage creep curve is therefore observed for these composites (Figure 3.19) although for some particulate reinforced systems the steady state creep region is limited and the creep curve is dominated by a monotonically increasing tertiary creep rate. Moreover, whereas the stress exponent might be around 4 for a metal the values for these composites are much higher (>15).

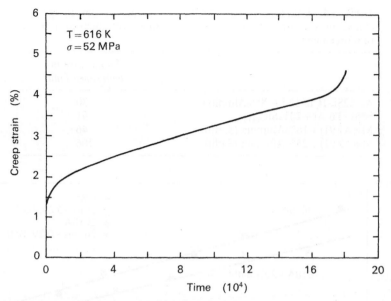

Figure 3.19 Three-stage creep curve for 6061 Al reinforced with 20wt.%SiC whiskers. (Source: Nieh, 1984.)

(d) *Fatigue resistance*

Fatigue is defined as the failure of a component under the repeated application of a stress smaller than that required to cause failure in a single application. In fatigue a crack is initiated and slowly grows under the action of the fluctuating stress until eventually failure occurs in a catastrophic manner with no gross distortion preceding the event.

Failure of MMCs under fluctuating or cyclic stresses has not been investigated to the same extent as the properties under so-called static loading discussed in the previous mechanical properties sections. In fatigue of MMCs the crack initiation sites are often internal defects such as large ceramic particles remaining from whisker production, unbonded clusters of reinforcement particles and undesirable brittle intermetallic compounds. Despite these crack initiation sites the reinforcement of a metal can lead to significant improvement in the fatigue resistance as shown by the increase in the stress needed to cause failure in 10^7 cycles (the 10^7 cycles endurance limit) presented in Table 3.4. However there are examples in the literature where the fatigue resistance has been degraded by reinforcing: Figure 3.20, which is a plot of alternating stress against number of cycles to failure (known as an *S–N* curve), has examples of the reinforcement producing (a) the more commonly encountered behaviour of a significant improvement and (b) a slight reduction in fatigue resistance.

Table 3.4 Fatigue data for MMCs in terms of percentage increase in the 10^7 endurance limit obtained on incorporating a reinforcement

	% increase in endurance limit
Al–12Si–1Cu–1Ni + 20%Alumina	30
6061–T6 Al + 20%SiC$_{(w)}$	91
Mg(AZ91) + 16%Alumina (Saffil)	46
Mg(AZ91) + 25%Alumina (Saffil)	106

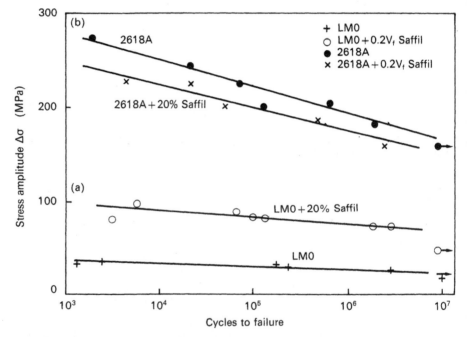

Figure 3.20 Fatigue (S–N) plots for matrix aluminium alloys and composites with 20% alumina (Saffil) fibres. (Source: Harris and Wilks, 1987.)

3.5 SOME COMMERCIAL MMCs

Two commercial systems will be discussed in more detail. They have been selected to illustrate the range of properties and applications of MMCs.

3.5.1 Multifilamentary superconductors

The superconducting constituent of these composites is the intermetallic compound Nb_3Sn, or occasionally V_3Ga. For an intermetallic superconductor to be commercially viable it must be possible to produce it in long lengths

with uniform properties and, for complex reasons associated with superconducting performance, it must be in the form of filaments or a thin tape. Both
forms are available but we shall concentrate on the former and a cross-section
of a multifilamentary superconducting composite is given in Figure 3.21.

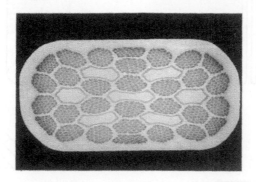

Figure 3.21 Multifilamentary superconducting composite with 41 070 filaments of
approximately 5 μm diameter: (a) cross-section; (b) matrix etched away to show the
filaments. (Courtesy J. A. Lee.)

Nb$_3$Sn is extremely brittle and cannot therefore be produced by conventional metal working processes, such as extrusion and wire drawing, of a
billet of the intermetallic. Instead a solid state processing route based on the
diffusion controlled reaction between niobium filaments and a copper–tin
(bronze) matrix is employed. This process is known as the *bronze route* and
is illustrated in Figure 3.22 (a)–(d). The first stage in the process is to drill
holes in a block of bronze and insert rods of niobium (Figure 3.22(a)). The
bronze block with the embedded niobium rods is swaged down to reduce
the cross-section of the niobium (Figure 3.22(b)). Sectioning of the swaged

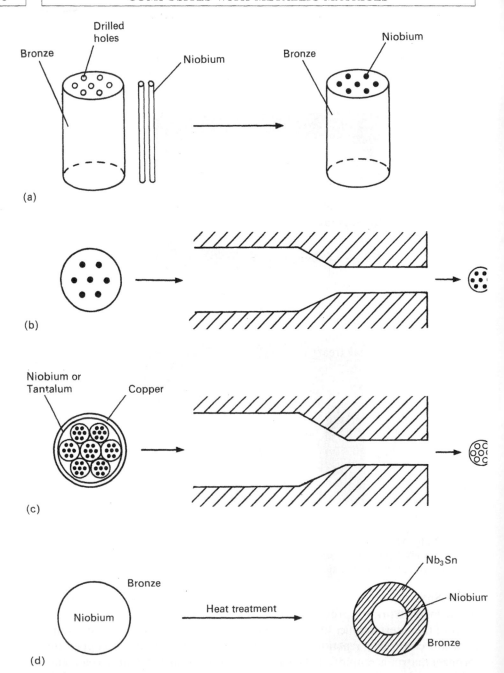

Figure 3.22 Schematic diagram of the production of multifilamentary supercon-ducting composite by the bronze route: (a) holes drilled in bronze block and niobium rods inserted; (b) swaging; (c) sectioning, rebundling, canning and final reduction; (d) heat treatment.

composite and rebundling are carried out to increase the number of niobium filaments in the cross-section. The rebundled composite is placed inside a copper can whose inner surface has a thin coating of tantalum or niobium. The presence of the copper in the final product increases the stability of the superconductor by minimizing local temperature rises. The canned, rebundled composite is drawn down to give a final niobium filament size of a few microns (Figure 3.22(c)). The composite wire is then heat treated in the temperature range 600–800 °C to allow diffusion of tin from the bronze into the niobium, and hence to form Nb_3Sn by a solid state reaction (Figure 3.22(d)). The niobium is not completely reacted and typically there will be a layer of Nb_3Sn of less than 2 μm thick on a core of unreacted niobium. The exact thickness will depend on the extent of interdiffusion and is therefore determined by the temperature (through its effect on D_d) and time of heat treatment in accordance with equation 2.4. The niobium or tantalum diffusion barrier prevents the tin from the bronze entering the copper. The Nb_3Sn has the A15 crystal structure and the layer is fine grained (grain size of less than 0.2 μm – see Figure 3.23). The superconducting properties are a function of the layer thickness and the grain size. In Figure 3.24 the optimum range for the thickness and grain size is superimposed on a plot of grain size against layer thickness for a series of multifilamentary superconductors where the heat treatment was varied.

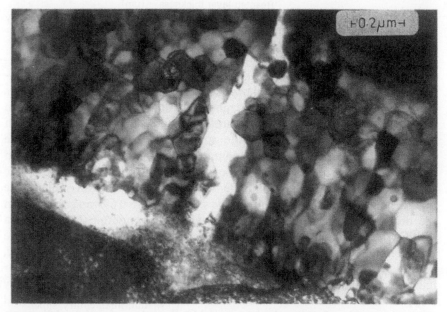

Figure 3.23 Transmission electron micrograph showing the fine grain size of the Nb_3Sn layer produced by the bronze route. Also shown is a crack in the Nb_3Sn layer which is stopped by the niobium core. (Source: West and Rawlings, 1979.)

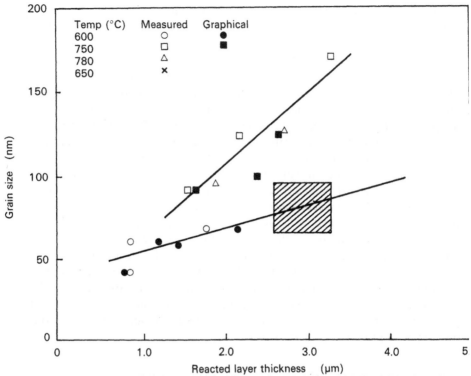

Figure 3.24   Plot of grain size against layer thickness for Nb$_3$Sn layer produced by the bronze route. Composites lying within the shaded area possess good superconducting properties. (Source: West and Rawlings, 1977.)

The major use of multifilamentary composites is as windings for superconducting magnets. The mechanical properties are important as the composite is stressed during assembly of the magnet and during service. The forces in superconducting magnets during operation are considerable and it has been found that the tensile stress on the composite can reach 100 MPa. The mechanical properties of multifilamentary superconductors are better than one might first expect for two reasons. First, the difference between the coefficients of thermal expansion of the bronze matrix and Nb$_3$Sn results in the intermetallic experiencing a compressive strain of up to 0.6% on cooling from the heat treatment temperature. Thus whereas monolithic Nb$_3$Sn has a ductility of less than 0.2%, the intermetallic layer does not fail in a composite until tensile strains of the order of $0.2 + 0.6 = 0.8\%$ are reached. Secondly, because of the toughness of the unreacted niobium core, the intergranular microcracks in the Nb$_3$Sn do not lead to immediate failure as they are initially stopped at the intermetallic-core interface (Figure 3.23). The mechanical properties are therefore largely determined by the volume fraction of filaments (Nb$_3$Sn + unreacted Nb), the layer thickness and the compressive strain.

3.5.2 Aluminium reinforced with silicon carbide particles

Aluminium and its alloys are light and are widely used in transport applications. There is considerable interest in producing low cost aluminium alloy based composites with improved properties over their monolithic counterparts. As a general rule both the material and fabrication costs are high for continuous fibre MMCs and hence many companies have concentrated in recent years on whisker, particulate or discontinuous fibre composites.

Silicon carbide is readily available as whiskers and particulates, the latter being less expensive and less of a hazard to health. For these reasons there are a number of manufacturers producing particulate reinforced aluminium alloys using a range of routes including casting, co-deposition and powder metallurgy. Some of the properties of these composites have already been presented, e.g., thermal expansion (Figure 3.8), Young's modulus (Figure 3.10) and strength (Figure 3.12), and not all of these will be discussed further.

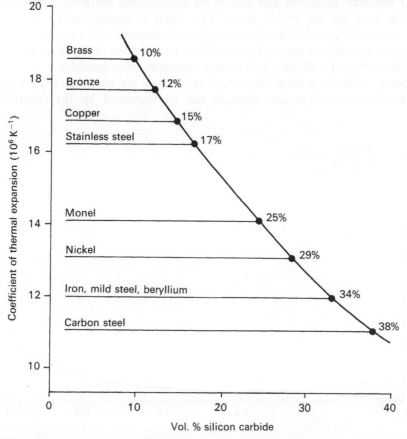

Figure 3.25 Graph of the coefficient of thermal expansion of aluminium–SiC$_{(P)}$ composites as a function of SiC content showing the matching with a range of metals (Source: Alcan).

The reader will recall that in section 3.4.1 it was pointed out that a ceramic reinforced metal has a lower coefficient of thermal expansion than the matrix metal. Consequently a series of aluminium alloy based composites are available with a range of values for the coefficient of thermal expansion, depending on the SiC content. As shown in Figure 3.25, this means an Al–SiC$_{(p)}$ composite (subscript p indicates particles) is available to match the expansion characteristics of many of the common metals and alloys.

It is interesting to compare the creep behaviour of aluminium reinforced with either SiC whiskers or particles. Whisker reinforcement leads to a three-stage creep curve as previously demonstrated in Figure 3.19. In contrast, particulate reinforced aluminium has a creep curve with a negligible steady state creep region and the curve is dominated by a marked monotonically increasing tertiary creep regime as shown in Figure 3.26. In the discussion on creep (section 3.4.2(c)) the reader will recall that it was stated that the stress exponent was greater for whisker and particulate reinforced metals than for the matrix metal. This point is exemplified by the data presented in Figure 3.27 for aluminium based composites. We can see that the two composites have similar high n-values (n is the slope of the plots) and that whiskers impart a slightly better creep resistance than particles.

Some of the aluminium alloys used as a matrix are heat treatable, i.e., the yield stress, hardness and strength can be increased by the controlled

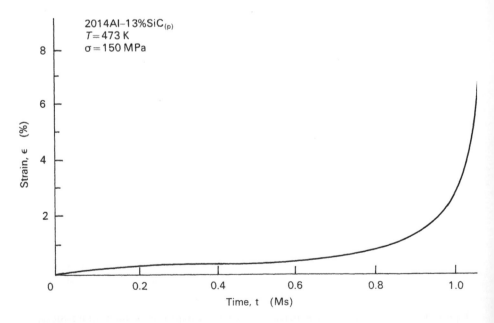

Figure 3.26 Creep curve for 2014Al–13%SiC$_{(p)}$ showing a monotonically increasing tertiary creep regime. (Source: McLean, 1990.)

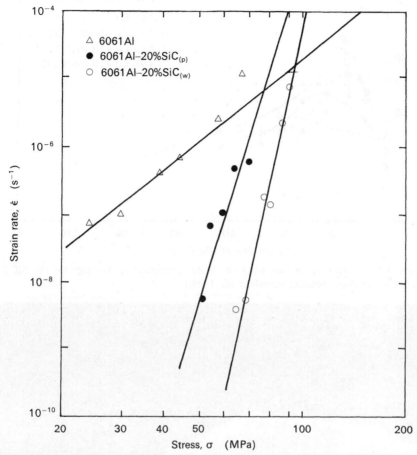

Figure 3.27 Effect of stress on the steady state creep rate for 6061Al and its composites. Note the higher stress sensitivity of the composites. (Source: Nieh, 1984.)

precipitation of a series of metastable phases. The details of the precipitation process depend critically on composition and heat treatment; generally the mechanical properties reach a maximum after a certain time at a given temperature and then fall. Commonly it is found that the precipitation process is enhanced in MMCs so that the peak in properties is reached after a shorter time of heat treatment. This behaviour is usually attributed to the dislocations produced by the thermal stresses in MMCs acting as nucleation sites for precipitation. A typical difference in response to heat treatment of an MMC and the monolithic matrix alloy is illustrated in Figure 3.28.

Many working processes can be used for forming aluminium–SiC$_{(p)}$ components. In some cases, due to the fine microstructure of the matrix,

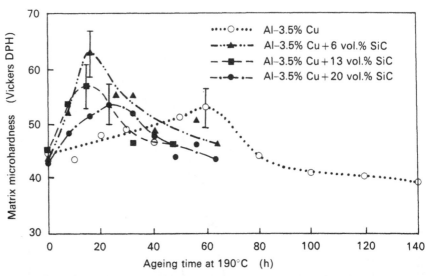

Figure 3.28 Ageing curves showing the faster precipitation kinetics for the SiC$_{(P)}$ reinforced samples. (Source: Suresh *et al.*, 1989.)

Figure 3.29 Aircraft panel produced by superplastic forming of an Al–SiC$_{(P)}$. (Courtesy British Aerospace.)

superplastic forming can be employed at a reasonable strain rate even though the SiC particles are not deforming superplastically. Figure 3.29 shows an aircraft panel produced by this method.

Welding is frequently used as the joining process for the assembly of metal structures. However conventional welding involves local melting and is therefore unsuitable for MMCs as the weld has a markedly different microstructure, and hence properties, from the parent composite. For this reason most MMC components are assembled by bolting. However for Al–SiC$_{(p)}$ MMCs *diffusion welding* and *friction welding*, both of which are solid state processes, show considerable promise.

3.6 SUMMARY

Metal matrix composites generally have higher values for specific modulus and specific strength than those for the unreinforced matrix, but this is achieved at the expense of toughness. The physical properties are also modified by the reinforcement and MMCs may be designed to have the appropriate values for, say, the coefficient of expansion and thermal conductivity, for a particular functional application.

Various fabrication methods have been described. It was seen that these usually involved high temperatures and consequently the matrix may react with the reinforcement to give an interfacial layer which can degrade properties. Interfacial layers may also be formed during high temperature service. However, provided there is no reaction, and no micro-damage because of differences in the coefficients of thermal expansion of the constituents, the high temperature static properties (Young's modulus and strength) and creep resistance of MMCs are good. Data on fatigue performance of MMCs are conflicting, but most reports suggest a significant improvement in fatigue resistance.

FURTHER READING

General

Begg, A. R. (ed.) (1992) *Metal Matrix Composites*, Arnold.

Bracke, P., Schurmans, H. and Verhoest, J. (1984) *Inorganic Fibres and Composite Materials*, EPO Applied Technology Series, Vol. 3, Pergamon Press.

Everett, R. K. and Arsenault, R. J. (eds) (1991) *Metal Matrix Composites. Vol. 1 – Processing and Interfaces; Vol. 2 – Mechanisms and Properties*, Academic Press.

Lee, J. A. and Mykkanen, D. L. (1987) *Metal and Polymer Matrix Composites*, Noyes Data Corporation, USA.

Lynch, C. T. and Kershaw, J. P. (1972) *Metal Matrix Composites*, CRC Press.

McLean, M. (1983) *Directionally Solidified Materials for High Temperature Service*, The Metals Society, London.

Taya, M. and Arsenault, R. J. (1989) *Metal Matrix Composites – Thermomechanical Behaviour*, Pergamon Press.

Terry, B. and Jones, G. (1990) *Metal Matrix Composites*, Elsevier Advanced Technology.

Specific

Ackermann, L., Moore, E. and Langman, C. A. J. (1985) *Proc. 5th. Int. Conf. on Composite Materials*, ICCM/V, TMS-AIME. Warrendale, USA.

Champion, A. R., Krueger, W. H., Hartman, H. S. and Dhingra, A. K. (1978) *Proc. 1978 Int. Conf. on Composite Materials*, ICCM/2, TMS-AIME, New York, p. 883.

Clyne, T. W. (1987) Fabrication and microstructure of metal matrix composites. *Proc. 6th Int. Conf. on Composite Materials*, Vol. 2. Elsevier, p. 1.

Clyne, T. W., and Flower, H. M. (1992) *Proc. 7th. World Titanium Conf*, San Diego, USA, TMS.

Dinwoodie, J., Moore, E. and Langman, C. A. J. (1985) *Proc. 5th Int. Conf. on Composite Materials*, ICCM/V, TMS-AIME, Warrendale, USA.

Feest, E. A. (1988) *Metals and Materials*, Institute of Metals, Vol. 4, p. 274.

Funk, W. and Blank, E. (1988) *Met. Trans. A*, **19A**, 987.

Harris, S. J. and Wilks, T. E. (1987) *Proc. 6th. Int. Conf. on Composite Materials*, Vol. 2, Elsevier, p. 113.

Huseby, I. and Shyne, J. (1973) *J. Powder Met.*, **9**, 91.

McLean, M. (1990) *Bull. Japan Inst. Metals*, **29**, 199.

Metcalfe, A. (1974). *Fibre Reinforced Titanium Alloys*, Composite Materials, Vol. 4, Academic Press, New York.

Nieh, T.(1984) *Metall. Trans.*, **15A**, 139.

Suresh, S., Christian, T. and Sugimura, Y. (1989) *Scripta Met.*, **23**, 1599.

Wei, W. (1992) *Metals and Materials*, **8**, 430.

West, A. W. and Rawlings, R. D. (1977) *J. Mat. Sci.*, **12**, 1862.

West, A. W. and Rawlings, R. D. (1979) *J. Mat. Sci.*, **14**, 1179.

Willis, T. C. (1988) *Metals and Materials*, **4**, 485.

PROBLEMS

3.1 Sketch typical creep curves for (a) a monolithic metal, (b) a continuous ceramic fibre reinforced metal and (c) a ceramic particulate reinforced metal.

The steady state creep rate of an aluminium alloy reinforced with 20%SiC particles when tested at 350 °C and at a stress of 50 MPa was $1.5 \times 10^{-7} \, \text{s}^{-1}$. Calculate the creep rate if the stress is increased to 60 MPa.

(The universal gas constant is 8.3 J/mol K and the stress exponent for creep is 17.)

3.2 Discuss the various liquid state processes for the production of metal matrix composites. In your answer make reference to the advantages and limitations of the processes.

SELF-ASSESSMENT QUESTIONS

Indicate whether statements 1 to 5 are true or false.

1. Powder metallurgy is commonly employed for the fabrication of MMCs, but only if the reinforcement is continuous fibre.

(A) true
(B) false

2. In the squeeze casting process molten metal is forced by mechanical pressure into a preform.

(A) true
(B) false

3. The electrical conductivity of an MMC is usually less than that of the matrix.

(A) true
(B) false

4. Although an MMC has a higher room temperature strength than the matrix the converse is true at elevated temperatures.

(A) true
(B) false

5. The superconducting properties of a multifilamentary superconductor are determined by the Nb_3 Sn layer thickness and grain size.

(A) true
(B) false

For each of the statements of questions 6 to 10, one or more of the completions given are correct. Mark the correct completions.

6. Rheocasting

(A) is a solid state technique,
(B) can only be employed for *in situ* composites,
(C) involves mixing the reinforcement with solid–liquid metal,
(D) is a modification of melt stirring,
(E) involves applying a mechanical pressure during casting.

7. *In situ* MMCs

   (A) are produced by squeeze casting,
   (B) are produced by unidirectional solidification,
   (C) are produced by spray co-deposition,
   (D) have an aligned microstructure,
   (E) usually have a two-phase eutectic microstructure.

8. The Young's modulus of an aligned continuous fibre metal matrix composite

   (A) increases with increasing volume fraction of fibre,
   (B) is independent of volume fraction of fibre,
   (C) is the same in the longitudinal and transverse directions,
   (D) is greater in the longitudinal direction,
   (E) is greater in the transverse direction.

9. The transverse tensile strength of an aligned continuous fibre composite

   (A) is obtained when testing normal to the fibre axis,
   (B) is obtained when testing parallel to the fibre axis,
   (C) is the lowest tensile stength,
   (D) is the highest tensile strength,
   (E) depends mainly on the properties of the matrix and of the fibre–matrix interface,
   (G) depends mainly on the properties of the fibres.

10. The creep curve of a metal reinforced with continuous ceramic fibres

   (A) asymptopically approaches a zero creep rate,
   (B) exhibits a marked monotonically increasing tertiary creep regime,
   (C) is a classical three-stage creep curve,
   (D) is identical to that of a continuous *in situ* composite,
   (E) is a consequence of significant creep of the ceramic fibres.

Each of the sentences in questions 11 to 14 consists of an assertion followed by a reason. Answer:

   (A) if both assertion and reason are true statements and the reason is a correct explanation of the assertion,
   (B) if both assertion and reason are true statements but the reason is not a true explanation of the assertion,
   (C) if the assertion is true but the reason is a false statement,
   (D) if the assertion is false but the reason is a true statement,
   (E) if both the assertion and reason are false statements.

11. The coefficient of thermal expansion of an MMC with a ceramic reinforcement is less than that of the matrix *because* the coefficient of thermal expansion of a ceramic is usually less than that of a metal.

12.   Generally reinforcing a metal degrades both ductility and toughness *because* there is an increase in Young's modulus.

13.   Unfortunately reinforcement of a metal markedly reduces the fatigue resistance *because* cracks are readily formed within the reinforcement.

14.   The specific strength and modulus of a silicon carbide particle reinforced aluminium alloy are superior to those of the matrix *because* silicon carbide is more dense than the common aluminium alloys.

15.   Fill in the missing words in this passage. Each dash represents a letter.

The superconducting constituent of a multifilamentary superconductor is the _ _ _ _ _ _ _ _ _ _ _ _ _ compound Nb_3Sn. Because of the _ _ _ _ _ _ _ _ _ _ of Nb_3Sn, multifilamentary superconductors have to be produced by a complex method known as the _ _ _ _ _ _ route. In this method rods of _ _ _ _ _ _ _ are embedded in a Cu–Sn alloy and swaged to reduce the cross-section. After sectioning, rebundling and further reductions in cross-section, the composite is _ _ _ _ _ _ _ _ _ _ to allow the _ _ _ _ _ _ _ _ _ of tin and the formation of Nb_3Sn. The Nb_3Sn is _ _ _ _ grained and has the A15 crystal structure.

ANSWERS

Problems

3.1 $3.3 \times 10^{-6} \, s^{-1}$.

Self-assessment

1. B; 2. A; 3. A; 4. B; 5. A; 6. C, D; 7. B, D, E; 8. A, D; 9. A, C, E; 10. A; 11. A; 12. B; 13. E; 14. B; 15. intermetallic, brittleness, bronze, niobium, heat treated, diffusion, fine

4 | Ceramic matrix composites

4.1 INTRODUCTION

As far as this chapter is concerned, included under the heading of 'ceramics' are technical ceramics, such as alumina and silicon nitride, glasses, glass-ceramics and carbon. Although the properties of these materials differ, e.g., as a rule glasses are not as strong and tough as glass-ceramics, they do have some common characteristics when compared with the other classes of materials (metals and polymers).

When discussing the characteristics of monolithic ceramics in Chapter 1, we learned that they had reasonably high strength and stiffness but were brittle (Table 1.1). Thus one of the main objectives in producing ceramic matrix composites is to increase the *toughness*. Naturally it is also hoped, and indeed often found, that there is a concomitant improvement in strength and stiffness. Figure 4.1 compares typical stress–strain curves for composites with that for a monolithic ceramic; the area under the stress–strain curve is the *energy of fracture* of the sample and is a measure of the toughness. It is clear from this figure that reinforcement with particulates and continuous fibres has lead to an increase in toughness but that the increase is more significant for the latter. Both the monolithic and the particulate reinforced composite fail in a catastrophic manner, which contrasts with the failure of the continuous fibre composite where a substantial load carrying capacity is maintained after failure has commenced. Therefore not only has the continuous fibre composite a better toughness but the failure mode is more desirable. However, fibres are a more expensive reinforcement than particles and, as we will see in section 4.3, the processing is more complex, therefore the improvement in toughness is associated with an extra cost burden. The toughening mechanisms operative in composites are described in detail in Chapter 11.

Ceramic matrix composite (CMC) development has lagged behind other composites for two main reasons. First, as we shall see in section 4.3, most

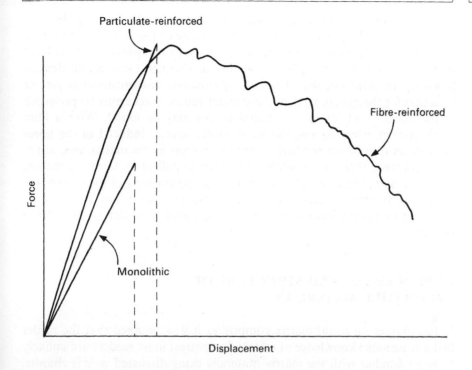

Figure 4.1 Schematic force–displacement curves for a monolithic ceramic and CMCs illustrating the greater energy of fracture of the CMCs.

of the processing routes for CMCs involve high temperatures and can only be employed with high temperature reinforcements. It follows that it was not until fibres and whiskers of high temperature ceramics, such as silicon carbide, were readily available was there much interest in CMCs. The high temperature properties of the reinforcement are also of importance during service. A major attribute of monolithic ceramics is that they maintain their properties to high temperatures and this characteristic is only retained in CMCs if the reinforcements also have good high temperature properties. Hence, there is only limited interest in toughening ceramics by the incorporation of reinforcements of materials, such as ductile metals, that lose their strength and stiffness at intermediate temperatures.

The second factor that has hindered the progress of CMCs is also concerned with the high temperatures usually employed for production. Differences in the coefficients of thermal expansion, α, between the matrix and the reinforcement lead to thermal stresses on cooling from the processing temperature as previously described for metal matrix composites (MMCs). However, whereas the thermal stresses can generally be relieved in MMCs by plastic deformation of the matrix, this is not possible for CMCs and cracking of the matrix can result. The nature of the cracking depends on the whether the reinforcement contracts more or less than the matrix on cooling as this determines the character (tensile or compressive) of the local thermal stresses. This is discussed further in Chapter 11; it is sufficient at this stage to point out that if α_R for a particulate reinforcement is greater than that for the matrix α_M then the circumferential cracks may be produced in the matrix, and for $\alpha_R < \alpha_M$ radial cracks may be found. With a fibre reinforcement, when $\alpha_R > \alpha_M$ the axial tensile stresses induced in the fibres produce an overall net residual compressive stress in the matrix and, as the fibres contract, there is a tendency for them to pull away from the matrix. The stress situation is reversed when $\alpha_R < \alpha_M$ and cracking of the matrix due to the axial tensile stresses may occur. Clearly there has to be some matching of the coefficients of thermal expansion in order to limit these problems.

4.2 PROCESSING AND STRUCTURE OF MONOLITHIC MATERIALS

In the chapter on metal matrix composites it was assumed that the reader had a reasonable knowledge of metals. In contrast most readers are unlikely to be so familiar with the matrix materials being discussed in this chapter. It is therefore necessary to look at the processing and structure of monolithic glasses, glass-ceramics and ceramics before we begin to study CMCs in detail.

4.2.1 Technical ceramics

Technical ceramics are mostly crystalline and their crystal structures are generally more complex than the simple metallic structures (see for example the crystal structure of alumina shown in Figure 2.8). The bonding is also variable; it may be ionic, covalent or have mixed ionic–covalent characteristics (Table 4.1.). The covalent bond is stronger than the ionic bond and is directional. A consequence of the bonding is that ceramics have high melting points as shown by the data of Table 4.1. The bonding, together with the complex crystal structure, makes ceramics brittle but relatively strong.

Because of their high melting points very few ceramics are produced by melting and casting. As far as modern technical ceramics such as alumina are concerned, most are produced via a powder route. The first stage in producing a component is to press the powder plus a *binder* into the required shape using a sufficiently high pressure so that a relatively dense and strong *green* compact, which can be handled, is formed. It is important that the green compact is of uniform density otherwise the properties will vary throughout the finished component. The uniformity of the density

Table 4.1 Bonding in some common ceramics (where a single bond type is given it comprises over 70% of the bonding) and its effect on melting point

Material	Bonding	Melting point ($°C$)
SiC	Covalent	2500[a]
Si_3N_4	Covalent	1900[a]
NaCl	Ionic	801
MgO	Ionic	2620
SiO_2	Covalent–ionic	1713
Al_2O_3	Covalent–ionic	2045

[a]sublimes

varies with component shape and direction of application of the pressure. The greatest uniformity of density is obtained by the application of pressure from all directions, which is known as *isostatic pressing*.

Although the strength of the green compact is sufficient for handling it is inadequate for service; to improve the strength the compact has to be heated to elevated temperatures in order to burn off the binder and to consolidate the powder further by *sintering*. It is possible to combine the forming and sintering into a single stage. This is more expensive than the two stage process but does lead to better properties. Simple pressing and isostatic pressing at an elevated temperature are termed *hot-pressing* and *hot isostatic pressing* (HIP) respectively. Whether produced by a single stage or two stage process the microstructure of the ceramic consists essentially of randomly orientated equiaxed grains and hence the properties are isotropic.

Wet processing is an alternative production route. First, a suspension of the ceramic powder in liquid, which is commonly water based, is made. This suspension is called a *slurry* or *slip*. The slurry may be employed in a number of methods, including tape and slip casting, to produce a finished product. As an illustration of these methods, *slip casting* will be briefly discussed. Slip casting depends on the ability of a mould, made from a porous material such as plaster of Paris, to absorb the liquid from the slip, so leaving an even layer of particles of the ceramic on the mould walls. The component is then allowed to dry until strong enough to be removed from the mould; it is then fired.

4.2.2 Glasses

The atoms in crystalline materials are arranged in an ordered manner over large distances; we say that there is long-range order. In contrast glasses have a more disordered, but not completely random, arrangement of atoms; glasses are said to have a short-range ordered structure (Figure 4.2(b)). The differences between the crystalline and glassy state are best understood

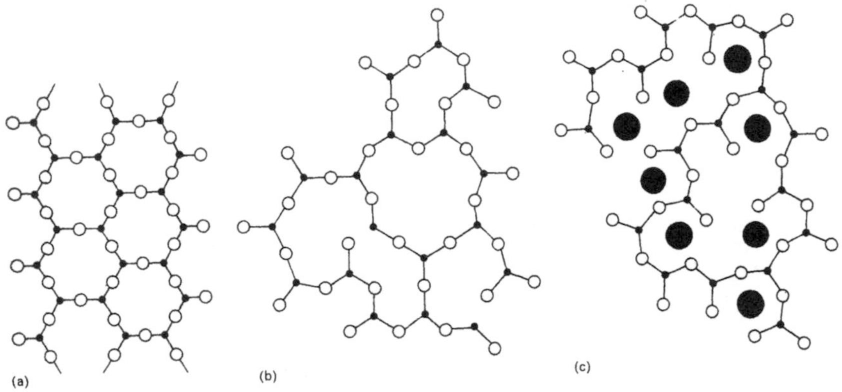

Figure 4.2 Two dimensional representation of: (a) crystalline structure of silica; (b) network structure of glassy silica; (c) soda-silica glass. Open circles indicate oxygen atoms, black dots silicon atoms and large shaded circles sodium atoms. (Source: Anderson *et al.*, 1990.)

by examining a material which can exist in either state. Silica (SiO_2) is such a material.

If silica is melted and cooled very slowly it will crystallise at a specific temperature T_M, called the freezing or melting point, in an identical manner to that of a metal. The specific volume as a function of temperature exhibits a discontinuity at the melting point (Figure 4.3(a)) and latent heat is evolved as shown by an exothermic peak in a differential thermal analysis plot (Figure 4.3(b)). From these changes we may deduce that there are significant differences in the atomic arrangements in the liquid and crystalline states. Silica can crystallize in a number of allotropic forms all of which can be regarded as a long-range ordered network of oxygen ions, in a cubic or hexagonal type lattice, with silicon ions in the tetrahedral spaces between them; this is shown schematically in two dimensions in Figure 4.2(a).

When silica is cooled more rapidly from the molten state there is insufficient time for the ions to attain the long-range order of the crystalline state and the specific volume versus temperature and the DTA plots are given by the dotted lines in Figure 4.3. The temperature T_g in this figure is the *glass transition temperature*, which is not a well-defined temperature like T_M but depends on the cooling rate. It should be noted that there is no change in slope of the specific volume curve at T_M and no latent heat evolved indicating that there is no change in structure at T_M, i.e., between T_M and T_g the material is a *supercooled liquid*. There is a change in slope on the specific volume curve at T_g, but no marked discontinuity. Similarly there is only a slight change in the base-line of the DTA plot and no marked peak indicative of latent heat at T_g. These observations suggest that the state below T_g, termed the glassy state, is very closely akin to the liquid state. This

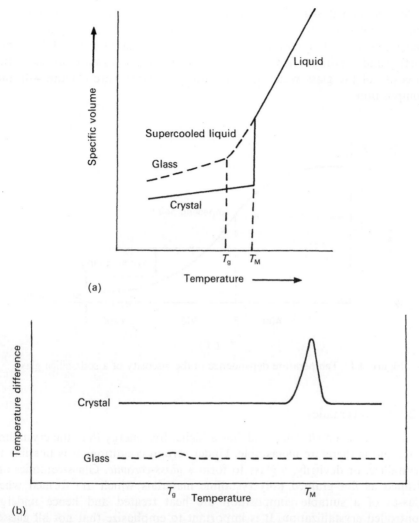

Figure 4.3 Plots showing the relationship between the liquid, crystalline and glassy states: (a) specific volume versus temperature; (b) differential thermal analysis results.

is indeed the case as we have already seen that the glassy state does not have the periodic lattice of the crystalline state but consists of a short-range ordered network (Figure 4.2(b)).

A consequence of the open network structure of glass (compare the structures in Figure 4.2(a) and (b)) is that glass can easily accommodate atoms of different species. These atoms may disrupt the continuity of the network or contribute to the network. The former are known as *network modifiers*, e.g., sodium (Figure 4.2(c)), and the latter as *network formers*, e.g., boron.

After molten glass has been produced from the appropriate raw materials, and usually some recycled glass, it can be formed into the required shape using any one of a number of methods including blow-moulding, spin casting and pressing. The important parameter in the forming process is the viscosity of the glass which is a function of temperature (Figure 4.4) and composition.

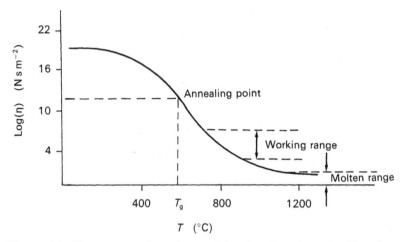

Figure 4.4 Temperature dependence of the viscosity of a soda-silica glass.

4.2.3 Glass-ceramics

A glass is called a *vitreous* solid, has a higher free energy than the crystalline state and is therefore metastable. Under certain conditions it is possible to crystallize, or devitrify, a glass to form a *glass-ceramic*. Glass-ceramics are defined as fine-grained polycrystalline materials which are formed when glasses of a suitable composition are heat treated and hence undergo controlled crystallization. It is important to emphasize that not all glasses can be crystallized to form a glass-ceramic – some glasses are too stable and difficult to crystallize whereas others crystallize too readily – only certain compositions are amenable to crystallization in a controlled manner.

A glass-ceramic is not fully crystalline; typically at least 50% and as much as 98% of the volume may be crystalline. The properties depend on the properties and volume fractions of the residual glass and the crystalline phase(s). A glass-ceramic has superior mechanical properties to its parent glass and may also exhibit unusual properties such as the extremely small coefficient of thermal expansion of certain compositions in the $Li_2O-Al_2O_3-SiO_2$ system.

One of the attractions of monolithic glass-ceramics is that the parent glass may be shaped using a number of relatively inexpensive glass forming techniques and then heat treated to crystallize it. This is the conventional production method but in some cases a powder route, as previously described for technical ceramics, is employed. The powdered parent glass is pressed into the desired shape and the green compact heat treated to sinter and crystallize. It is generally found that the sintering route yields better mechanical properties.

4.3 PROCESSING OF CMCs

4.3.1 Conventional mixing and pressing

This is simply an extension of the powder route for producing technical ceramics and some glass-ceramics. A powder of the matrix constituent is mixed with the toughening constituent, in particulate or whisker form, together with a binder. The mixture is then pressed and fired or hot pressed.

Difficulty can be experienced in obtaining a homogeneous mixture of the two constituents and high proportions of the toughening phase cannot easily be achieved. Additional problems may arise with whiskers. Whiskers tend to form strong aggregates and significantly reduce the packing efficiency. Also damage to the whiskers can occur during the mixing and pressing operations, particularly when cold pressing.

4.3.2 Techniques involving slurries

Because of the difficulties encountered in obtaining homogeneous materials by conventional mixing, especially with whiskers, wet processing is sometimes favoured. A simplified flow sheet for this form of processing is given in Figure 4.5. It is essential that the constituents should remain *deflocculated*, that is, well dispersed throughout the slurry, and this is achieved by control of the pH of the aqueous solution. The dispersion is further improved by agitation of the slurry, usually by ultrasonic vibration. At this stage the composite can be formed by slip casting. Alternatively heat is applied to evaporate the water and the dried, well-mixed constituents cold pressed and sintered or hot pressed.

Certain ceramics can be produced from slurries by shaping using a range of methods including casting and extrusion followed by setting at ambient or slightly elevated temperatures. Examples are plaster of Paris, which is obtained from gypsum $CaSO_4.2H_2O$, and $AlPO_4$ formed from phosphoric acid and aluminium hydroxide solutions. The slurries of these types of ceramic are amenable to the incorporation of particles, whiskers or chopped fibres and the normally employed shaping techniques may be used to produce the composite component. However, these ceramics are intrinsically poor mechanically and furthermore the production route results in much

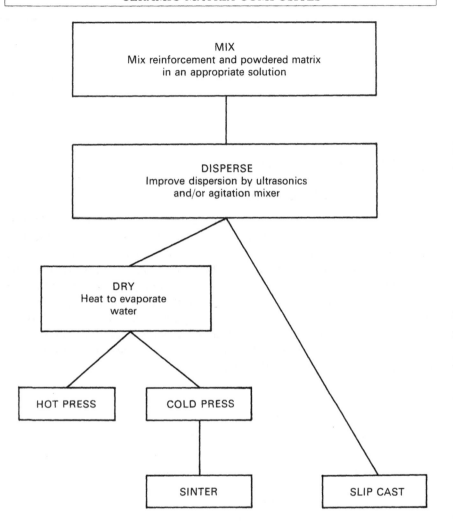

Figure 4.5 Simplified flow sheet for mixing as a slurry prior to shaping.

porosity; it is therefore unlikely that high performance composites will be produced by this method.

So far the methods involving slurries have been suitable for reinforcements by particles, whiskers and chopped fibres but not continuous fibres. We now turn out attention to the most commonly employed slurry method for continuous fibre reinforced composites. This was developed about twenty years ago in the United Kingdom for the production of glass matrix composites but it is now also widely used for glass-ceramic matrix composites. The intimate mixing of continuous fibres and the glass (parent glass in the case of a glass-ceramic) is achieved by drawing bundles of the fibres, called tows, through a slurry of powdered glass in water and a water soluble resin binder (Figure 4.6). The tows, impregnated with the slurry, are wound on to a mandrel to form a monolayer tape. The tape is cut into plies which are stacked into the required stacking sequence, e.g., unidirectional, cross-plied, etc., prior to burnout of the binder. This is followed by hot pressing to consolidate the matrix. In glass-ceramic composite production some crystallization occurs during the hot pressing stage but an additional heat treatment may be required to complete devitrification. The micrograph of Figure 4.7 show the high fibre content and the good alignment obtainable by this method.

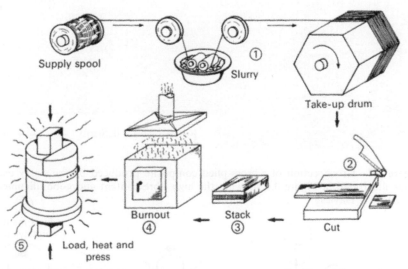

Figure 4.6 Processing using a slurry and hot pressing for continuous fibre reinforced glass or glass-ceramic. (Source: Prewo, 1989.)

4.3.3 Liquid state processing

The reader will recall that melt infiltration techniques are well established for metal matrix composites (section 3.2.2). Modifications of these techniques for CMCs have met with only limited success so far because of (a) reactions with the reinforcement due to the high melting temperatures of refractory ceramics and the reactivity of molten glasses, and (b) the low rates of infiltration resulting from the high viscosities. The most successful of the melt techniques is *matrix transfer moulding* which was originally developed for glass matrix composites but can also be used for glass-ceramic matrix composites. The advantage of matrix transfer moulding is that it permits the fabrication of components, such as tubes, which are difficult to produce by many of the other methods. Tube production is illustrated in Figure 4.8. A

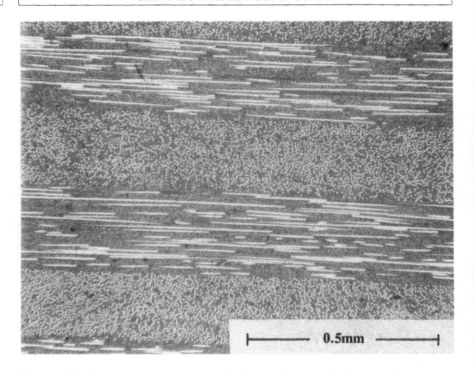

Figure 4.7 Cross-section of a cross-plied composite produced by the slurry technique illustrated in Figure 4.6 showing the high fibre content and good alignment. (Courtesy J. Davies.)

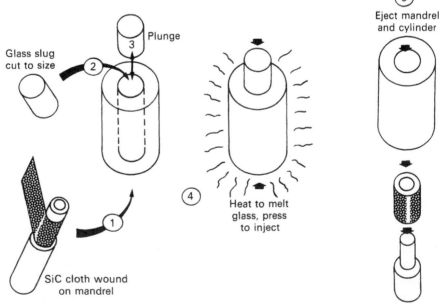

Figure 4.8 Matrix transfer moulding of a thin wall tube. (Source: Prewo, 1989.)

preform and a glass slug are inserted into a cylindrical mould. Application of heat and pressure forces the fluid glass into the pores in the preform and, after cooling, the composite tube is ejected from the mould.

Certain ceramics can be produced by the *pyrolysis* of a polymer. These ceramic-producing polymers may be used in place of the slurry in the production route for continuous fibre composites illustrated in Figure 4.6, but more commonly they are employed for the liquid impregnation of a preform. After impregnation a heat treatment is required to pyrolyse the polymer. The pyrolysis temperature is relatively low, typically 600–1000 °C, which reduces any degradation of the reinforcement. However, the ceramic yield is low, generally less than 60% and hence multiple polymer impregnation – pyrolysis cycles are needed to achieved high density material. As will be seen later in this chapter (section 4.4.3), this is a standard method for producing carbon–carbon composites but is also applicable to other systems, e.g., pyrolysis of polysilastyrene or dodecamethylcyclohexasilane yields SiC.

4.3.4 Sol–gel processing

A *sol* is a dispersion of small particles of less than 100 nm which is usually obtained by precipitation resulting from a reaction in solution, for example, the precipitation of zirconium hydroxide particles from a solution of $ZrOCl_2$ by ammonia according to

$$ZrOCl_2 + NH_3 + 3H_2O = 2NH_4Cl + Zr(OH)_4.$$

A *gel* is a sol that has lost some liquid and hence has an increased viscosity. Currently there is much interest in sol–gel processing of ceramics and this will undoubtedly extend into the field of CMCs.

With the exception of the production of zirconia-toughened alumina (section 4.4.1), sol–gel processing of CMCs is still in the experimental stage. If a sol is poured over a preform it will infiltrate because of the fluidity of the sol (Figure 4.9(a)). The sol is then dried in a subsequent heat treatment. The processing temperature is normally low, thus reducing the risk of damage to the preform, and complex shapes can be produced. However there are the disadvantages of high shrinkage and low yield and consequently repeated infiltration is necessary to increase the density of the matrix. Furthermore for some materials higher temperatures than those needed just for drying are required to produce the desired ceramic, e.g., the $Zr(OH)_4$ would be *calcined* at about 550 °C to give zirconia, ZrO_2.

A good dispersion of particulate or whisker reinforcement can be made by mixing in the sol or gel state. This is then followed by drying, calcining if necessary, and pressing plus sintering or more commonly hot pressing (Figure 4.9(b)). The mixing is better than that obtained via conventional mixing or slurries for particulate reinforcements but of course the final stage

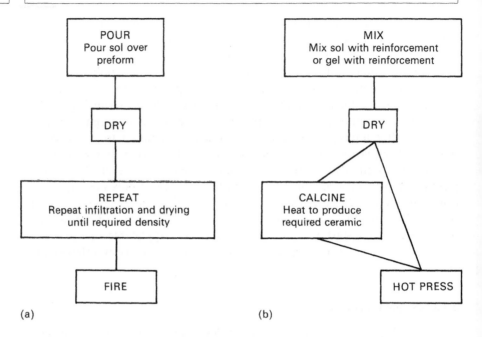

Figure 4.9 Sol-gel processing: (a) infiltration of a preform; (b) mixing reinforcement in a sol or a gel.

of sintering or hot pressing removes the benefit of the low processing temperatures usually associated with sol–gel processing.

4.3.5 Vapour deposition techniques

A number of methods involving deposition from a vapour have been employed in CMC production including *chemical vapour deposition* (CVD), evaporation and plasma assisted processes, e.g., *ion plating* and *sputtering*. Choice of techniques depends on many factors such as the composition of the matrix and the reinforcement, the form of the reinforcement and the desired properties and no technique is superior in every instance. As a general rule CVD gives better control of composition and has a faster deposition rate (except for evaporation) but often requires higher temperatures than other vapour deposition techniques. We will concentrate on CVD as it is the more widely used technique.

In CVD chemical reactions in the gaseous state lead to the deposition of a solid on to the surface of a heated substrate. In composite technology CVD is used to produce fibres (e.g., boron and SiC fibres – see Chapter 2), to coat fibres and to infiltrate porous preforms to form the matrix. In the latter case the process is also called *chemical vapour infiltration*, CVI, and it is CVI that we will now discuss.

The basic stages of a CVI process are shown in Figure 4.10. The reactants may be solids, liquids or gases at room temperature. Liquids or solids are heated to raise their vapour pressures and hence increase to an acceptable level the amount of reactants in the gaseous state. The gaseous reactants are 'picked-up' by a flowing gas, known as the *carrier gas*, and transported to the reactor. The gaseous reactant infiltrates the heated substrate positioned in the reactor; a chemical reaction occurs in the gaseous state and deposition of the matrix takes place. Although the rate of deposition is faster than for most other vapour deposition processes it is still relatively slow, the maximum rate being only of the order of 2500 μm/h. The effluent, which consists of the carrier gas, partially decomposed reactants and some reaction products, has to be continuously removed.

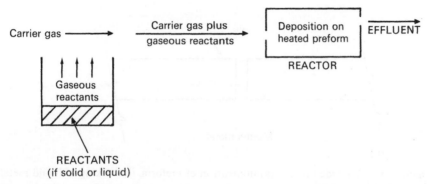

Figure 4.10 Simplified diagram of the stages in a chemical vapour infiltration (CVI) process.

The best established CVI process is for the production of carbon–carbon composites and this will be discussed in more detail later in this chapter (section 4.4.3). However it must be appreciated that it has been employed for the production of a wide range of ceramic matrices including carbides (e.g., B_4C, SiC, TaC and TiC), nitrides (e.g., BN and Si_3N_4), TiB_2 and Al_2O_3. Two typical reactions are given below:

$$TiCl_4(g) + 2BCl_3(g) + 5H_2(g) = TiB_2(s) + 10HCl(g),$$

$$SiCl_4(g) + CH_4(g) = SiC(s) + 4HCl(g).$$

4.3.6 Lanxide process and *in situ* techniques

The *Lanxide* process involves the formation of a ceramic matrix by the reaction between a molten metal and a gas, e.g., molten aluminium reacting with oxygen to form alumina. Growth of the ceramic occurs outwards from the original metal surface and through a preform as shown in Figure 4.11(a). In fact a preform is not a prerequisite and particulate reinforced composites

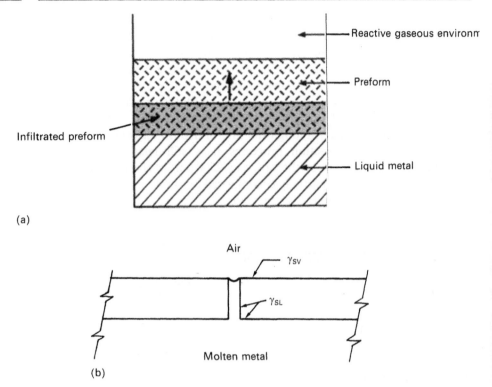

Figure 4.11 Lanxide process: (a) infiltration of preform; (b) wicking of liquid metal along grain boundaries.

may also be produced by simply placing the particles above the liquid metal. In both cases the only requirements are that the fibres/particles do not react with the gas and are wetted by the ceramic.

The rate of growth is constant as it is controlled by the rate of the chemical reaction and not transport of the reacting species. Transport of the liquid metal is thought to occur rapidly by a wicking process along grain boundaries in the ceramic matrix when $\gamma_{SV} > 2\gamma_{SL}$, where γ_{SV} and γ_{SL} are the grain boundary and liquid metal–ceramic interfacial energies respectively (Figure 4.11(b)). The rates of growth are low, typically 1 mm/h, nevertheless reasonable size components (~ 20 cm thick) of complex shape can be fabricated.

A number of novel techniques are under investigation whereby the composite is formed *in situ* via a chemical reaction. A possible reaction is

$$2AlN + B_2O_3 \rightarrow Al_2O_3 + 2BN.$$

Such reactions have the potential to give good homogeneous distributions of the toughening phase, and the raw materials may be less costly than the products, e.g., BN is expensive.

If the reaction is strongly *exothermic* it may be self-propagating, i.e., the reaction has to be initiated only in one part of the compact of the reactants for it to extend throughout the compact. High but transient temperatures are involved and hence metastable structures may be produced. The following is an example of a self-propagating reaction that could produce an interesting composite:

$$4Al + 3TiO_2 + 3C \rightarrow 2Al_2O_3 + 3TiC.$$

4.4 DETAILED REVIEW OF SELECTED CMCs

Zirconia-toughened and SiC whisker toughened alumina, continuous SiC fibre reinforced glass-ceramics and carbon–carbon composites have been selected for further review. These composites have been chosen because they represent three different matrix materials, cover the main forms of reinforcement (particulate, whisker and continuous fibre), differ in method of production and illustrate the wide range of properties available with CMCs.

4.4.1 Alumina matrix composites

Alumina is probably the most familiar of the modern technical ceramics and is used for a variety of applications ranging from substrates for electrical devices to femoral spheres in hip prostheses. It was therefore an obvious choice to be the matrix of a composite.

(a) SiC whisker reinforced

SiC reinforced alumina is usually made by some variation of the slurry method shown in the simplified flow sheet of Figure 4.5. The whisker content is typically 25% or less and the whiskers tend to lie perpendicular to the pressing axis. The matrix consists of small, equiaxed, randomly orientated grains of α-alumina. Cutting tools for the wood and metal industries are manufactured from these composites and are said to give significant productivity benefits over conventional carbide cutting tools.

SiC is slightly less dense than alumina so reinforcing with SiC whiskers leads to a small decrease in density, e.g., the densities of alumina and a composite with 25% SiC whiskers are about 3.9 and 3.7 Mg/m^3 respectively. The Young's modulus is improved by the presence of the stiff whiskers and approaches 400 GPa for a composite with 25% SiC (Table 4.2). The decrease in density and increase in stiffness on the incorporation of the whiskers obviously results in higher specific modulus values.

Although these improvements in stiffness and specific modulus are desirable they can only be considered a minor benefit in comparison with the

Table 4.2 Properties of SiC whisker reinforced alumina produced for cutting tools. (Source: Sandvik, Sweden)

%SiC	0	7	15	25
Density (Mg/m^3)	3.9	3.8	3.8	3.7
Young's modulus (GPa)	340	340	350	390
Specific modulus [(GPa)/(Mg/m^3)]	87	89	92	105
Bend strength (MPa)	300	650	700	900
Weilbull modulus	6	8	10	13
Fracture toughness (MPa m$^{1/2}$)	4.5	5.5	6.0	8.0
Coeff. thermal expansion × 10^6 (K^{-1})	8.0	8.0	7.0	6.0

significant increase in the strength and toughness. There is considerable scatter in the mechanical property values reported in the literature, nevertheless Figure 4.12 demonstrates that the best strength and toughness values from composites are some 50% greater than the corresponding values for alumina. There is some evidence that the stress intensity for crack growth increases with crack length; this suggests that bridging of the crack faces by the strong SiC whiskers, which has been observed by scanning electron microscopy, may be giving some wake toughening. Other likely operative toughening mechanisms are crack deflection and pull-out. (The reader is referred to Chapter 11 for a general review of toughening mechanisms.) Finally, it is interesting to note that some cutting tool manufacturers are quoting Weibull modulus (see equation 8.16) values for composites which are equal to, or greater than, those normally associated with monolithic alumina (Table 4.2); this reflects the fact that the composites are more defect tolerant and that the defects are controlled in the manufacturing process.

For all practical purposes the temperature dependence of the toughness and strength of the whisker reinforced composites is similar to that of alumina; in other words the room temperature improvement in properties is maintained as the temperature is increased (Figure 4.13). The results of Figure 4.13 illustrate the high temperature behaviour over a short period of time; when the stress is applied for long periods the mechanical performance is determined by the creep resistance. The data of Figure 4.14 demonstrate the substantial improvement in creep resistance at 1500 °C obtained as a result of the incorporation of 15% SiC whiskers. The stress exponent, i.e., the slope of the curves of Figure 4.14, is 1 to 2 for fine-grained alumina, which is consistent with creep being controlled by a diffusional mechanism. A similar creep mechanism is thought to be operative in composites tested at low stress at 1200 °C and 1300 °C, whereas at 1500 °C, and at high stresses at the lower temperatures, the stress exponent is nearer to 5 which may be a consequence of cavitation at grain boundary–interface junctions.

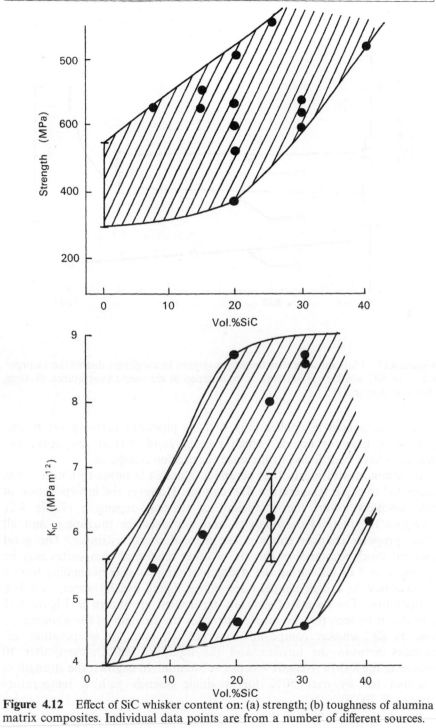

Figure 4.12 Effect of SiC whisker content on: (a) strength; (b) toughness of alumina matrix composites. Individual data points are from a number of different sources.

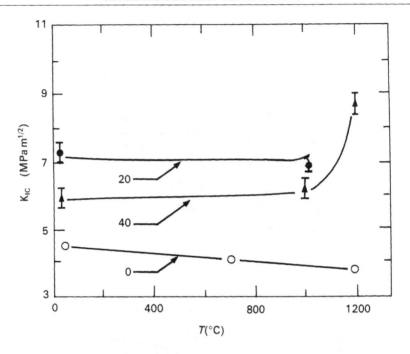

Figure 4.13 Plots showing that the improvement in toughness due to the incorporation of SiC whiskers in alumina is maintained at elevated temperatures. (Source: Wei and Becher, 1985.)

Alumina is used in applications such as pipework carrying solids suspended in gases or liquids, which require good erosion resistance. SiC whiskers have been found to increase the erosion resistance.

In common with most other ceramics, alumina is prone to failure when subjected to rapid temperature fluctuations. However the incorporation of SiC whiskers lowers the coefficient of thermal expansion (Table 4.2), increases the thermal conductivity and improves the toughness, and all these property changes enhance the *thermal shock resistance*. The good thermal shock resistance of the SiC whisker reinforced composites may be demonstrated by measuring the strength of samples after quenching from a furnace held at a high temperature into a liquid, usually water, at a low temperature. The results of such an experiment are shown in Figure 4.15 and it can be seen that for single quenches the strength of the alumina–20 vol. % SiC whisker composite does not fall even for temperature differences between the furnace and the liquid of 900 °C; even after 10 quenches the loss in strength was not substantial. In contrast the strength of alumina fell by over 50% for a single quench with a temperature change of 700 °C.

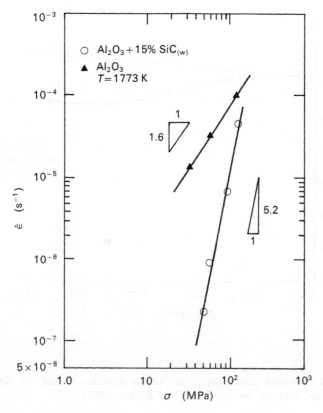

Figure 4.14 Log–log plot of strain rate versus stress showing that the creep rate at a given stress is less for the SiC reinforced alumina. (Source: Chokshi and Porter, 1985.)

(b) Zirconia-toughened alumina

The best established particulate-reinforced alumina is *zirconia-toughened alumina*, ZTA. ZTA contains 10–20 vol. % of fine zirconia, ZrO_2, particles (Figure 4.16). The equilibrium condition of the zirconia at elevated temperatures is tetragonal (t) and at low temperatures monoclinic (m). On cooling ZTA from the high temperatures required for fabrication the t→m transformation may occur in the particles and is accompanied by an increase in volume of about 3%. The transformation is what is known as an *athermal* transformation; a characteristic of an athermal transformation is that it is not time dependent, i.e., if it takes place it proceeds very rapidly. If the transformation takes place in the zirconia particles in ZTA during production, then the 3% volume change accompanying the transformation produces microcracks in the alumina matrix. Control of the extent of the microcracking augments the toughness by means of the *microcrack*

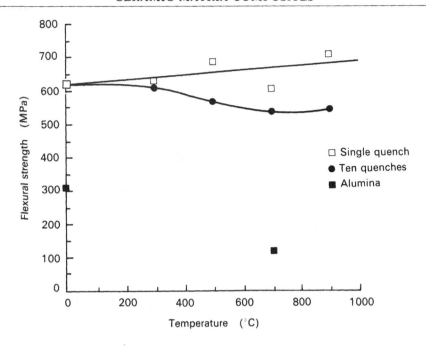

Figure 4.15 Thermal shock behaviour of an alumina–20 vol. % SiC whisker composite and alumina. (Source: Tiegs and Becher, 1987.)

Figure 4.16 Scanning electron micrographs of zirconia-toughened alumina (ZTA). (Courtesy I. Thompson.)

toughening mechanism discussed in Chapter 11. However, as we can see from Figure 4.17, although microcracking may lead to an improvement in toughness there is an adverse effect on strength.

If some *stabilizing oxide*, say 3 mol. %Y_2O_3, is added to the zirconia then it is possible to suppress the t→m transformation on cooling from the fabrication temperature.

When the particles are retained in the tetragonal state at room temperature they are said to be in a *metastable* state. Whether the transformation takes place or not depends not only on the amount of stabilizing oxide but also on the size of the particles. The presence of the alumina matrix around the particles makes it difficult for the volume expansion associated with the transformation to be accommodated; we say that the elastic constraint of the matrix hinders the transformation. The constraint is such that small particles are less likely to transform than large particles.

Thus with a suitable combination of amount of stabilizing oxide and particle size we are able to obtain a dispersion of metastable tetragonal particles. However under the influence of the stress field at a crack tip these

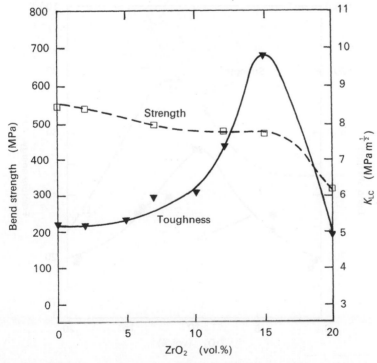

Figure 4.17 Strength and toughness of ZTA as a function of the volume fraction of zirconia particles when the $t \rightarrow m$ transformation has produced microcracks. (Source: Claussen, 1976.)

particles will transform athermally to the monoclinic state. This leads to *transformation toughening* and the toughening increment ΔK_{TT} which can be achieved by this mechanism is proportional to the following parameters:

$$\Delta K_{TT} \propto V_{zirc} \Delta \varepsilon E_m r_o^{1/2},$$

where V_{zirc} is the volume fraction of the metastable zirconia particles in the composite, $\Delta \varepsilon$ is volume strain accompanying the transformation, E_m is Young's modulus of the alumina matrix and r_o is the width of the process zone around a crack containing transformed particles (Figure 11.25).

The strength and toughness of ZTA as a function of zirconia content when the transformation toughening mechanism is operative is shown in Figure 4.18. It is important to note that, in contrast to ZTA toughened by microcracking, transformation toughening can lead to an improvement in both toughness and strength and consequently is the preferred toughening mechanism.

Care must be taken when using a component manufactured from ZTA in liquid environments. It has been found that certain liquids cause the metastable tetragonal particles at the surface to transform to the monoclinic.

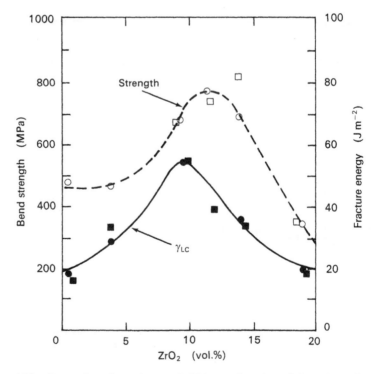

Figure 4.18 Strength and toughness of ZTA as a function of the volume fraction of zirconia particles when the stress induced transformation toughening mechanism is operative. (Source: Becher, 1981.)

Microcracking due to the transformation enables the liquid to penetrate further into the material and to propagate the transformation to greater depths. The loss of the capacity for transformation toughening, together with the large cracks formed by the linking of the microcracks, produces a marked degradation in mechanical performance as illustrated by the strength data for ZTA in a 20 vol. % HCl solution (Figure 4.19). Also shown in Figure 4.19 is that the yttria content of the HCl solution increases as a function of time in contact with the ZTA. These results clearly demonstrate that the instability of the zirconia particles is due to the HCl leaching out the stabilizing oxide, yttria, from the zirconia particles. It may be that zirconia particles stabilized with other oxides may not be so prone to liquid attack but, at the present, yttria is the most commonly employed stabilizing oxide and hence this must be considered as a weakness of ZTA.

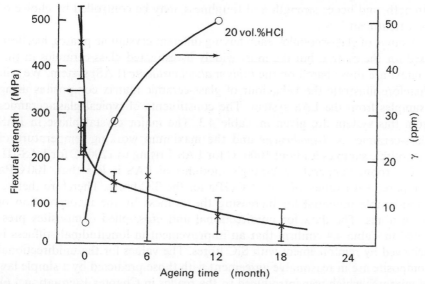

Figure 4.19 Data showing the loss in strength of ZTA in 20 vol. % HCl is associated with an increase in the yttria content of the HCl. (Source: Thompson and Rawlings, 1992.)

4.4.2 Glass-ceramic matrix composites

Development of this class of composite material has centred on continuous fibre reinforcement. Early work, which followed on from studies of glass matrix composites, used carbon fibres but nowadays effort is almost soley concentrated on reinforcing with the SiC yarns, 'Tyranno' and 'Nicalon' discussed in an earlier chapter (section 2.5.5).

The normal production route is the slurry based method illustrated in Figure 4.6. The parent glass composition, the particle size of the powder and the hot pressing temperature have to be carefully selected if the required microstructure, and hence properties, are to be obtained. It is essential that sintering is complete before crystallization becomes well established; crystallization gives an effective increase in viscosity which hinders sintering and results in porosity and poor strength.

The importance of the matrix–fibre interface in determining the mechanical properties of a composite has been mentioned previously, as have the methods for measuring the interfacial strength (Chapter 2). In the case of glass-ceramic matrix composites the interfacial bond strength is a function of the chemistry of the matrix and fibre (as one would expect), but also of the heat treatment given to sinter and crystallize. The variation in the interfacial shear stress, as measured by the micro-indentation technique, with time and temperature of heat treatment for a SiC fibre reinforced glass-ceramic is shown in Figure 2.28. These graphs demonstrate how the interfacial strength, and hence strength and toughness, may be controlled by choice of heat treatment.

A range of glass-ceramics, each having different crystalline phases, has been used for the matrix, but the most widely investigated glass-ceramics in this context are those based on the *lithium aluminosilicate* (LAS) system. We will therefore illustrate the behaviour of glass-ceramic matrix composites using examples from the LAS system. The constituents of typical glass-ceramics from this system are given in Table 4.3. The major crystal phase in LAS glass-ceramics is *β-spodumene* and the maximum working temperature of monolithic samples is about 1000 °C for LAS-I rising to 1200 °C for LAS-III.

The room temperature Young's modulus of LAS is less than 100 GPa compared with values of over 200 GPa for the SiC fibres. Therefore there is considerable potential for increasing the stiffness by the incorporation of these fibres. The data for unidirectional and cross-plied composites presented in Table 4.4 confirm that an improvement in longitudinal stiffness is achieved by reinforcement with SiC fibres. The values for the unidirectional composite are in reasonable agreement with those predicted by a simple law of mixtures which was introduced to the reader in Chapter 1 (equation 1.6). The improvement in stiffness is, of course, not so marked for composites with a more complex arrangement of the reinforcement, e.g., for a cross-plied composite the Young's modulus is only 118 GPa compared with 133 GPa for the unidirectional.

Table 4.3 Typical compositions of lithium aluminosilicate (LAS) glass-ceramics

	Basic Constituents	*Major additions*	*Minor additions*
LAS-I	$Li_2O-Al_2O_3-MgO-SiO_2$		ZnO,ZrO_2,BaO
LAS-II	$Li_2O-Al_2O_3-MgO-SiO_2$	Nb_2O_5	ZnO,ZrO_2,BaO
LAS-III	$Li_2O-Al_2O_3-MgO-SiO_2$	Nb_3O_5	ZrO_2

Table 4.4 Young's modulus of LAS–SiC composites: (a) Prewo (1986); (b) Brennen and Prewo (1982); (c) Fareed *et al.* (1987)

	Vol.%SiC		*Young's modulus (GPa)*		*Ref.*
			Experimental	*Predicted*[a]	
LAS	0		86		(a)
LAS–I	46	(unidirectional)	133	143	(a)
LAS–II	46	(unidirectional)	130	143	(a)
LAS–II	44	(unidirectional)	136	141	(a)
LAS–I	~50	(cross-plied)	118		(b)
LAS–III	~40	(3-D braid)	79–111		(c)

[a]Predicted from simple law of mixtures (equation 1.6) with $E_f = 210\,GPa$.

The strain to failure of SiC fibres is greater than that of the glass-ceramic matrix. Consequently on loading a glass-ceramic matrix composite, matrix cracking occurs before failure of the fibres. The composite is designed to have sufficient fibres to carry the load transferred from the matrix and to continue to sustain an increasing load. The onset of cracking is indicated in a mechanical test by a change in slope of the load–displacement curve (see points M on Figures 4.20 and 4.21 (a)). The composite continues to carry an increasing load until the failure strain of the weakest fibres is reached (point F), thereafter fibre fracture and pull-out continue under a decreasing load. This is a generalized description of the various stages of deformation during testing of a SiC fibre reinforced glass-ceramic and whether all stages are exhibited by a composite depends on the matrix and the testing mode. For example extensive matrix microcracking is observed in notched bend tests on composites with an LAS-I matrix (Figure 4.20) and in LAS-II

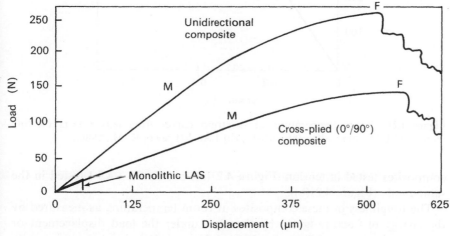

Figure 4.20 Room temperature load–deflection curves from single-edge notched bend specimens of SiC reinforced LAS-I. (Source: Brennan and Prewo, 1982.)

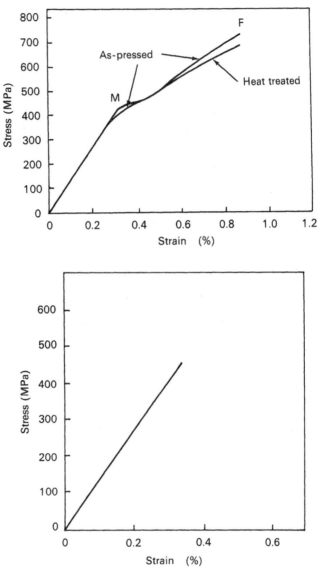

Figure 4.21 Room temperature stress–strain curves from tensile tests on SiC reinforced LAS composites: (a) LAS-11; (b) LAS-I (Prewo, K. M., 1986.)

composites tested in tension (Figure 4.21(a)) but not in LAS-I tested in the same mode (Figure 4.21(b)).

The toughness of these composites at room temperature, as measured by the energy of fracture (given by the area under the load–displacement or stress–strain curves of Figures 4.20 and 21) and the critical stress intensity factor (Table 4.5) is significantly greater than that of the monolithic

Table 4.5 Room temperaure toughness of LAS–SiC composites. (Source: Brennan and Prewo, 1982)

		$K_{IC}\,(MPa\,m^{1/2})$
LAS		1.5
LAS–I + ∼50%SiC	(unidirectional)	17.0
LAS–II + ∼50%SiC	(cross-plied)	10.0

glass-ceramic. In these, and in other tough continuous SiC reinforced glass-ceramic composites, a considerable amount of fibre *debonding* and *pull-out* has been observed as illustrated by Figure 4.22. It will be shown later (Chapter 11) that the ratio of the maximum energy of pull-out ($W_{P(max)}$) to the maximum energy of debonding ($W_{D(max)}$) is given by

$$\frac{W_{P(max)}}{W_{D(max)}} = \frac{3E_f}{\hat{\sigma}_{Tf}}.$$

Substituting typical values for the Young's modulus, E_f, and strength, $\hat{\sigma}_{Tf}$, for SiC fibres indicates that pull-out is the more important mechanism as the ratio is several hundreds. Another mechanism which undoubtedly contributes to the toughness is *wake toughening* due to the strong SiC fibres bridging the crack faces and thereby hindering crack opening.

|—————— 10mm ——————|

Figure 4.22 Tensile fracture of continuous fibre composite showing a 'brush-like' failure indicative of debonding and pull-out. (Courtesy J. Davies.)

A relatively weak fibre–matrix interface is necessary to obtain the debonding and pull-out required for good toughness. We have already seen that the interfacial strength is a function of heat treatment (Figure 2.26) and matrix composition. It has been established that in some glass-ceramic matrix systems, including LAS, a carbon rich interfacial layer develops during processing of the composite and this is responsible for the low interfacial strength and the marked improvement in toughness. In the niobium containing LAS-II and LAS-III matrix composites there is also a second reaction layer of fine NbC particles.

At the strain at which the weakest fibres fail the matrix is carrying a negligible load. It follows that the longitudinal tensile strength, $\hat{\sigma}_{1T}$, of a unidirectional composite is approximately (for further details see section 8.3.1)

$$\hat{\sigma}_{1T} = \hat{\sigma}_{Tf} V_f$$

The tensile strength, $\hat{\sigma}_{Tf}$, of SiC fibres is high, say 2000 MPa, and as the volume fraction of fibres is typically 0.5 the strength of the unidirectional composite should be about 1000 MPa. Strength values much better than that from the monolithic matrix, and of the order of 1000 MPa, have been obtained from a number of glass-ceramic matrix composites. Some values for LAS composites reinforced with about 45% SiC are presented in Table 4.6; it can be seen that the strength follows a similar trend to the Young's modulus in that the cross-plied composite has inferior properties to the unidirectional. However, both the strength and stiffness of the cross-plied are less anisotropic than the unidirectional composite. In fact, because of the low fibre–matrix interfacial strength associated with the carbon rich layer, the strength of a unidirectional composite falls rapidly as the tensile axis deviates from the fibre axis and the transverse strength can be less than 5% of the longitudinal strength.

Now let us turn our attention to the mechanical properties at elevated temperatures. In an inert atmosphere, such as argon, the room-temperature improvement in stiffness, strength and toughness is maintained up to

Table 4.6 Room temperature strength of LAS–40 to 50 vol. % SiC composites. (Source: Minford and Prewo, 1986)

		Test	Strength (MPa)
LAS		Bend	180
LAS–I–SiC	(unidirectional)	Bend	620
LAS–I–SiC	(cross-plied)	Bend	350
LAS–II–SiC	(unidirectional)	Bend	830
LAS–II–SiC	(unidirectional)	Tensile	520
LAS–III–SiC	(unidirectional)	Tensile	620
LAS–II–SiC	(cross-plied)	Tensile	370
LAS–III—SiC	(cross-plied)	Tensile	260

temperatures in excess of 1000 °C. However, there is a major problem with SiC fibre reinforced LAS and other glass-ceramic matrix composites when loaded at elevated temperatures in air. As mentioned earlier, the design of these composites is such that microcracking of the matrix occurs prior to failure. An undesirable consequence is that the environment is able to penetrate through the microcracks and reach the fibres. In the case of an oxygen containing environment, the oxygen reacts with the carbon rich layer leading to a dramatic degradation in mechanical performance. The failure mode becomes more brittle with less fibre pull-out and this is accompanied by a loss of strength. The temperature at which the mechanical properties begin to deteriorate depends on the composite and the test conditions but is typically around 800 °C as illustrated by the plots of strength as a function of temperature in Figure 4.23. The degradation proceeds rapidly at elevated temperatures and occurs even at low oxygen partial pressures as shown by the strength data at 900 °C for an LAS-II matrix composite (Table 4.7).

Compared with other structural materials there is a relative dearth of fatigue data for ceramics and it is therefore not surprising that little is known of the fatigue behaviour of glass-ceramic matrix composites. It

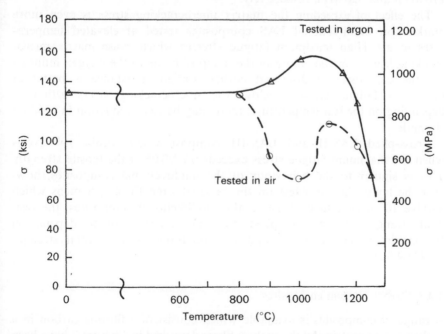

Figure 4.23 Data illustrating the adverse effect of an oxygen containing environment on the strengh of unidirectional SiC fibre reinforced LAS-III. (Source: Prewo, 1989.)

Table 4.7 Effect of oxygen partial pressure on the strength of an LAS–II–SiC composite tested at 900 °C. (Source: Mah *et al.*, 1985)

Oxygen (N/m^2)	Partial pressure (Atm)	Strength (MPa)
2×10^4	0.2	521
1×10^4	0.1	586
1×10^3	0.01	664
1×10^2	0.001	1010

appears that at room temperature the fatigue performance is quite good; in tension–tension fatigue the maximum fatigue stress has to exceed about 0.7 of the tensile fracture stress to cause failure within 10^5 cycles in any of the unidirectional LAS matrix composites (Table 4.8). Residual strength measurements on specimens which survive 10^5 cycles suggest that significant damage, and hence a reduction in residual strength, occurs mainly when the maximum fatigue stress is greater than the stress required for matrix microcracking. This is evident from the data of Table 4.8 bearing in mind that LAS-I exhibits a linear stress–strain curve to failure, i.e., no microcracking, whereas the matrix cracking stress for LAS-II and LAS-III is about 270 MPa and 320 MPa respectively.

The effect of exceeding the matrix microcracking stress is even more marked on unidirectional LAS composites tested at elevated temperatures in air. High maximum fatigue stresses which cause matrix microcracking lead to brittle failure of the composite due to the oxygen infiltrating the composite as discussed earlier. Fatigue resistance is better at maximum fatigue stresses below the microcracking stress although, depending on the test temperature, there may be some reduction in residual strength.

Cross-plied LAS-II and LAS-III composites fail within 10^5 cycles when the maximum fatigue stress exceeds 0.7 ± 0.05 of the tensile strength. This is similar to the behaviour of the unidirectional composites, however the lay-up has an effect on the residual strength of specimens which survive 10^5 cycles. In contrast to the unidirectional composites, the residual strength of the cross-plied materials appears not to depend on the maximum fatigue stress and is comparable to the initial strength (Table 4.8).

4.4.3 Carbon–carbon composites

A range of composites is available which consist of a fibrous carbon in a carbonaceous matrix. All the carbon fibres described in Chapter 2 have been employed in carbon–carbon composites in a number of forms, e.g., chopped fibres and continuous fibre mats. The carbon matrix is equally as variable as

Table 4.8 Room temperature fatigue behaviour of SiC fibre reinforced LAS. (Sources: Prewo, 1987; Minford and Prewo, 1986)

Material	Tensile strength, $\hat{\sigma}_{1T}$ (MPa)	Max. fatigue stress, σ_{max} (MPa)	$\dfrac{\sigma_{max}}{\hat{\sigma}_{1T}}$	Fatigue cycles	Residual strength (MPa)
LAS–I (unidirectional)					
	261	207	0.79	1.9×10^2	
	261	207	0.79	2.2×10^3	
	261	172–138	0.66–0.53	10^{5a}	286–226
LAS–II (unidirectional)					
	550	355	0.65	10^{5a}	485
	550	310	0.56	10^{5a}	525
	550	275	0.50	10^{5a}	485
	550	225	0.41	10^{5a}	620
LAS–III (unidirectional)					
	575	456	0.79	5×10^1	
	575	421	0.73	5×10^2	
	575	357	0.62	10^{5a}	462
	575	315	0.55	10^{5a}	480
	575	280	0.49	10^{5a}	602
	575	223	0.39	10^{5a}	538
LAS–II (cross-plied)					
	325	361	1.10	6×10^1	
	325	285	0.88	1.2×10^2	
	325	280	0.86	1.5×10^2	
	325	280	0.86	3×10^2	
	325	262	0.81	10^2	
	325	256	0.79	3×10^4	
	325	245–210	0.75–0.65	10^{5a}	385–315
LAS–III (cross-plied)					
	269	181	0.67	10^4	
	269	179	0.67	3×10^3	
	269	179	0.67	5×10^3	
	269	186–163	0.69–0.61	10^{5a}	291–227

[a]Specimen had not fractured after 10^5 cycles.

it may be highly crystalline graphite or glassy depending on production details. Finally the porosity content can be negligible or as high as 80%. Bearing in mind the possible combinations of fibre type, morphology and proportion, matrix structure and proportion, and porosity content, it is not surprising that the properties differ greatly from composite to composite.

(a) *Porous carbon–carbon composites*

The major application of carbon–carbon composites with porosity contents in the range 70 to 90% is as insulation at high temperatures. These composites are also known as *carbon bonded carbon fibre*, CBCF, and can only be employed under vacuum or in an inert atmosphere as oxidation occurs at temperatures above about 400 °C.

The fibres used for CBCF are low modulus fibres produced from a rayon precursor and are typically a couple of millimetres in length. A slurry is made with the fibres together with some ground recycled CBCF, termed *rework*. Phenolic resin, a binder and water are added to the slurry and mixed. The mixture is pumped into a mould and the water extracted under vacuum. It is at this stage that some orientation of the fibres takes place perpendicular to the direction of application of the vacuum to give an ill-defined two-dimensional mat effect (Figures 4.24 and 10.21). The vacuum moulded components go through various drying procedures before a carbonization heat treatment at an intermediate temperature ($\sim$950 °C) which gives a carbon yield from the phenolic resin of about 50%. The final stage is a high temperature heat treatment which is terminated at a low pressure in order to remove gaseous impurities. The final product is extremely pure; the carbon content is greater than 99.9%.

The properties are determined mainly by the porosity content and, because of the orientation of the fibres, are anisotropic (Figure 4.25). A number of properties for a CBCF of about 90% porosity are presented in Table 4.9. It can be seen that the strength is low compared with that of other CMCs but is adequate for insulating applications. Of particular interest is the anisotropy of the thermal conductivity in an insulating component. Let us consider the insulation for a furnace. The CBCF insulation is produced such that the low thermal conductivity direction is perpendicular to the furnace wall, i.e., parallel to the direction in which heat flow is to be minimized. The high thermal conductivity of the CBCF perpendicular to this direction assists in the achievement of a uniform temperature in the furnace.

(b) *Dense carbon–carbon composites*

Continuous and discontinuous fibres are used for the reinforcement in these dense composites. Discontinuous fibres tend to be used to fabricate large components, to produce more isotropic materials and to improve inter-laminar strengths. The discontinuous fibres are normally made from rayon or pitch precursors. Continuous fibre reinforcement is generally employed to exploit the good mechanical properties of the better quality fibres and/or to produce a material with a desired degree of anisotropy. Applications of these composites include disc brakes for racing cars and aircraft, gas turbine

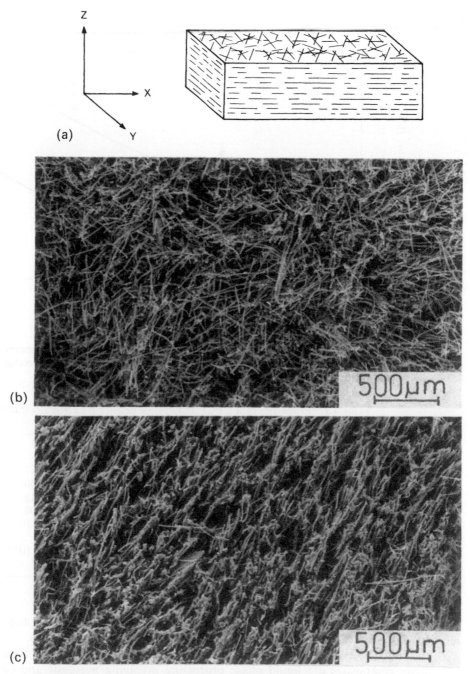

Figure 4.24 Orientation of the fibres in carbon bonded carbon fibre material: (a) schematic representation with the axis system used to denote planes and directions; (b) micrograph in a plane perpendicular to the direction of application of vacuum, i.e., *xy* plane; (c) micrograph in a plane parallel to the direction of application of vacuum, e.g. *zx* with *x* direction at 55° to the bottom of the micrograph. (Courtesy I. J. Davies.)

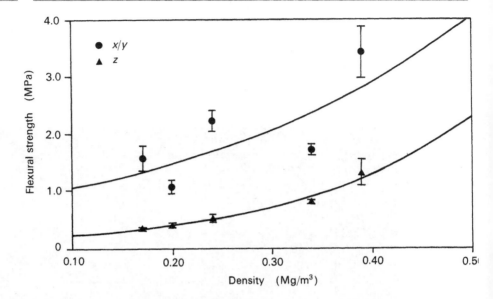

Figure 4.25 Strength (measured in three-point bend) of carbon bonded carbon fibre as a function of density and orientation. z and x/y denote the direction of the tensile stress in the bend test (see Figure 4.24 for axis system used).

Table 4.9 Properties of a CBCF materal of density 0.17 Mg/m³. (Source: Calcarb Ltd, UK and I. Davies (Ph.D. thesis, London 1992))

	x/y^a	z^a
Young's modulus (MPa)	105.6	9.57
Compressive strength (MPa)	0.78	0.60
Flexural strength (MPa)	1.03	0.15
Tensile Strength (MPa)	0.48	0.08
Coeff. of thermal expansion 0–1000 °C (K^{-1})	3.0×10^{-6}	2.8×10^{-6}
Electrical Resistivity (Ω m)	1.1×10^{-3}	4.1×10^{-3}

[a]measurement direction (see Figure 4.24(a) for axis system used).

components (e.g., exhaust nozzle flaps and seals), nose cones and leading edges for missiles, and biomedical implants such as bone plates.

The two main techniques used to produce the matrix of these dense composites are liquid phase processing involving pyrolysis and carbonization, and chemical vapour infiltration. Any liquid which flows sufficiently to permit impregnation and which gives a high carbon yield on pyrolysis of at least 40% can be used as the matrix precursor in liquid phase processing. In practice two precursors types dominate, namely *thermosetting resins* and *pitches*.

An advantage in using thermosetting resins, such as *phenolic, furan* and *polyimide,* in the development stage of carbon–carbon composites was the knowledge available from processing of polymer matrix composites from these resins. Typically these resins polymerize at low temperatures ($\sim 250\,°C$) to form a cross-linked amorphous material. Higher temperatures of the order of $1000\,°C$ are required to carbonize to a glassy, isotropic carbon. The carbon yield is usually in the range 45 to 65% but in some cases may reach 80% (Table 4.10). Multiple impregnation and carbonization cycles are necessary to obtain a dense composite. A cycle may take up to three days as the gases formed during pyrolysis have to be released slowly in order to avoid structural damage, therefore the complete fabrication process is time-consuming. The glassy carbon matrix does not transform readily to graphite and temperatures in excess of $2500\,°C$, together with stress (applied or residual), are needed for graphitization to occur at a reasonable rate.

The pitches, such as petroleum pitch, used in carbon–carbon composite fabrication are a mixture of *polynuclear aromatic hydrocarbons* which are thermoplastic in nature. The yield from pitches at atmospheric pressure is about 50%, i.e., similar to that obtained from many thermosetting resin precursors. The main structural feature on heating a pitch is the formation of small volumes (0.1 µm diameter) of a highly orientated structure similar to that of a liquid crystal. These are known as *mesophases* and on prolonged heating they coalesce to form larger regions with an ordered structure. The lamellar arrangement of the ordered structure is favourable for the formation of graphite when heated at temperatures of the order of $2500\,°C$. A consequence of the differences in the structures of the matrix from pitch and thermosetting resin precursors is that composites with the former precursor have higher densities (Figure 4.26).

The yield from a pitch at atmospheric pressure is around 50%, however, unlike a resin, the yield from a pitch is greatly improved on increasing the pressure. Yields approaching 90% have been achieved at pressures in excess of about 1000 atm. (100 MPa) (Figure 4.27). There is clearly an advantage in processing at high pressures and *hot isostatic pressure impregnation carbonization* (HIPIC), where, as the name suggests, an isostatic pressure is applied during heating, is now an available method of composite production.

Table 4.10 Carbon yields on pyrolysis of thermosetting resins and petroleum pitch

Material	% Yield
Phenolic resin	45
Furan resin	60
Polyimide	80
Poly-phenylene	80
Pitch	50
Pitch (100atm)	80

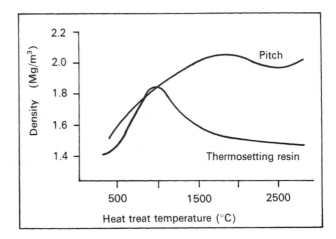

Figure 4.26 Graph of density as a function of heat treatment temperature demonstrating the higher density of pitch precursor carbon–carbon composites. (Source: Burns and McAllister, 1976.)

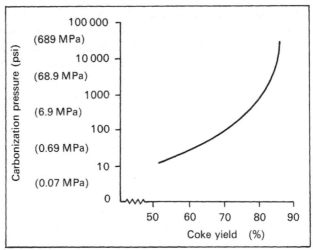

Figure 4.27 Plot showing the improvement in yield from petroleum pitch precursor with increase in pressure. (Source: Latchman *et al.*, 1978.)

With suitable combinations of temperature and pressure, carbon may be deposited in a desirable form by CVI involving the thermal decomposition of a *hydrocarbon*. Although propane, benzene and other hydrocarbons have been used, traditionally *methane*, which decomposes according to the following simplified equation, is the reactant:

$$CH_4(g) = C(s) + 2H_2(g).$$

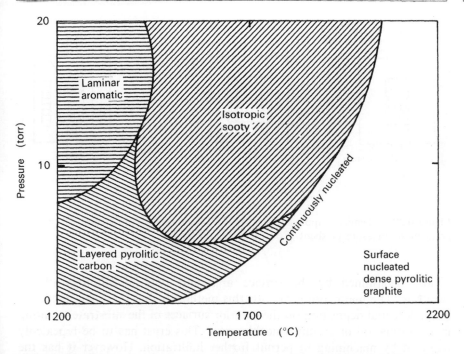

Figure 4.28 Effect of pressure and temperature on the structure of CVD carbon. (Source: Kotlensky, 1973.)

Figure 4.28 shows the dependence of the structure of CVD carbon on the deposition temperature and pressure. At low pressures and temperatures high density layered *pyrolitic* carbon is formed. Pyrolitic carbon is a microcrystalline carbon with a grain size that is so small that it appears to be non-crystalline when examined by many conventional techniques such as X-ray diffraction. Dense pyrolitic graphite is deposited at higher temperatures by a surface nucleated process. Other deposition conditions covered by the temperature–pressure range of this figure lead to the production of unsatisfactory structural carbon forms, for example soot.

In CVI the structure of the carbon and the effectiveness of the infiltration process depend critically not only on the temperature and pressure but also on the characteristics of the substrate such as surface roughness, surface area, pore morphology and size. Concerning the latter, there must be an open, permeable pore structure and the pore size must be small for successful densification in a reasonable time scale.

There are essentially three main CVI methods for carbon production, namely isothermal, temperature gradient and pressure gradient methods (Figure 4.29). In the *isothermal method* the gas and the substrate are maintained at a constant temperature of around 1100 °C. The infiltration is carried out at a reduced pressure of 0.6–6 kPa (5–50 torr) and the flow is

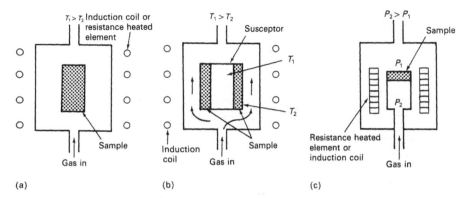

Figure 4.29 Chemical vapour infiltration (CVI) methods: (a) isothermal; (b) thermal gradient; (c) pressure gradient. (Source: Savage, 1988.)

mainly determined by the surface area and pore morphology of the substrate. The major problem with this method is that there is a tendency for preferential deposition on the exterior surfaces of the substrate resulting in the formation of an impermeable crust. This crust has to be repeatedly removed by machining to permit further infiltration. However it has the advantages that it is relatively simple and a number of substrates may be processed at the same time.

In the *thermal gradient method* the hottest part of the substrate is the interior and the outer surface is cooler. This is achieved by supporting the substrate on a susceptor which is inductively heated (Figure 4.29(b)). Surface crusting is eliminated as the deposition rate is less at the cooler outer surface of the substrate and deposition proceeds radially outwards from the interior. The normal processing conditions are atmospheric pressure and an interior temperature of 1100 °C. The thermal gradients have to be carefully controlled and consequently the method is only used for single work-pieces.

The final method, the *pressure gradient method*, is also only suitable for single work-pieces but is not as extensively used as the other CVI methods. In this method the precursor gas is forced into the interior of the substrate in order to obtain a more homogeneous deposit than by the isothermal method.

Having discussed the various fabrication processes let us now turn our attention to the mechanical properties of dense carbon–carbon composites. Stress–deflection curves from flexural tests on composites fabricated by liquid phase processing using a thermosetting resin are given in Figure 4.30. It can be seen that the mechanical properties are sensitive to the heat treatment. At low heat treatment temperatures of below 2500 °C a glassy carbon matrix is formed which bonds strongly to the fibres. This results in a brittle composite as matrix cracks propagate readily across the strong

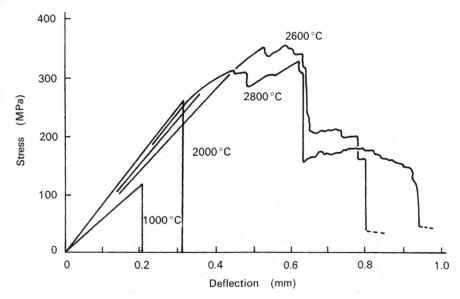

Figure 4.30 Stress–deflection curves obtained from 4-point bend tests on thermo-setting precursor matrix C–C composites showing the improved fracture behaviour associated with high temperature heat treatment. (Source: Tanaka *et al.*, 1978.)

interface and through the fibres. On the other hand, higher processing temperatures are associated with a more desirable failure mode and increased toughness. This is a consequence of graphitization which reduces the fibre–matrix bond strength and facilitates debonding and pull-out.

As a general rule the fracture behaviour of composites fabricated using a pitch precursor, or by CVI, is similar to that of the high temperature treated, thermosetting resin precursor matrix material.

Of course the mechanical behaviour is not solely determined by the processing of the matrix but is also strongly dependent on the proportion and form of the fibre reinforcement. This is illustrated by the schematic stress–strain curves for one- (1-D), two- (2-D) and three- (3-D) dimensional woven carbon fibre reinforced composites presented in Figure 4.31. The 1-D composite is strong but brittle whereas the 3-D has much improved toughness which is achieved at the expense of the strength; the 2-D has properties intermediate to those of the 1-D and 3-D. The low toughness of the 1–D, and to a certain extent the 2-D, composites is attributed to the poor interlaminar properties due to the brittleness of the matrix. The interlaminar properties may be enhanced by the incorporation of fine graphite particles in the matrix to give crack blunting and deflection. This is an example of an *hybrid* composite and it can have an interlaminar toughness which is double that of the particle-free matrix composite.

Fatigue studies have shown that the dependence of the number of cycles to failure on the dynamic stress is similar to that observed for carbon

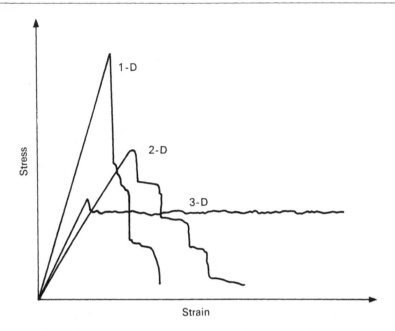

Figure 4.31 Schematic stress–strain curves illustrating the effect of the form of the substrate on strength and toughness. (Source: Fritz *et al.*, 1979.)

fibre-reinforced polymers (Figure 4.32). However, after long periods under a fluctuating stress it appears that local microcracking of the matrix may occur to such an extent that small, unretained matrix particles are formed which can be lost from the composite as a fine dust.

The forte of carbon–carbon composites is that they maintain their mechanical properties to extremely high temperatures. Their performance at elevated temperatures is particularly impressive compared with other high temperature materials when viewed in terms of specific properties as illustrated by the specific strength data of Figure 4.33. (Note that the adjective 'dense' is used to differentiate between these composites and the porous CBCF materials; the density of dense carbon–carbon composite is in fact low, typically in the range 1.4–1.7 Mg/m^3.) However, as for CBCF material, dense carbon–carbon composites suffer from a lack of oxidation resistance. Protective coatings may be applied to components and these currently operate reasonably well up to about 1400 °C, although coating systems capable of offering protection up to 1800 °C are being developed.

For a coating to be successful it must satisfy the following criteria: (i) be mechanically, chemically and thermally compatible with the composite, (ii) adhere to the composite, (iii) prevent diffusion of oxygen from the environment through to the composite, and (iv) prevent diffusion of carbon

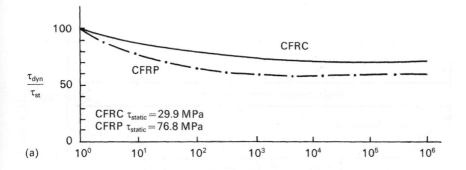

(a)

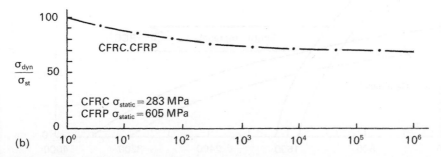

(b)

Figure 4.32 Comparison of the fatigue performance of carbon–carbon composites (CFRC) and carbon fibre reinforced polymers (CFRP): (a) torsion; (b) flexural. (Source: Huttner *et al.*, 1982.)

from the composite through to the environment. Most protective systems are complex as these criteria are stringent and difficult to satisfy by means of a single coating. To illustrate the problems let us consider thermal compatibility. Figure 4.34 shows that the expansion coefficient of most of the ceramics which offer some potential for an oxidation resistant coating is much greater than that of carbon–carbon composites. Large differences in the expansion behaviour of the composite and the coating during cooling from the coating temperature or the service temperature lead to cracking of the coating and loss of oxidation protection. For this reason SiC or Si_3N_4, both of which have relatively low coefficients of thermal expansion, are commonly employed for the *primary oxidation barrier coat*. Nevertheless, even these coatings are prone to some cracking and hence a *secondary protection system* is also required. The secondary system is usually based on incorporating particles of a glass former (e.g., Si or B) into the matrix and/or having an additional glass coating. Either secondary system results in the cracks in the primary coat being filled up at elevated temperatures with a glassy phase.

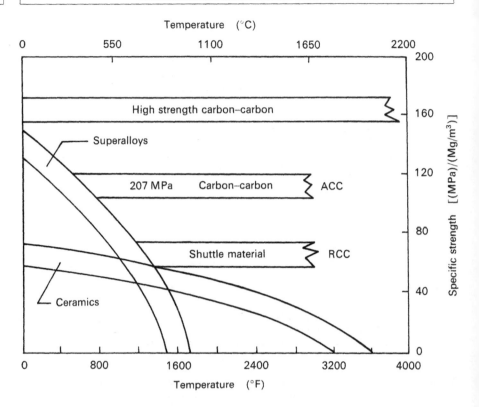

Figure 4.33 Specific density as a function of temperature for carbon–carbon composites and other high temperature materials: shuttle material – produced from low modulus rayon precursor fibres; ACC – made using woven carbon cloth; high strength carbon–carbon – made with unidirectional carbon fibres interplied with woven cloth. (Source: Buckley, 1988.)

4.5 SUMMARY

The fabrication routes for ceramic matrix composites (CMCs) are varied and include mixing of the reinforcement with the powdered matrix and pressing, techniques involving slurries, and vapour deposition methods. A common feature of the fabrication routes is that usually high temperatures are employed at some stage in the process. For this reason, and also because the reinforcement should not degrade the high temperature capability of the matrix, the reinforcements used normally have good high temperature properties.

Monolithic ceramics are brittle and one of the main objectives is to produce composites with improved toughness. We have seen that significant

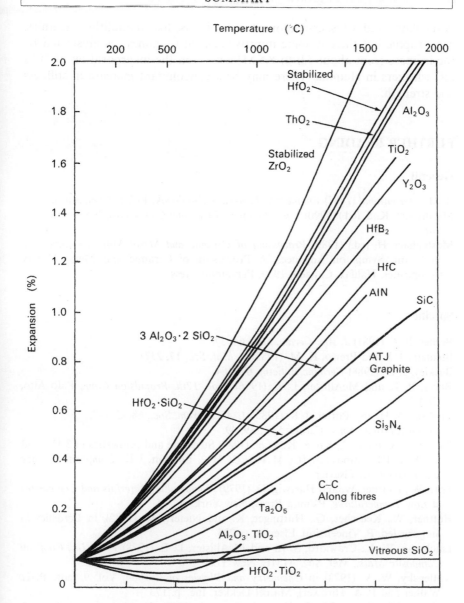

Figure 4.34 Thermal expansion characteristics of candidate ceramics for primary oxidation barrier coatings on carbon–carbon composites. (Source: Strife an Sheehan, 1988.)

improvements in toughness, and some load carrying capacity after failure has commenced, can be achieved with continuous fibre reinforcements, e.g., the toughness of LAS glass-ceramic and LAS–50% SiC (unidirectional) composite is 1.5 and 17.0 MPa m$^{1/2}$ respectively. Failure of particulate- and whisker-reinforced ceramics is, like monolithic ceramics, catastrophic but there is some improvement in toughness as illustrated by the increase from about 5.0 to 8.5 MPa m$^{1/2}$ on the incorporation of 30% SiC whiskers in alumina. There may be a concomitant increase in stiffness and strength.

FURTHER READING

General

ASM International (1987) Engineered Materials Handbook, Vol. 1, Composites.
Mazdiyasni, K. S. (ed.) (1990) *Fibre Reinforced Ceramic Composites*, Noyes Publication.
Mostaghaci, H. (ed.) (1989) *Processing of Ceramic and Metal Matrix Composites*. Proc. Int. Symp. on Advances in Processing of Ceramic and Metal Matrix Composites, Halifax, Canada, 1989, Pergamon Press.

Specific

Becher, P. F. (1981) *J. Am. Ceram. Soc.*, **64**, 37.
Brennan, J. J. and Prewo, K. M. (1982) *J. Mat. Sci.*, **17**, 2371.
Buckley, J. D. (1988) *Ceramic Bulletin*, **67**, 364.
Burns, R. L. and McAllister, L. E. (1976) *Proc. 12th. Propulsion Conf.*, Palo Alto, USA.
Chokshi, A. H. and Porter, J. R. (1985) *J. Am. Ceram. Soc.*, **68**, C–144.
Claussen (1976) *J. Amer. Ceram. Soc.*, **59**, 49.
Fareed, A. S., Fang, P., Ko, F. K., *et al.* (1987) Structure and properties of 3-D braid SiC/LAS–III composites. *ICCM VI/ECCM 2*, London, UK; *Composite Science and Technology*, Elsevier.
Fritz, W., Huttner, W. and Hartwig, G. (1979) *Nonmetallic Materials and Composites at Low Temperatures*, Plenum Press, New York, p. 245.
Huttner, W., Keuscher, G., Huttinger, K. and Nietert, M. (1982), in *Ceramics in Surgery*, (ed. P. Vincenzini), Elsevier, p. 225.
Latchman, W. L., Crawford, S. A. and McAllister, L. E. (1978) *Proc. Int. Conf. on Composite Mats.*, Met. Soc. AIME, New York.
Kotlensky, W. V. (1973) in *Chemistry and Physics of Carbon*, Vol. 9, (eds. P. L. Walker and P. A. Thrower), Marcel Dekker, Inc., p. 173.
Mah, T., Mendiratta, M. G., Katz, A. P. *et al.* (1985) *J. Amer. Ceram. Soc.*, **68**, C–248.
Minford, E. and Prewo, K. M. (1986), in *Materials Science Research*, Vol 20, (eds. R. E. Tressler, G. L. Messing, C. G. Pantano and R. E. Newnham), Plenum Press, p. 561.
Prewo, K. M. (1986) *J. Mat. Sci.*, **21**, 3590.
Prewo, K. M. (1987) *J. Mat. Sci.*, **22**, 2695.
Prewo, K. M. (1989) Fibre reinforced glasses and glass-ceramics, in *Glasses and Glass-Ceramics* (eds. M. H. Lewis), Chapman and Hall, London, p. 336.

Savage, G. (1988) *Metals and Materials*, **4**, 544.
Strife, J. R. and Sheehan, J. E. (1988) *Ceramic Bulletin*, **67**, 369.
Tanaka, H. *et al.* (1978) *Trans. Japan. Soc. Comp. Mat.*, **4**, 37.
Tiegs, T. N. and Becher, P. F. (1987) *J. Amer. Ceram. Soc.*, **70**, C–109.
Thompson, I. and Rawlings, R. D. (1992) *J. Mat. Sci.*, **27**, 2823, 2831.
Wei, C. C. and Becher, P. F. (1985) *Am. Ceram. Soc. Bull.*, **64**, 298.

PROBLEMS

4.1   Discuss the effect of environment on the performance of the following composite materials:

   (i) zirconia toughened alumina,
   (ii) SiC fibre reinforced LAS, and
   (iii) dense carbon–carbon.

4.2   What ceramic matrix composite would you select, giving your reasoning, for the following applications:

   (i) cutting tool,
   (ii) insulation for an inert atmosphere furnace, and
   (iii) leading edge of a missile.

SELF-ASSESSMENT QUESTIONS

Indicate whether statements 1 to 6 are true or false.

1.   Most monolithic technical ceramics are processed in a similar manner to metals, namely melting and casting.

   (A) true
   (B) false

2.   Most of the processing routes for ceramic matrix composites involve high temperatures and can only be employed with reinforcements capable of withstanding such temperatures.

   (A) true
   (B) false

3.   Chemical vapour infiltration is simply chemical vapour deposition applied to the formation of a matrix in a porous preform.

   (A) true
   (B) false

4. During the pressing stage of the production of SiC whisker reinforced alumina the whiskers tend to become orientated perpendicular to the pressing axis.

  (A) true
  (B) false

5. Stabilizing oxides, such as yttria, are added to the zirconia in zirconia-toughened alumina in order to enhance the transformation to the monoclinic state.

  (A) true
  (B) false

6. The carbon yield from pitch is greatly improved on increasing the pressure.

  (A) true
  (B) false

For each of the statements of questions 7 to 12, one or more of the completions given are correct. Mark the correct completions.

7. A slurry is

  (A) a sol,
  (B) a sol that has lost some liquid,
  (C) a dispersion in a liquid of small particles of less than 100 nm,
  (D) a suspension of large (typically 1–50 µm) particles in a liquid,
  (E) a gel.

8. Production of a CMC by impregnation of a preform with a ceramic-producing polymer

  (A) requires a heat treatment which may be at relatively low temperature in the range 600 °C to 1000 °C,
  (B) involves pyrolysis of the polymer,
  (C) is extremely rapid as the ceramic yield is high,
  (D) requires multiple impregnations as the ceramic yield is low,
  (E) is used to produce some carbon–carbon composites,
  (F) is used to produce zirconia-toughened alumina.

9. Reinforcing alumina with SiC whiskers

  (A) enhances the thermal shock resistance,
  (B) lowers the coefficient of thermal expansion,
  (C) decreases the thermal conductivity,
  (D) increases the density,
  (E) improves the toughness.

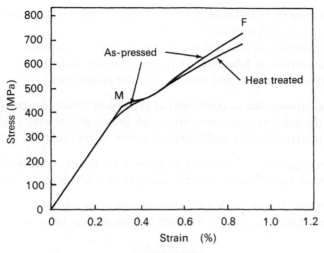

Figure 4.35

10. The room temperature tensile stress–strain curves in Figure 4.35 are for LAS-SiC fibre composites. The break-down of linearity in the curves at the point marked M is due to

(A) plastic deformation associated with dislocation motion,
(B) fracture of the fibres,
(C) matrix microcracking,
(D) viscous flow of a glassy phase,
(E) fibre pull-out.

11. Chemical vapour deposition of a carbon matrix

(A) is also commonly known as gas carbonization,
(B) is also commonly known as chemical vapour infiltration,
(C) produces a brittle composite with a glassy matrix,
(D) is independent of deposition pressure,
(E) is dependent on deposition temperature.

12. Oxidation protection of carbon–carbon composites

(A) is only necessary at temperatures in excess of 1400 °C,
(B) currently is only satisfactory up to temperatures of 1400 °C,
(C) normally involves an alumina or zirconia coat,
(D) normally involves a silicon nitride or carbide coat,
(E) may consist of a primary oxidation barrier coat and secondary protection.

Each of the sentences in questions 13 to 19 consists of an assertion followed by a reason. Answer:

(A) if both assertion and reason are true statements and the reason is a correct explanation of the assertion,

(B) if both assertion and reason are true statements but the reason is not a true explanation of the assertion,

(C) if the assertion is true but the reason is a false statement,

(D) if the assertion is false but the reason is a true statement,

(E) if both the assertion and reason are false statements.

13. Thermal stresses due to coefficient of expansion differences between the matrix and the reinforcement are important in ceramic matrix composites *because* the matrix is brittle and therefore prone to cracking.

14. Melt techniques are commonly employed for the processing of CMCs *because* ceramic melts flow easily and rarely react with the reinforcement.

15. Reinforcing alumina with SiC whiskers increases both the stiffness and the specific modulus *because* the fracture toughness may be increased by up to 50%.

16. All the zirconia particles in a zirconia-toughened alumina always exist in the monoclinic crystal form at room temperature *because* the tetragonal to monoclinic transformation is athermal.

17. Carbon bonded carbon fibre is widely used as an insulating material *because* the matrix precursor is phenolic resin and rework is a constituent.

18. The toughness of a carbon–carbon composite produced using a thermosetting resin precursor is improved by heat treatment at temperatures in excess of about 2500 °C *because* a large quantity of hydrogen gas is given off from the matrix at these temperatures.

19. The strength of 3-D carbon–carbon composites is high, but usually they are extremely brittle *because* the interlaminar properties of the matrix may be improved by the incorporation of fine particles of graphite.

20. Fill in the missing words in this passage. Each dash represents a letter.

Glass-ceramics are polycrystalline materials which are formed when _ _ _ _ _ _ _ of a suitable composition are heat treated and hence undergo controlled _ _ _ _ _ _ _ _ _ _ _ _ _ _ _ _. Currently there is considerable interest in reinforcing glass-ceramics with continuous fibres of _ _ _ _ _ _ _ _ _ _ _ _ _ _. The usual production route involves drawing the fibres through a _ _ _ _ _ _ to produce a monolayer tape which is stacked in the desired sequence and _ _ _ _ _ _ _ _ _ _ to consolidate into a composite. The most widely investigated glass-ceramic matrix composites have matrices based on the _ _ _ _ _ _ _ _ _ _ _ _ _ _ _ _ _ _ _ _ _ _ (LAS) system. The room temperature mechanical properties of these LAS matrix composites are good, especially the toughness. The good toughness is attributed to a _ _ _ fibre–matrix interfacial strength due to the presence of a _ _ _ _ _ _ rich interfacial layer. However the mechanical performance is degraded at elevated temperature as a result of _ _ _ _ _ _ microcracking allowing oxygen from the environment to penetrate the composite.

ANSWERS

Self-assessment

1. B; 2. A; 3. A; 4. A; 5. B; 6. A; 7. D; 8. A, B, D, E; 9. A, B, E;
10. C; 11. B, E; 12. B, D, E; 13. A; 14. E; 15. B; 16. D; 17. B;
18. C; 19. D; 20. glasses, crystallization, silicon nitride, slurry, hot pressed, lithium aluminosilicate, low, carbon, matrix.

5 | Polymer matrix composites

5.1 INTRODUCTION

The most common matrix materials for composites are polymeric. The reasons for this are twofold. First, as we saw in Chapter 1, in general the mechanical properties of polymers are inadequate for many structural purposes. In particular their strength and stiffness are low compared with metals and ceramics. This meant that there was a considerable benefit to be gained by reinforcing polymers and that the reinforcement, initially at least, did not have to have exceptional properties.

Secondly, as we will see in section 5.3, the processing of polymer matrix composites (PMCs) need not involve high pressures and does not require high temperatures. It follows that problems associated with the degradation of the reinforcement during manufacture are less significant for PMCs than for composites with other matrices. Also the equipment required for PMCs may be simpler. For these reasons polymer matrix composites developed rapidly and soon became accepted for structural applications. Today glass-reinforced polymers are still by far the most used composite material in terms of volume with the exception of concrete.

A simple classification of polymers is given in Figure 5.1; the three classes, thermosets, thermoplastics and rubbers, are all important as far as matrices of PMCs are concerned. Within any class there are many different polymers, e.g., epoxy, polyester, polyimide and phenolic are all thermosets. Even a given polymer, such as polyester, exists in many forms; there are a large number of formulations, curing agents and fillers which result in an extensive range of properties for polyesters. Indeed, polyesters, and other polymers, are often marketed according to their properties by the employment of descriptive terms including 'general purpose', 'chemically resistant' and 'heat resistant'. Bearing in mind also the variety of materials used for reinforcement, and the possible arrangements of the reinforcement (random chopped fibres, unidirectional, numerous weaves,

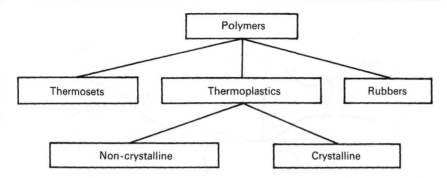

Figure 5.1 Simple classification of the polymers used for matrices of composites.

etc.) it is clear that the range of properties exhibited by PMCs is quite remarkable.

The main disadvantages of PMCs are their low maximum working temperatures, high coefficients of thermal expansion and hence dimensional instability, and sensitivity to radiation and moisture. The absorption of water from the environment may have many harmful effects which degrade mechanical performance, including swelling, formation of internal stresses and lowering of the glass transition temperature. There are of course exceptions, for example carbon fibre-reinforced polymers may be designed to have very low coefficients of thermal expansion and epoxies are radiation resistant.

5.2 POLYMER MATRICES

5.2.1 Thermosetting

It has been estimated that over three-quarters of all matrices of PMCs are thermosetting polymers. *Thermosetting polymers*, or *thermosets*, are resins which readily cross-link during curing. *Curing* involves the application of heat and pressure or the addition of a catalyst known as a curing agent or hardener.

Cross-linking is schematically illustrated in Figure 5.2(a); the bonds of the link are, like the bonds in the polymer chain, covalent. These strong bonds of the cross-links have the effect of pulling the chains together. This restricts the movement of the polymer chains and so increases the glass transition temperature to above room temperature. Consequently thermosets are brittle at room temperature and have low fracture toughness values, K_{IC}, of typically 0.5 to 1.0 MPa m$^{1/2}$. Also, because of the cross-links, thermosets cannot be reshaped by reheating: thermosets just degrade on reheating, and in some cases may burn, but do not soften sufficiently for reshaping.

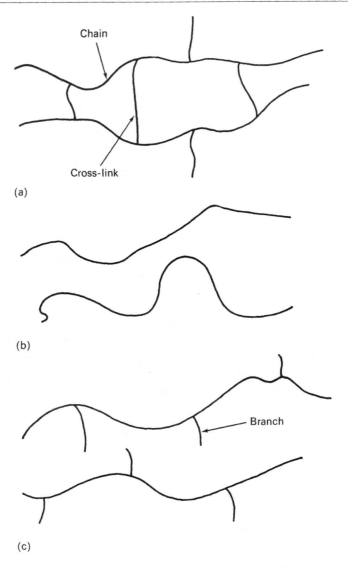

Figure 5.2 Arrangements of polymer chains: (a) cross-linked; (b) linear; (c) branched.

On the other hand many advantages accrue from the presence of the cross-links as demonstrated by the property data presented in Table 5.1. The stiffness is increased as the weak van der Waals bonding between polymer chains is replaced by the strong cross-links. Furthermore, thermosets may be used at higher temperatures as they have higher softening temperatures and better creep properties than thermoplastics. Finally, they are more resistant to chemical attack than most thermoplastics.

Table 5.1 Comparison of typical ranges of property values for thermosets and thermoplastics

	Thermosets	Thermoplastics
Young's modulus (GPa)	1.3–6.0	1.0–4.8
Tensile strength (MPa)	20–180	40–190
Fracture toughness		
K_{Ic} (MPa m$^{1/2}$)	0.5–1.0	1.5–6.0
G_{Ic} (kJ/m^2)	0.02–0.2	0.7–6.5
Maximum service temperature (°C)	50–450	25–230

Let us look at some of the thermosets in more detail, starting with *polyester* resins which dominate the market. A common resin, first developed in 1942, it consists of unsaturated (i.e., double bonds exist between certain carbon atoms) linear polyesters dissolved in styrene. Styrene is the cross-linking monomer and curing is effected by use of an organic peroxide initiator which generates free radicals leading to the formation of the three-dimensional, cross-linked network shown in Figure 5.3. Polyester resins are relatively inexpensive and have low viscosities, which is beneficial in many fabrication processes. However the shrinkage which occurs on curing is high (4–8%).

Epoxy resins are more expensive and are more viscous than polyester resins making impregnation of woven fabrics more difficult. Although curing may require temperatures up to 180 °C, epoxies have a major advantage in that they are usually cured in two or more stages. This allows preforms to be pre-impregnated with the epoxy in a partially cured state. The *pre-preg* may be stored, often at sub-zero temperatures, for a reasonable length of time before being moulded into the final shape and then cured. The shrinkage on curing is smaller than for polyesters, being typically 1–5%.

As for polyesters a wide range of formulations for epoxy resins is available. Generally they start as linear low molecular weight polymers (Figures 5.2(b) and 5.4(a), (b)). Many materials are able to act as curing agents for epoxy resins, e.g., polyamides and polyamines. During curing, active groups on the curing agent react with the epoxy, also known as epoxide, or the hydroxyl groups to produce the cross-links so forming the network structure. In the example given in Figure 5.4(c), the diamine hardener provides four active hydrogen atoms each of which is capable of reacting with an epoxide group. The mechanical properties depend on the particular resin system and the curing; generally epoxies are stiffer and stronger, but more brittle than polyesters as can be seen from the typical values presented in Table 5.2. Epoxies also maintain their properties to higher temperatures than polyesters.

Phenolics are the oldest of the thermosets discussed here but, nevertheless, because of their low cost and good balance of properties they still find many applications. Indeed because of their good fire resistance their usage has

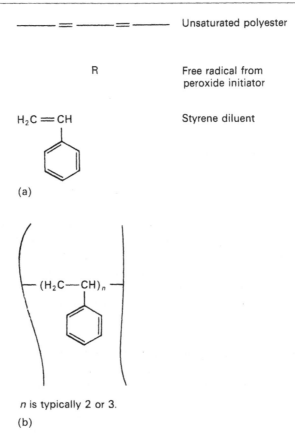

—— ≡ —— ≡ —— Unsaturated polyester

R   Free radical from
                        peroxide initiator

$H_2C = CH$ Styrene diluent

(a)

$-(H_2C—CH)_n-$

n is typically 2 or 3.

(b)

Figure 5.3 Polyester: (a) constituents of the resin; (b) cured resin with cross-linked network.

increased again lately because of the introduction of more stringent fire, smoke and toxicity regulations. Phenolic resins are produced by reacting phenol and formaldehyde; the characteristics of the resin product depending on the proportions of the reactants and the catalyst employed (Figure 5.5). Some resins, termed *resoles* or one-stage resins, contain the necessary constituents – including a catalyst – to be cured by heating, e.g., during hot moulding, without the need for any additional curing agent. Others, known as two-stage resins or *novalacs*, require the addition of an agent prior to fabrication. An undesirable feature of phenolic resins is that volatile by-products are evolved during curing, hence high pressures are often necessary in composite production.

Polyimides are more expensive and less widely used than polyesters or epoxies but can withstand relatively high service temperatures. The poly-imide chain, which is rigid and heat resistant owing to the presence of ring structures (Figure 5.6), results in a high stiffness, a low coefficient of thermal expansion and a service temperature as high as 425 °C for several hours and

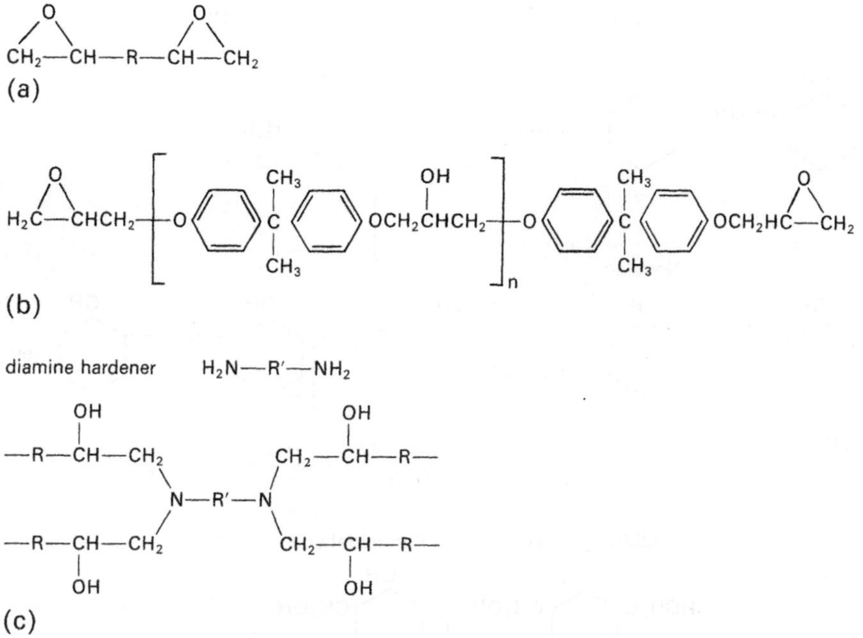

(a)

(b)

(c)

Figure 5.4 Epoxy (a) general structure – R can have various forms; (b) example of an epoxy structure, i.e., a specific form for R; (c) curing using a diamine hardener.

Table 5.2 Some typical properties of thermosets

	Epoxy	Polyester	Phenolics	Polyimides
Density (Mg/m^3)	1.1–1.4	1.1–1.5	1.3	1.2–1.9
Young's modulus (GPa)	2.1–6.0	1.3–4.5	4.4	3–3.1
Tensile strength (MPa)	35–90	45–85	50–60	80–190
Fracture toughness				
K_{Ic} ($MPa\,m^{1/2}$)	0.6–1.0	0.5		
G_{Ic} (kJ/m^2)	0.02			0.3–0.39
Thermal expansion	55–110	100–200	45–110	14–90
($10^{-6}\,K^{-1}$)				
Glass trans. temp. (°C)	120–190			

over 500 °C for a few minutes (Table 5.2). However, like the other thermosets the polyimides are brittle.

5.2.2 Thermoplastics

Thermoplastics readily flow under stress at elevated temperatures, so allowing them to be fabricated into the required component, and become solid and retain their shape when cooled to room temperature. These polymers may be repeatedly heated, fabricated and cooled and consequently

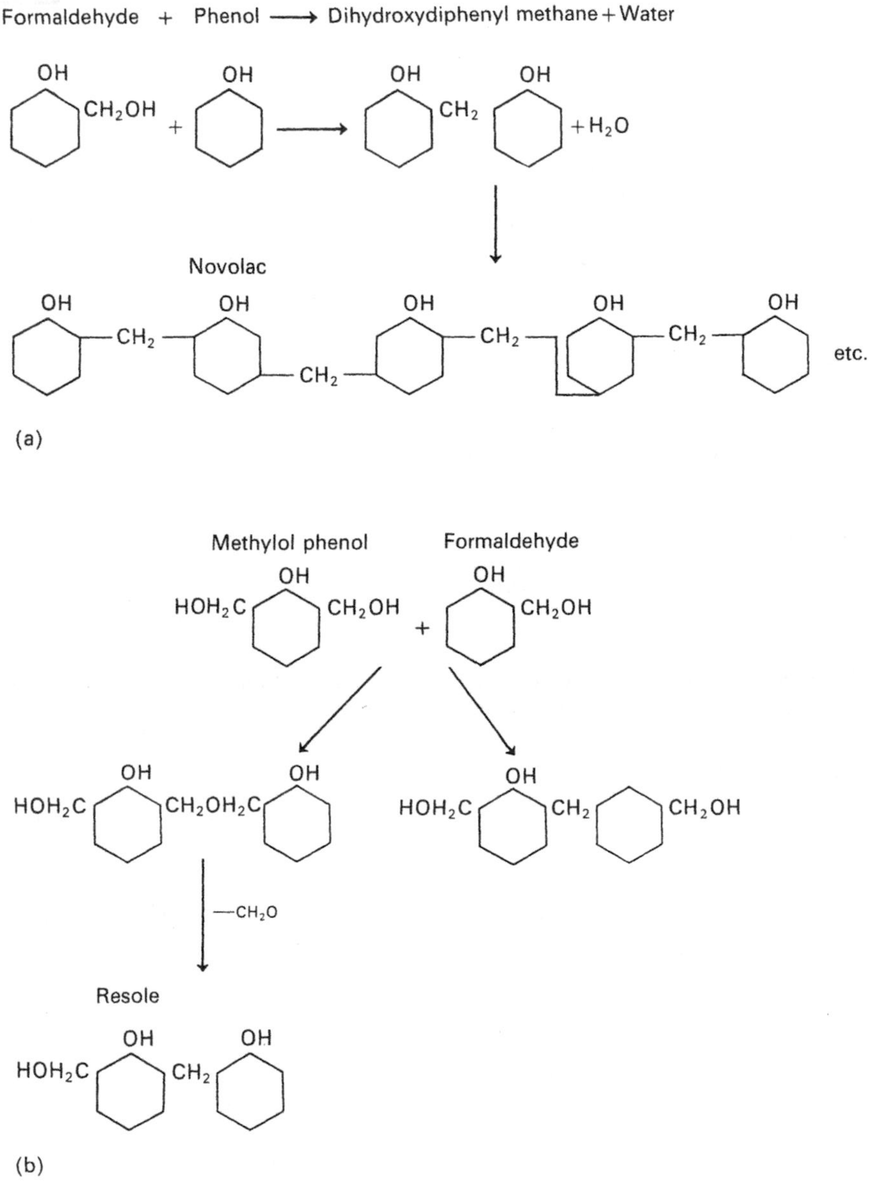

Formaldehyde + Phenol ⟶ Dihydroxydiphenyl methane + Water

(a)

(b)

Figure 5.5 Formation of phenolic resins: (a) novolacs; (b) resoles. (Source: Bill-meyer, 1971.)

scrap may be recycled, though there is evidence that this slightly degrades the properties probably because of a reduction in molecular weight. Well known thermoplastics include acrylic, nylon (polyamide), polystyrene, poly-ethylene and polyetheretherketone. Some typical properties are given in

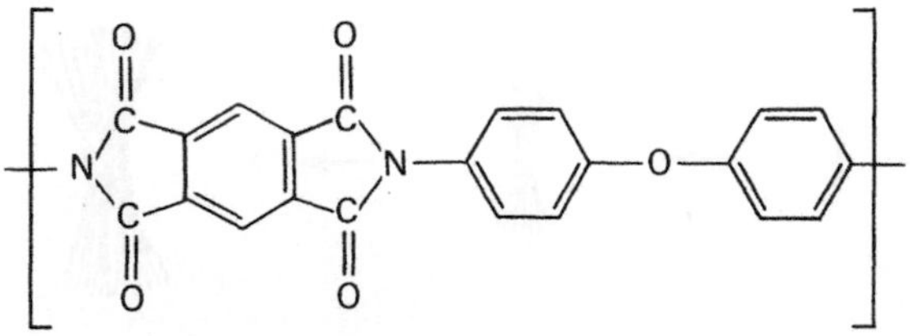

Figure 5.6 Polyimide structure.

Table 5.3 Properties of thermoplastics

	Acrylic (PMMA)	Nylon (6.6)	Polycarbonate	Polypropylene
Density (Mg/m^3)	1.2	1.1	1.1–1.2	0.9
Young's modulus (GPa)	3.0	1.4–2.8	2.2–2.4	1.9–1.4
Tensile strength (MPa)		60–70	45–70	25–38
Ductility (%)		30–100	90–110	100–600
Fracture toughness K_{Ic} (MPa m$^{1/2}$)	1.5			
Thermal expansion (10^{-6} K^{-1})	50–90	90		
Glass trans. temp. (°C)	90–105		150	
Melting point (°C)		261		175

Tables 5.1 and 5.3 and it can be seen that thermoplastics have superior toughnesses to thermosets.

We have seen that the properties of the thermosets were determined by the molecular structure and the same is true for the thermoplastics. Thermoplastics are linear polymers (Figure 5.2(b)): they do not cross-link to form a rigid network although the chains may be *branched* (Figure 5.2(c)). A branched chain is still discrete unlike the chains in a cross-linked network. The bonding between the chains in a thermoplastic is due to weak van der Waals forces which are easily broken by the combined action of thermal activation and applied stress. This is why the thermoplastics flow at elevated temperatures.

Provided the chains of a thermoplastic are relatively simple without bulky side-groups they are likely to fold back on themselves to form crystalline regions. The crystallization process on cooling a polymer melt proceeds as follows. First small plate-like lamellae of the folded crystalline structure are formed. Each lamella rapidly breaks up and splits into a sheaf of tape-like crystalline *fibrils*, separated by amorphous material. Continued splitting splays the sheaf outwards until the intimately mixed fibrils plus amorphous material has a spherical form. This is called a *spherulite* (Figure 5.7). The

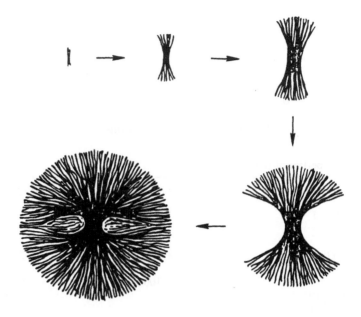

Figure 5.7 Development of a spherulite. (Source: Bassett, 1981.)

dimensions of spherulites can vary from submicrons to several millimetres; generally the slower the cooling the larger the spherulite size. Most thermoplastics are partially crystalline but the extent of crystallization depends on the details of the molecular structure and varies considerably. For example, linear *high density polyethylene* (HDPE), because of the simplicity and regularity of the chains, is able to attain up to 90% crystallization. In contrast *low density polyethylene* (LDPE) has a branched structure which interferes with the regular packing required for crystallization and hence crystallinity only reaches about 60%. At the other extreme from HDPE, *polymethylmethacrylate* (PMMA), which has irregularly spaced, bulky benzene side-groups is amorphous.

Let us turn out attention to a more detailed study of some selected examples of thermoplastics, commencing with *polycarbonate*. Polycarbonate is an amorphous polymer characterized by the –OCOO– unit and rings in its chain as illustrated in Figure 5.8. It has good impact resistance and also has reasonable high temperature properties being serviceable at temperatures up to about 140 °C. Other notable features are that it is transparent and can be easily processed using conventional moulding and extrusion methods.

Nylon is an accepted generic term for synthetic *polyamides*. They are formed either by the condensation of diamines with dibasic acids or by a polymerization process involving amino acids. By varying the reactants different molecular structures may be produced giving a wide range of

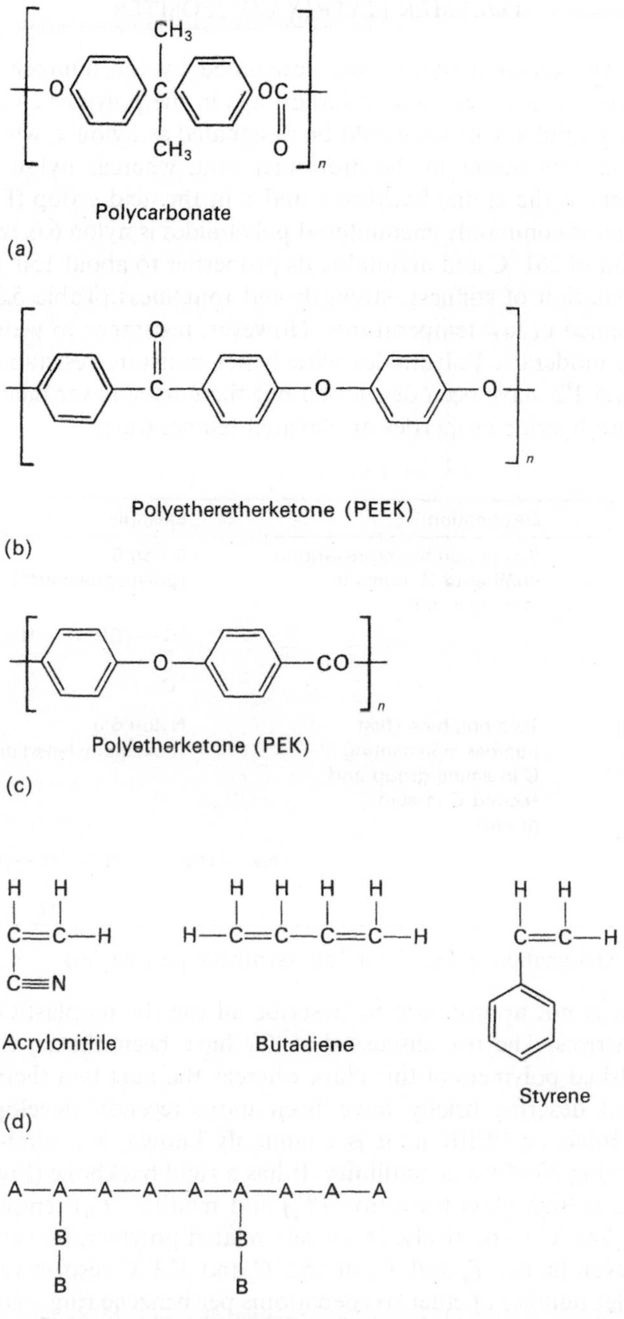

Figure 5.8 Structure of some thermoplastics: (a) polycarbonate; (b) polyetherether-ketone (PEEK); (c) polyetherketone (PEK); (d) the monomers of acrylonitrile–butadiene–styrene (ABS); (e) grafted copolymer structure of styrene acrylonitrile (SAN).

properties. The different nylons are designated by the number of carbon atoms attached to the acid and amine groups in the polymer chain. Thus a straight-chain aliphatic nylon would be designated as nylon x, where x is the number of carbon atoms in the monomer unit, whereas nylon $x.y$ has x carbon atoms in the amine backbone and y in the acid group (Figure 5.9). One of the most commonly encountered polyamides is nylon 6.6, which has a melting point of 261 °C and maintains its properties to about 150 °C. It has a good combination of stiffness, strength and toughness (Table 5.3) and the latter is retained at low temperatures. However, resistance to water absorption is only moderate. Polyamides with better moisture resistance, such as nylon 1.1 and 1.2, have been developed but these have lower melting points and therefore inferior properties at elevated temperatures.

Reactants	Designation	Example
Amino acids	Single number representing number of C atoms in monomer unit	Nylon 6 (polycaprolactum) $$[C\!-\!(CH_2)_5\!-\!NH]_n$$ $$\overset{\|}{O}$$
Diamines and dibasic acids	Two numbers (first number representing C in amine group and second C in acid group)	Nylon 6.6 (hexamethylenediamine) $$[NH\!-\!(CH_2)_6\!-\!NH\!-\!\underset{\underset{O}{\|}}{C}\!-\!(CH_2)_4\!-\!\underset{\underset{O}{\|}}{C}]_n$$

Figure 5.9 Designation system for nylons (synthetic polyamides).

Clearly it is not appropriate to describe all the thermoplastics currently used as matrices. The two discussed so far have been chosen to illustrate well-established polymers of this class whereas the next two thermoplastics that we will describe briefly have been more recently developed. *Polyetheretherketone*, or PEEK as it is commonly known, is a semi-crystalline polymer having 20–40% crystallinity. It has a rigid backbone (Figure 5.8(b)) which gives it high glass transition (T_g) and melting (T_M) temperatures of 143 °C and 343 °C respectively. (A closely related polymer, polyetherketone, PEK, has even higher T_g and T_M of 162 °C and 373 °C respectively because of the smaller number of ether oxygen atoms per benzene ring – compare the chain structures of Figures 5.8(b) and 5.8(c).) A consequence of the high T_g and T_M of PEEK is that mechanical properties are retained up to 120 °C and it can be employed at temperatures as high as 230 °C. In addition PEEK has a good toughness with a critical stress intensity factor of the order of 6 MPa m$^{1/2}$ and is resistant to many solvents.

It is possible to blend two or more polymers to obtain a multi-phase product with enhanced properties. In fact such polymers can be considered to be composite materials in their own right as well as serving as the matrix for PMCs. A thermoplastic example of such a multi-phase polymer is *acrylonitrile–butadiene–styrene* (ABS) which, as the name suggests, is a *terpolymer* (i.e., it contains three monomeric units – see Figure 5.8(d)). The constituent phases of ABS are a single-phase copolymer, *styrene acrylonitrile* (SAN) which is a grafted copolymer (Figure 5.8(e)), and a lightly cross-linked *butadiene elastomer*. The elastomer phase forms as a fine dispersion of particles which are grafted on to a layer of SAN making them compatible with the SAN matrix. The deformable butadiene particles blunt cracks and thereby increase the toughness. The proportion of the two phases can be adjusted according to need; increasing the butadiene elastomer content improves the toughness but reduces strength.

5.2.3 Rubbers

Natural rubber is obtained from the latex from the tree *Hevea Brasiliensis* and is over 98% polyisoprene. *Polyisoprene* exists in two forms and it is the *cis* form that is the main constituent of natural rubber. Nowadays a wide range of synthetic rubbers are available and these dominate the market. A number of synthetic rubbers are derived from *butadiene* (Figure 5.8(d)), such as *polybutadiene*, *styrene–butadiene* (SBR) and *nitrile–butadiene* (NBR) (Figure 5.10), and of these the copolymers SBR and NBR are the most important.

To achieve properties suitable for structural purposes most rubbers have to be *'vulcanized'*, i.e., the long molecules of the rubber have to be cross-linked. The cross-linking agent in vulcanization is commonly sulphur and the stiffness and strength increases with the number of cross-links.

5.3 PROCESSING OF PMCs

We will see in the following section that an extensive range of well established processing methods is available for PMCs. These vary from simple labour intensive methods suitable for one-offs to automated methods for rapidly producing large numbers of complex components. The method of production and PMC selected by a manufacturer will depend on factors such as cost, shape of component, number of components and required performance.

5.3.1 Hand methods

In *hand lay-up* the reinforcement is put down to line a mould previously treated with a *release agent* to prevent sticking and perhaps a *gel coat* to

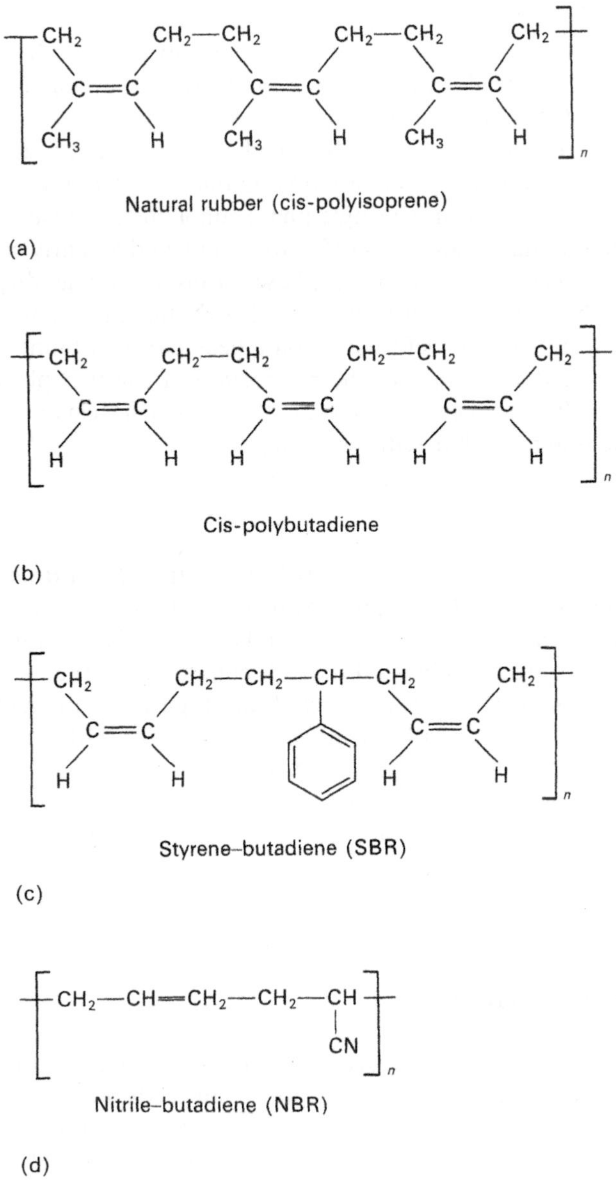

Natural rubber (cis-polyisoprene)

(a)

Cis-polybutadiene

(b)

Styrene–butadiene (SBR)

(c)

Nitrile–butadiene (NBR)

(d)

Figure 5.10 Structure of rubbers: (a) natural rubber (cis-polyisoprene); (b) poly-butadiene; (c) styrene–butadiene (SBR); (d) nitrile–butadiene (NBR).

give a decorative and protective surface. The reinforcement can be in many forms including woven rovings (WR) and chopped strand mat (CSM). The liquid thermosetting resin is mixed with a curing agent and applied with a brush or roller taking care to work it into the reinforcement. The most commonly employed resins are polyesters and curing is usually at room temperature. A prime consideration is the viscosity and the working time of the resin.

Hand lay-up requires little capital equipment but is labour intensive. It is particularly suited for one-offs or short production runs and can be used for large components such as hulls of boats and swimming pools. The main disadvantages of this method are the low reinforcement content of about 30 vol. % and the difficulty in removing all of the trapped air, hence the mechanical properties are not good.

The *spray-up* method has most of the advantages previously describe for the hand lay-up method, except that continuous fibre reinforcement is not possible, with the additional benefit of a greater rate of production. Chopped fibres, resin and catalyst are fed into a gun which sprays the mixture on to a mould prepared in a similar way, i.e., with a release agent and gel coat, to that for hand lay-up. The sprayed composite has to be rolled to remove entrapped air and to give a smooth surface finish.

A modification of these techniques is to spray or lay-up on the back of a preformed thermoplastic component. This modified process is used for components such as sinks and baths as the composite backing gives strength and rigidity to the component.

5.3.2 Moulding methods

(a) Matched-die moulding

This method is widely used for long production runs for components ranging in size from small domestic items to doors and cab panels for large commercial vehicles. The material to be shaped is pressed between heated matched dies as illustrated in Figure 5.11. The pressure required depends on the flow characteristics of the feed material and may be as high as 50 MPa but is usually less than 10 MPa. The feed material flows into the contours of the mould and when the temperature is high enough it rapidly cures. The time for the complete moulding process obviously depends on the feed material but also on the dimensions of the component and whether preheating of the feed has been employed to shorten the time; times typically range from several seconds to several minutes. Good moulded detail and dimensional accuracy are possible although the cost of a complex tool steel die can be considerable.

Two forms of feed which are particularly suited to matched-die moulding are sheet moulding compound and dough moulding compound. *Sheet moulding compound*, often abbreviated to SMC, is a prepared sheet of resin-fibre blend which contains all the necessary additives such as curing agent, release agent and pigment. It reduces the number of components to be stored, is clean to use and gives good consistency in the finished component. As all the constituents are pre-mixed SMC has a shelf life of

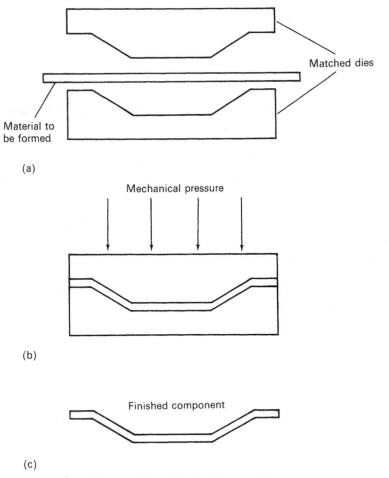

Figure 5.11   Matched die moulding.

around three to six months at room temperature. *Dough moulding compound* (DMC) is also a blend of all the necessary constituents but only short fibres are used. The resulting fibrous mixture has the consistency of dough or putty and can be readily made into accurately measured quantities for feeding the press. The shelf life of DMC is less than that of SMC.

(b) *Forming methods employing gas pressure*

These forming methods are sometimes known as *bag moulding* processes and we will discuss three such methods, namely vacuum forming, autoclave moulding and pressure bagging.

Unlike matched-die moulding only one die is required for *vacuum forming*, also known as *vacuum bagging*, and the pressure is not obtained mechanically but is atmospheric. The principle of the method is shown in Figure 5.12(a), where we can see that the heated preimpregnated reinforcement, which is sealed to the mould by a flexible sheet or membrane, is forced into

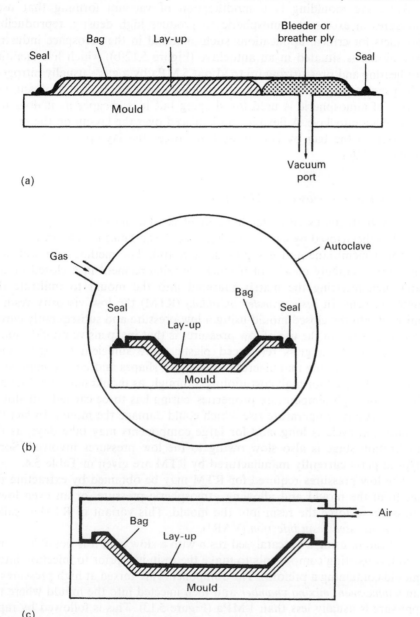

(a)

(b)

(c)

Figure 5.12 Open mould forming methods employing gas pressure (bag-moulding methods): (a) vacuum forming (vacuum bagging); (b) autoclave moulding; (c) pressure bagging.

the mould as air is removed. It should be noted that the mould may be either male of female. The vacuum is maintained until the curing process has reached completion.

Autoclave moulding is a modification of vacuum forming that uses pressures in excess of atmospheric to produce high density, reproducible products for critical applications such as found in the aerospace industry. The mould is situated in an autoclave (Figure 5.12(b)) which has facilities for heating and pressurizing up to about 5 MPa by a gas – usually nitrogen. The pressure bag process works on a similar principle in that a pressure in excess of atmospheric is used for shaping but it is cheaper as it does not require an autoclave. A flexible bag is placed over the lay-up on the mould. Inflation of the bag by compressed air forces the lay-up into the mould (Figure 5.12(c)).

(c) Low pressure, closed mould systems

The methods discussed so far have involved starting with an open mould which is then closed by means of a matched die and mechanical pressure (see (a)) or a membrane and gas pressure (see (b)). The methods we will now describe essentially consist of placing the reinforcement in a closed mould and then inserting the matrix material into the mould to infiltrate the reinforcement. In *resin transfer moulding* (RTM) the low viscosity resin is injected into the closed mould using a low pressure and subsequently cured. A consequence of the use of low pressures is that inexpensive moulds, made for example from glass reinforced plastic, have sufficient strength. Such moulds facilitate the manufacture of complex shapes and large components without the need for high cost tooling although, as mould material does not have good high temperature properties, curing has to be carried out slowly to restrict any temperature rise which could damage the mould. In fact the production cycle is long, and for large components may take days, as the infiltration stage is also slow owing to the low pressures involved. Some typical parts currently manufactured by RTM are given in Table 5.4.

The low pressures required for RTM may be obtained by extracting the air from the mould and allowing atmospheric pressure, or an even lower pressure, to force the resin into the mould. This variant of RTM is called *vacuum-assisted resin injection* (VARI).

Instead of using a precatalysed resin with a slow cure it is possible to mix two fast reacting components to make the resin just prior to injection into a mould containing a preform. The components are mixed at high pressures in an *impingement mixing chamber* and then injected into the mould where the pressure is usually less than 1 MPa (Figure 5.13). This is followed by rapid curing so the cycle time for this process, which is known as *reinforced reaction injection moulding* (RRIM) is far less than that for VARI and is typically 1–2 min.

Table 5.4 Typical parts currently manufactured by resin transfer moulding, RTM. (Source: Stark and Breitigam, 1987)

Use	Part
Industrial	solar collectors, electrostatic precipitator plates, fan blades, business machine cabinetry, water tanks
Recreational	canoe paddles, television antennae, snowmobiles
Construction	seating, bathtubs, roof sections, bus shelters
Aerospace	wing ribs, cockpit hatch covers, escape doors
Automobile	crash members, leaf springs, car bodies

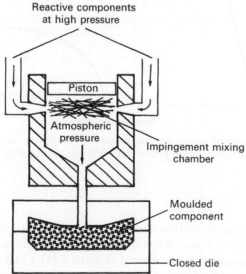

Figure 5.13 Diagram of reinforced reaction injection moulding.

It is important to appreciate the relative merits of the different processing methods and to know under what circumstances a particular method is likely to be selected for manufacture. It is therefore appropriate that we recap briefly on the main features of some of the methods described so far. Hand lay-up can be used to produce complex and/or large components in small quantities. The properties obtained are variable. Hand lay-up capital costs are low but it is very labour intensive and slow, therefore these methods are used in region A of Figure 5.14. The equipment for matched die moulding methods (section (a)) is expensive but components can be produced rapidly. These, and related methods are especially suited for the production of a large number of components the complexity of which is limited by the need to used steel dies (region C). RTM processes lie between these two extremes (region B); they are employed for relatively small runs on simple components and for longer runs on more complex components.

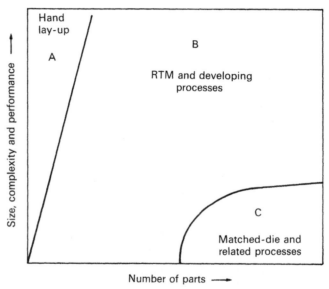

Figure 5.14 Process diagram for the automotive industry showing the ranges of manufacturing variables appropriate for different processing methods. (Source: Johnson, 1987.)

(*d*) *Pultrusion*

Rods of uniform cross-section can be produced in long lengths by pultrusion. Continuous rovings of the reinforcement are impregnated with resin by being passed through a bath of resin (Figure 5.15). (The reader may recall that in CMC production continuous fibres are impregnated in a similar manner by pulling fibres through a slurry – see section 4.3.2 and Figure 4.6). The impregnated fibres are then pulled through a heated die which compacts and shapes to the required profile in a manner reminiscent of extrusion. However, since the process relies on a pulling action the name *pultrusion* has been devised. Curing takes place in the heated die but is sometimes completed in an oven. Pultrusion is a continuous process and,

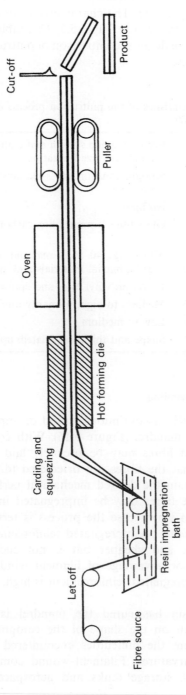

Figure 5.15 Schematic diagram of pultrusion process.

depending on size and complexity of the section, rates of several metres per minute may be achieved. The characteristics of the pultrusion process and products are summarized in Table 5.5. This table includes some assessment of costs but a more detailed comparison of pultrusion with other processes is given in Table 5.6.

Table 5.5 Characteristics of the pultrusion process and products. (Source: Martin and Sumerak, 1987)

Size	Forming guide system and equipment pulling capacity influences size limitation
Shape	Straight constant cross-sections; some curved sections possible
Length	No limit
Reinforcement (type)	Glass fibre, aramid fibre, carbon fibre and thermoplastic
Reinforcement (content, wt. %)	All rovings, 40–80%; mat and roving, 25–50%; woven roving, 55%; biaxial materials and mat, 40–70%
Resin systems	Polyester, vinyl ester and epoxy
Strength	Medium to high, primarily unidirectional
Labour intensity	Low to medium
Production rate	Shape and thickness related; up to several metres per minute

5.3.3 Filament winding

In *filament winding* a continuous strand of impregnated fibres, or tape, is wound on to a mandrel (Figure 5.16). With computer controlled systems the impregnated fibres may be precisely laid down in a predetermined manner such that the fibres are orientated (different incident angles and patterns) to obtain the desired mechanical performance from the finished component. The fibres may be impregnated in a similar manner to that in pultrusion, in which case the process is termed *wet winding*. Alternatively, in *dry winding* preimpregnated reinforcements are used. Wet winding is more flexible and cheaper but is not such a clean process as dry winding. A major advantage of filament winding is that the rate of lay down of the impregnated reinforcement is high, typically in the range 50 to 350 kg/h.

After the resin has cured the mandrel is withdrawn and this imposes some limit on the shape of the component. Another shape limitation arises from the difficulties encountered in producing components with reverse curvatures. Filament-wound components include pipework, pressure vessels, storage tanks and aerospace parts such as helicopter blades.

Table 5.6 Comparison of costs and process efficiency of PMC production processes. (Source: Quinn, 1989)

Process	Typical cycle time	Equipment capital (£1000)	Mould capital (£)	Product value per cycle (£)	Product value per hour (£)	Process efficiency[a]
Compression	3 min	50	5000–20000	1–5	20–100	365–1428
Autoclave	8 h	150	1000	10–100	1.25–12	8.3–79
Filament	4 h	20–100	1000	10–100	2.5–25	119–250
Injection (VARI)	10–60 min	5–10	300–1000	1–10	6–10	1200–1000
Pressure bag	1 h	5	100–500	1–4	1–4	200–800
Spray	3 h	5	100–500	5–25	2–8	400–1600
Hand lay-up	5 h	0	100–500	5–25	1.5	10000
Pultrusion	0.5–3 m/min	50–100	2000–10000	3/m	90–540	2884–5400

[a]Process efficiency $= \dfrac{\text{Product value per hour} \times 10^6}{\text{total capital}}$.

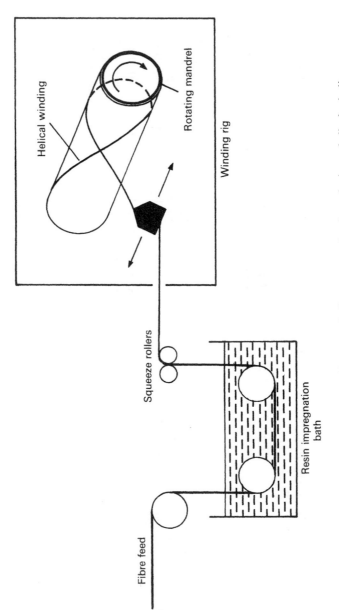

Figure 5.16 Schematic diagram of filament winding process (wet) producing a helical winding.

5.4 SOME COMMERCIAL PMCs

In this section we will study in more detail an example of a thermoset and a thermoplastic matrix composite together with rubber matrix composites. The thermoset chosen is epoxy and the thermoplastic, polyetheretherketone (PEEK). Neither are the most widely used polymer of their class but represent high performance materials.

5.4.1 Fibre-reinforced epoxies

Glass fibres are the most common reinforcement for PMCs, but carbon and aramid fibres are being used more frequently. Glass fibre-reinforced polyesters (GRPs) have by far the largest proportion of the market with epoxies being second. Although *glass fibre-reinforced epoxies* (GREs) may be a poor second to their polyester counterparts in terms of quantity, as we can see from the data of Table 5.7 this is not the case as far as properties are concerned. As a consequence of their higher cost and superior properties GREs are generally employed in low volume, high technology applications.

Table 5.7 Comparison of the room temperature properties of unidirectional glass roving fibre reinforced epoxy and polyester (fibre content 50–80vol.%)

	Epoxy	*Polyester*
Density (Mg/m^3)	1.6–2.0	1.6–2.0
Tensile modulus (GPa)	30–55	12–40
Flexural modulus (GPa)		10–35
Tensile strength (MPa)	600–1165	140–690
Flexural strength (MPa)	1000–1500	205–690
Compressive strength (MPa)	150–825	140–410
Interlaminar shear (MPa)	30–75	

The superior mechanical properties of GREs demonstrated in Table 5.7 are a result of the good strength and stiffness of the epoxy matrix and the strong bonding of glass fibres to the epoxy. Epoxies bond more strongly to glass fibres than either of the other widely used thermosets, the polyesters and phenolics. Nevertheless coupling agents are still used to improve wetting and bonding to the epoxy matrix and a wide range of coupling agents have been found to be effective including the silanes. The good bonding of the epoxies to the glass reinforcement leads to high interlaminar shear strengths. Both E-glass and S-glass are used with epoxies, with the incorporation of the higher strength S-glass (also known as R-glass) producing the better mechanical properties (Table 5.8).

The glass fibres in a composite can be degraded by water and this results in a loss of strength. The water reaches the fibres via defects in the matrix,

Table 5.8 Comparison of unidirectional mechanical properties of epoxies reinforced with various fibres. (Source: Dow Chemical Company)

Fibre	Strength (MPa)		Young's Modulus (GPa)	Density (Mg/m³)
	Tensile	Compressive		
E-glass	1165	490	50	1.99
S-glass	1750	495	60	1.99
Carbon (AS4)	1480	1225	145	1.55
Carbon (HMS)	1275	1020	205	1.63
Aramid	1310	290	85	1.38

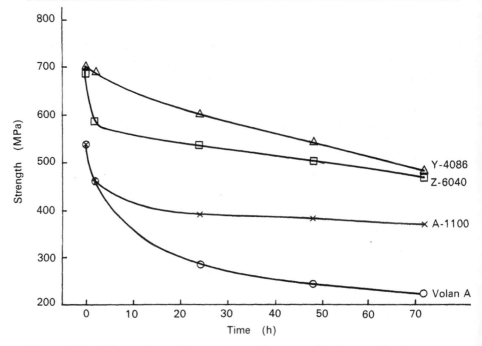

Figure 5.17 Effect of coupling agent on the strength of epoxy laminates as a function of the time exposed to boiling water. Volan A, methacrylatochromic chloride; A-1100, triethoxysilyl propylamine; Z-6040, glycidoxypropyl trimethoxysilane; Y-4086, 3,4-epoxycyclohexylethyl trimethoxysilane. (Source: Lotz and Milletari, 1963.)

down poor fibre–matrix interfaces or by diffusion through the matrix. There is evidence that coupling agents may not only affect the dry strength of GREs but also reduce the loss of strength in wet environments. As shown in Figure 5.17 the dry flexural strength of epoxy laminates varies from 540 to 705 MPa and after exposure to boiling water for 72 h these have fallen to 225 to 470 MPa depending on the coupling agent. The good performance associated with the silane coupling agents has been explained as follows. As described in Chapter 2 (section 2.7.2) the silane is attached to the resin through a resin compatible R group and to the glass by M where M is typically Si, Fe or Al. In the presence of water the covalent M–O bond hydrolyses as shown in Figure 5.18(a). Under the action of a shear stress the fibre can slide relative to the matrix without permanent bond failure and hence damage to the integrity of the interface (Figure 5.18(b)).

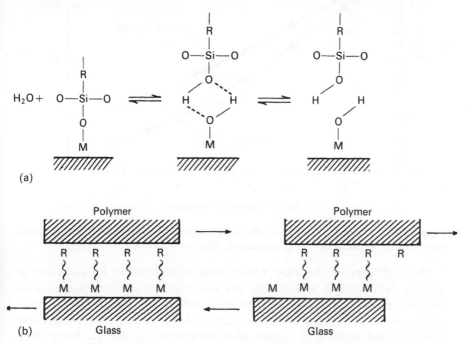

(a)

(b)

Figure 5.18 The beneficial effect of a silane coupling agent on interfacial behaviour in the presence of water according to Plueddemann (Plueddemann, 1974): (a) hydrolysis of the covalent M–O bond; (b) shear displacement at the polymer–glass interface without permanent bond rupture. (Source: Hull, 1981.)

Changes in the matrix due to moisture, as well those at the matrix–glass fibre interface, also modify the mechanical properties of the composite. There are a number of sites in the cross-linked epoxy matrix for the attachment of water molecules by hydrogen bonding. The water acts as a plasticizer and reduces the glass transition temperature T_g significantly (several tens of degrees) as shown by the data for six cured epoxy resins in Figure 5.19. The reduction in T_g may be beneficial with respect to properties at ambient temperature, but can have a marked detrimental effect on the performance at elevated temperatures. It is worth reminding the reader at this stage that epoxy resins are available with a variety of compositions and structures and consequently have a range of glass transition temperatures and respond differently to water. For example, it has been found that the curing agent plays a role in determining the kinetics of water uptake and the degradation in mechanical properties.

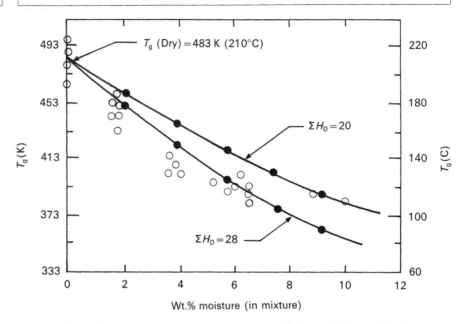

Figure 5.19 The effect of water on the glass transition temperature (T_g) of six cured epoxy resins; •, calculated; ○, experimental. (Source: Kaelble *et al.*, 1988.)

Although glass is the most widely used reinforcement for polymers in general, both carbon and aramid are employed to obtain epoxy matrix composites with different characteristics which are required for certain applications. As we saw earlier in Table 5.8, the tensile and compressive strengths and stiffness of carbon fibre-reinforced epoxies are better than those of GREs, however the performance in shear of these two epoxy based composites is similar. The carbon reinforced materials have lower densities so their specific properties are far superior, but it should be remembered that a higher cost is involved. The lightweight aramid fibres result in even lower density composites which gives good specific properties even though the tensile strength and stiffness values are less than those obtained using carbon fibres. The compressive properties are relatively poor. The forte of the aramid–epoxy system is that it has the best impact resistance of the epoxy matrix composites (Figure 5.20). Because of their good specific properties aramid and carbon fibre-reinforced epoxies are increasingly used for applications in the aerospace industry.

5.4.2 PEEK matrix composites

It has been previously pointed out that thermosetting polymers are the most widely used for the matrices of PMCs. Nevertheless there is growing interest in thermoplastics because of their essentially unlimited shelf life, ease of fabrication, possibility of reusing offcuts and scrap, and improved per-

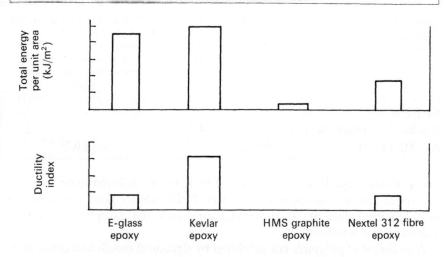

Figure 5.20 Comparison of the impact characteristics of fibre reinforced epoxies. (Source: Weeton *et al.*, 1987.)

formance in certain areas such as toughness. *Polyetheretherketone*, PEEK, is a relatively costly thermoplastic with good mechanical properties. As such it is most likely to be employed in high performance applications; for example, carbon fibre-reinforced PEEK is a competitor with carbon reinforced epoxies and Al–Cu and Al–Li alloys in the aircraft industry.

PEEK is a semi-crystalline polymer whose mechanical behaviour depends on the *degree of crystallization* and size and distribution of the *spherulites*. These microstructural features are determined by thermal history and the reinforcement. Concerning thermal history, by cooling extremely rapidly from the melt at a rate greater than 1000 °C/min an amorphous matrix may be obtained, whereas under more usual cooling conditions the degree of crystallinity is 20–40%. The spherulites are nucleated in the PEEK matrix and at the fibre surfaces. The extent of the nucleation at the fibre–matrix interface varies with the reinforcement; even for a given reinforcement such as carbon fibre the nucleation is sensitive to the type of fibre used.

One of the major attractions of thermoplastics is their good toughness in comparison with thermosets (Table 5.1). PEEK is tough even for a thermoplastic and the toughness is not degraded when PEEK is reinforced (Table 5.9).

We have seen that the mechanical properties of epoxies can be affected by moisture, however in contrast PEEK, like many thermoplastic matrices, does not absorb a significant amount of water. Thus even in wet conditions a PEEK matrix composite may be used at temperatures as high as 120 °C. For aerospace applications materials also require a resistance to other liquids such as fuel, hydraulic fluid, lubricants and solvents. Whereas the amorphous thermoplastics polysulphone and polyetherimide degrade in

Table 5.9 Comparison of the toughness of monolithic PEEK and PEEK matrix composites. (Sources: (a) Friedrich *et al.*, 1986; (b) Alger and Dyson, 1990)

	PEEK	30vol.% glass fibre	20 or 30vol.% glass fibre or 30vol.% carbon fibre
G_{IC} (kJ/m^2)	6.6		
Notched Izod impact (kJ/m^2)	12.8[b]	14.8[b]	
K_{IC} (MPa m$^{1/2}$)	6.0[a]		6.6 ± 0.7[a]

many of these liquids, semi-crystalline PEEK has been found to be resistant to typical aircraft environments. In fact PEEK shows good resistance to most environments at room temperature with the exception of strong sulphuric acid solutions.

A weakness of polymers not exhibited by structural metals and ceramics is their inflammability. The volatile decomposition products of polymers are often rich in hydrogen which supports the combustion process. However PEEK, and the other *aromatic* (i.e., one or more benzene ring in the molecule) thermoplastics, have a low ratio of hydrogen to carbon and consequently do not produce large amounts of combustible volatiles. PEEK is therefore difficult to ignite. A measure of inflammability is the *limiting oxygen index* (LOI) which is the percentage of oxygen in a nitrogen–oxygen mixture which will just support combustion. LOI values for some aromatic thermoplastics are presented in Table 5.10. The high LOI value of 35% for PEEK is extremely desirable; air only contains 22% oxygen and therefore PEEK and its composites are self-extinguishing in a normal atmosphere.

Table 5.10 Limiting oxygen index (LOI) for aromatic thermoplastics. (Clegg and Collyer, 1985)

	LOI (%)
Polycarbonate	25
Polysulphone	30
Polyphenylene sulphide	44
Poyethersulphone	36
Polyetheretherketone	35

Not only are PEEK matrix composites difficult to set on fire but, owing to the high T_g and T_M of PEEK, mechanical performance is maintained to high temperatures. The data for PEEK reinforced with continuous carbon fibres and 30% discontinuous glass fibres presented in Figure 5.21 demonstrate that even at 150 °C the strength and stiffness of the poorer performance glass reinforced composite were still over 60% of the room temperature values.

The strength and stiffness at elevated temperatures that we have just discussed refers to the behaviour of the composites when subjected to stress

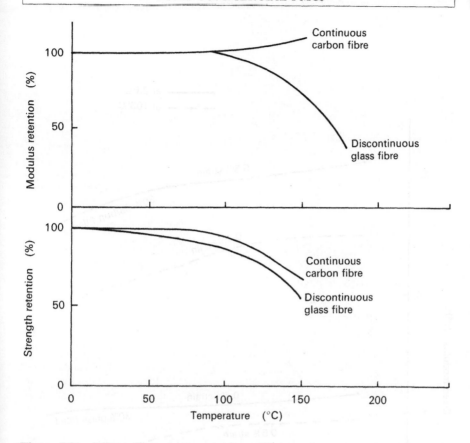

Figure 5.21 Effect of temperature on the stiffness and strength of PEEK matrix composites. (Source: Hancox, 1989.)

for a short time. If a composite is under stress for long periods at elevated temperatures, or indeed room temperature, then for many polymers, we must consider the creep properties. As a general rule thermoplastics readily creep, although this tendency decreases with increasing crystallinity. The creep resistance of PEEK, in common with other thermoplastics, is significantly improved by the incorporation of a reinforcement. This fact is illustrated in Figure 5.22 which is a plot of *creep modulus* against time; from this plot it can be seen that at 100 °C unfilled PEEK under a stress of 3 MPa will deform 2% in 10^4 s while 30% carbon filled PEEK held at the same temperature for the same time, but with a much larger stress of nearly 17 MPa, only undergoes 0.4% strain.

Finally fatigue crack growth in PEEK and its composites has been investigated and found to obey the well-known *Paris–Erdogan* equation

$$\frac{da}{dN} = A\Delta K_I^m,$$

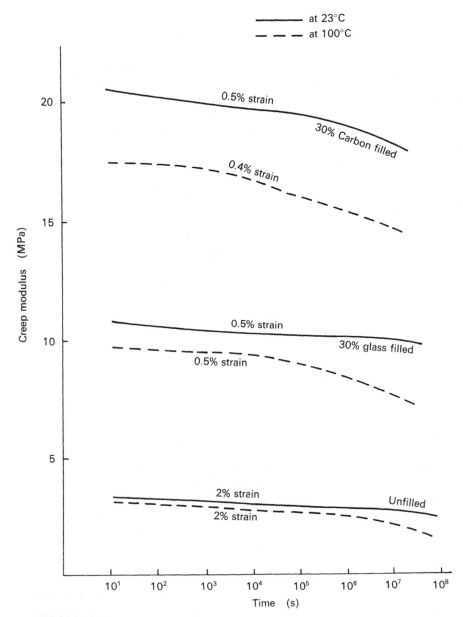

Figure 5.22 Plot of creep modulus against time showing the improved creep resistance obtained when PEEK is reinforced with carbon or glass. (Source: Alger and Dyson, 1990.)

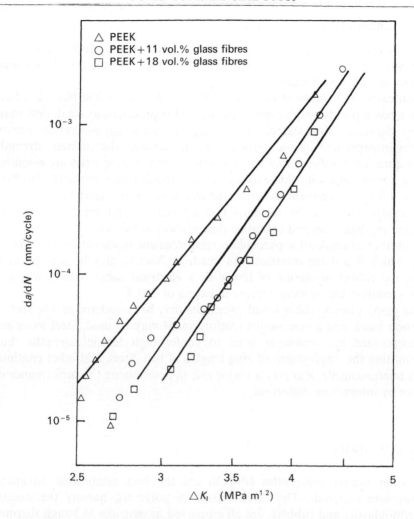

Figure 5.23 Linear relationship between $\log(da/dN)$ and $\log(\Delta K_I)$ in accordance with Paris–Erdogan equation for PEEK and glass-reinforced PEEK. The composites have the better fatigue crack growth resistance. (Source: Friedrich et al., 1986.)

where da/dN is the crack growth rate per cycle, ΔK_I the stress intensity factor range, and A and m are constants. Figure 5.23 shows that short glass fibre reinforcement decreases the fatigue crack growth rate at a given ΔK_I; for example at $\Delta K_I = 4$ MPa m$^{1/2}$ the crack growth rate in the composites is less than half that of monolithic PEEK. Short carbon fibres are more effective and at the same ΔK_I value da/dN is one to two orders of magnitude less than in the unreinforced PEEK.

5.4.3 Rubber matrix composites

Reinforced rubbers are used in a variety of applications including conveying systems, e.g., conveyer belts; transport, e.g., tyres and inflatable boats; and electrical, e.g., fire-resistant cable sheathing.

An early, and the most common, reinforcement is *carbon black*. Carbon black has a poorly graphitized structure and is produced as small (less than 1 μm diameter) spherical particles, although some agglomeration occurs when incorporated into a rubber. Carbon increases the stiffness, strength and abrasion resistance and these improvements in properties are essential for a number of applications, such as in tyre treads where typically 65wt.%C is used for reinforcement. A further benefit arises from the ability of carbon to absorb most of the ultraviolet component of natural light thereby decreasing light-initiated oxidative degradation of rubber.

Another example of a particulate reinforcement is *aluminium tri-hydroxide*, which is a flame retardent. As much as 70wt.% may be incorporated into the rubber insulation of heavy duty electrical cables which, in some cases, enables use at temperatures in excess of 100 °C.

In many cases a rubber based composite may be considered as a hybrid as carbon black and a continuous reinforcement may be used. Steel wires are incorporated into conveyer belts to confer high tensile strengths thus permitting the employment of long lengths of belt. Steel, and other continuous reinforcements, also play a major role in determining the performance of tyres by minimizing distortion.

5.5 SUMMARY

Polymer matrix composites (PMCs) are the best established advanced composite materials. The three classes of polymers, namely thermosets, thermoplastics and rubbers, are all employed as matrices although thermosetting polymers dominate the market. The structure and properties of the three classes of polymers were described. As far as mechanical properties of polymers were concerned we learnt that their stiffness and strength were low but could be dramatically improved by reinforcement, e.g., the Young's modulus and tensile strength of a polyester is typically 3 GPA and 65 MPa respectively whereas the corresponding values for glass fibre-reinforced material are approximately 25 GPa and 400 MPa.

The extensive range of processing methods available for PMCs were reviewed. We saw that these vary from simple labour intensive methods, such as hand lay-up, to automated processes requiring heavy capital investment and suitable for producing large numbers of components, e.g., pultrusion and filament winding. The temperatures involved in processing PMCs are low compared with those employed during the fabrication of metal and ceramic matrix composites, hence reinforcements with low temperature capabilities, such as organic and glass fibres, may be used.

FURTHER READING

General

Phillips, L. N. (ed.) (1989) *Design with Advanced Composite Materials*, Design Council and Springer-Verlag.
Richardson, T. (1987) *Composites: A Design Guide*, Industrial Press Inc., New York.

Specific

Alger, M. S. M. and Dyson, R. W. (1990) Thermoplastic Composites, in *Engineering Polymers*, (ed. R. W. Dyson), Blackie, p. 1.
Anderson, J. C., Leaver, K. D., Rawlings, R. D. and Alexander, J. M. (1990) *Materials Science*, 4th edn., Chapman and Hall.
Bassett, D. C. (1981) *Principles of Polymer Morphology*, Cambridge University Press.
Billmeyer, F. W. (1971) *Textbook of Polymer Science*, Wiley, p. 469.
Clegg, D. W. and Collyer, A. A. (1985) *High Performance Plastics*, **2** (2), 1.
Dow Chemical Company *Electrical Laminates*, *NEMA FR-4m*, Technical Bulletin 296-396-484.
Friedrich, K., Walter, R., Voss, H., and Karger-Kocsis, J. (1986) *Composites*, **17**, 205.
Hancox, N. L. (1989) High-Performance Composites with Thermoplastic Matrices, in *Concise Encyclopedia of Composite Materials*, (ed. A. Kelly), Pergammon Press, p. 134.
Hull, D. (1981) *An Introduction to Composite Materials*, Cambridge Solid State Science Series, Cambridge University Press.
Johnson, C. F. (1987) Resin transfer molding, in *Composites, Vol. 1, Engineered Materials Handbook*, ASM International, p. 564.
Kaelble, D. H., Moacanin, J. and Gupta, A. (1988) Physical and mechanical properties of cured resins, in *Epoxy Resins – Chemistry and Technology*, 2nd edn., (ed. C. L. May), Marcell Dekker.
Lotz, E. and Milletari, S. (1963) *SPE Annual Tech. Conf.*, Los Angeles, USA, paper 215A.
Martin, J. D. and Sumerak, J. E. (1987) Pultrusion, in *Composites, Vol. 1, Engineered Materials Handbook*, ASM International, p. 533.
Plueddemann, E. P. (ed.) (1974) *Interfaces in Polymer Matrix Composites*, Academic Press, New York.
Quinn, J. A. (1989) *Met. and Mat.*, **5**, 270.
Stark, E. B. and Breitigam, W. V. (1987) Resin transfer moulding materials, in *Composites, Vol 1, Engineered Materials Handbook*, ASM International, p. 168.
Weeton, J. W., Peters, D. M. and Thomas, K. L. (1987) *Engineers' Guide to Composite Materials*, ASM.

PROBLEMS

5.1. A component is to be constructed from glass fibre-reinforced plastic. Initial information had led to the final choice of matrix being between epoxy and PEEK. Further details of the working environment of the component 'were then supplied. These were that the component was likely to be used in a moist environment at temperatures as high as 100 °C. Which matrix would you select? Give the reasons for your selection.

5.2. A component was constructed from PEEK reinforced with 18 vol. % discontinuous glass fibres. It was found to have a crack of 2 mm in length. Fracture toughness calculations determined that catastrophic failure would take place when the crack had grown to 14 mm. In service the component would be subjected to fluctuating stresses, at a frequency of 5 Hz, which gave a stress intensity factor range at the crack tip of 3.5 MPa m$^{1/2}$ irrespective of crack length. Estimate whether the component would survive 20 days in service. Use the data of Figure 5.23.

SELF-ASSESSMENT QUESTIONS

Indicate whether statements 1 to 7 are true or false.

1. Polyester and epoxy are examples of thermosets.

   (A) true
   (B) false

2. Phenolic resins known as resoles or one-stage resins require the addition of a curing agent prior to fabrication.

   (A) true
   (B) false

3. Polycarbonate is an accepted generic term for synthetic polyamides.

   (A) true
   (B) false

4. Hand lay-up requires little capital equipment but is labour intensive.

   (A) true
   (B) false

5. Figure 5.24 is a diagram of matched-die moulding.

   (A) true
   (B) false

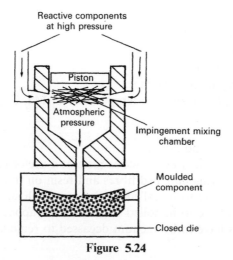

Reactive components
at high pressure

Piston

Atmospheric
pressure

Impingement mixing
chamber

Moulded
component

Closed die

Figure 5.24

6. It has been estimated that over three-quarters of all matrices of PMCs are thermosets and of these the majority is polyester.

(A) true
(B) false

7. Although many benefits accrue from reinforcing PEEK, unfortunately the creep properties are significantly degraded with the result that PEEK matrix composites cannot be used at temperatures in excess of 40 °C.

(A) true
(B) false

For each of the statements of questions 8 to 14, one or more of the completions given are correct. Mark the correct completions.

8. Thermosetting polymers

(A) are also known as thermoplastic polymers,
(B) are also known as rubbers,
(C) readily cross-link during curing,
(D) cannot be cured,
(E) cannot be reheated and reshaped.

9. Thermoplastics

(A) are cross-linked polymers,
(B) are linear polymers,
(C) may have branched chains,
(D) readily flow at elevated temperatures,
(E) cannot be reheated and shaped.

10. Polyetheretherketone

   (A) is also known as PEEK,
   (B) is a type of nylon,
   (C) is a thermoplastic,
   (D) is a semi-crystalline polymer,
   (E) has a high glass-transition temperature of 143 °C.

11. Spray-up

   (A) is faster than hand lay-up,
   (B) involves spraying a non-viscous resin at a preform,
   (C) involves spraying chopped fibres and resin on to a mould,
   (D) composites have to be rolled to give a smooth surface finish,
   (E) composites have to be rolled to remove entrapped air,
   (F) composites have to be vacuum degassed to remove entrapped air.

12. Pultrusion

   (A) is a slow labour intensive production method,
   (B) is particularly suited to the production of large, complex, planar shapes,
   (C) is used for the production of rods of uniform cross-section,
   (D) involves pushing fibres into a closed mould containing resin,
   (E) involves pulling resin impregnated fibres through a heated die.

13. The mechanical properties of glass-reinforced epoxies are affected by moisture because

   (A) the epoxy matrix becomes more brittle,
   (B) the water acts as a plasticizer,
   (C) the glass transition temperature is reduced,
   (D) the epoxy transforms to a weaker polyester,
   (E) the glass fibres can be degraded.

14. Fatigue crack growth rate in PEEK matrix composites

   (A) is slower than in monolithic PEEK for the same test conditions,
   (B) is faster than in monolithic PEEK for the same test conditions,
   (C) obeys the law of mixtures,
   (D) obeys the Paris–Erdogan equation,
   (E) is proportional to the stress intensity factor range raised to a power,
   (F) is proportional to the square root of the creep modulus.

Each of the sentences in questions 15 to 20 consists of an assertion followed by a reason. Answer:

   (A) if both assertion and reason are true statements and the reason is a correct explanation of the assertion,
   (B) if both assertion and reason are true statements but the reason is not a true explanation of the assertion,

(C) if the assertion is true but the reason is a false statement,
(D) if the assertion is false but the reason is a true statement,
(E) if both the assertion and reason are false statements.

15. Thermosets are usually stiffer than thermoplastics *because* of the strong covalent bonding of the cross-links in the thermosets.

16. Many thermoplastics are partially crystalline *because* thermoplastics cross-link to form a rigid network.

17. Acrylonitrile–butadiene–styrene (ABS) is a terpolymer *because* it is a multi-phase material.

18. The moulds used in resin transfer moulding may be complex, large and made from relatively inexpensive materials such as glass reinforced plastics *because* only low pressures are employed in the process.

19. Vacuum forming requires matched-dies *because*, as in vacuum-assisted resin injection, extremely high pressures are involved.

20. The longitudinal stiffness of continuous fibre reinforced epoxy is about 100 GPa irrespective of whether the reinforcement is glass, carbon or aramid *because* the glass-fibre reinforced epoxy has the highest density.

21. Fill in the missing words in this passage. Each dash represents a letter.

_ _ _ _ _ _ _ _ _ _ _ _ _ _ _ _ _ _ _ _, commonly known as PEEK is a _ _ _ _ - _ _ _ _ _ _ _ _ _ _ _ _ thermoplastic. Its mechanical properties are determined by the degree of _ _ _ _ _ _ _ _ _ _ _ _ _ _ and the size of the _ _ _ _ _ _ _ _ _ _ _. PEEK is tough and does not absorb significant amounts of _ _ _ _ _. In fact PEEK shows good resistance to most environments at room temperature with the exception of strong sulphuric acid solutions. PEEK is an _ _ _ _ _ _ _ _ thermoplastic and has a low ratio of hydrogen to oxygen and consequently, on heating, does not produce large amounts of combustible volatiles. PEEK is therefore difficult to ignite and has a high _ _ _ _ _ _ _ _ _ _ _ _ _ _ _ _ _ (LOI) of 35%.

ANSWERS

Problems

5.2 No, fails in 6.7 h.

Self-assessment

1. A; 2. B; 3. B; 4. A; 5. B; 6. A; 7.B; 8. C, E; 9. B, C, D; 10. A, C, D, E; 11. A, C, D, E; 12. C, E; 13. B, C, E; 14. A, D, E; 15. A; 16. C; 17. B; 18. A; 19. E; 20. D; 21. polyetheretherketone, semi-crystalline, crystallization, spherulites, water, aromatic, limiting oxygen index.

6	# Stiffness, strength and related topics

6.1 INTRODUCTION

Often we need to ensure that a component does not break, i.e., it has adequate strength, or that it does not suffer excessive deformation under load, i.e., it has adequate stiffness. For these reasons we shall discuss methods of calculating the stiffness and strength of composites in the following chapters. In particular we shall be concerned with the behaviour of flat, plate-like elements, as composite components are frequently fabricated in this way. Since most of these methods are derived from the basic principles of mechanics and stress analysis, the latter are reviewed below before going on to consider composites. Full details of these methods can be found in standard texts on such topics.

6.2 LOADS AND DEFORMATIONS

6.2.1 Loads

We consider the loads acting on a structure or component to be of two basic types; *forces* and *couples*. Forces can be further subdivided into *tensile*, *compressive* and *shear*, and couples into *bending moments* and *twisting moments* (or *torques*). All of these can act in isolation or in combination, depending on the situation.

6.2.2 Structural entities

When carrying out a structural analysis it is convenient to represent the real structure by one, or more, idealized elements. Each element will almost certainly be a simplification of the actual situation, in that we will make assumptions about the way the loads are distributed within it. These simplifications are often based on the dimensions of the part.

A straight member whose length is much greater that its cross-sectional dimensions, and is subjected to a tension or compression along its length, is referred to as a *rod*. A similar member, but subjected to a bending moment is known as a *beam*, and when loaded by a torque is called a *shaft*.

When we have an element whose length and breadth are much greater than its thickness, and which is subjected to in-plane forces, we refer to it as a *membrane*. If we subject it to bending moments we call it a *plate*.

Cylindrical components can be treated as membranes provided the diameter is constant and much larger than the wall thickness. If the diameter of the cylinder varies along its length we would call it a *shell*; in this case its behaviour is similar to, but more complicated than, a plate.

6.2.3 Deformations

Each type of load referred to above produces a characteristic deformation of the body on which it acts. These are shown in Figure 6.1. We also frequently refer to states of *plane stress* and *plane strain*. In plane stress, stresses normal to the plane in which the stresses act are considered to be zero, e.g. a plate in simple bending. In plane strain, strains normal to the plane in which the stresses act are considered to be zero, e.g. the cross-section of a bar held between rigid walls.

6.3 STRESS AND STRAIN

6.3.1 Stress

To allow us to deal with components of differing size we work with normalized loads. Thus tensile and compressive forces are divided by cross-sectional area to give the corresponding *direct stress* (σ) and shear force by a surface area to give a *shear stress* (τ) (see Figure 6.2). Stress therefore has units of force/area, i.e., N/m^2 or Pascal (Pa). A bending moment alone will give rise to a direct stress which varies linearly through the depth of the component from tensile on one surface, to compressive on the other; a transverse force (normal to the longitudinal axis of the beam) will cause shear stresses as well as direct stresses (see Figures 6.2 and 2.24). A torque will give rise to shear stresses. Again, these will vary in a linear fashion but can only be easily calculated for simple components (such as a circular section rod – see Figure 6.2). For plates it is usual to work with loads (forces and couples) per unit width, as shown in Figure 6.3.

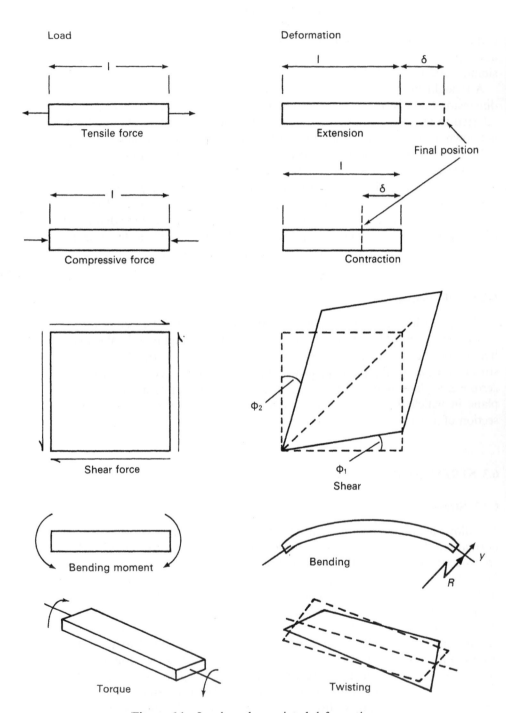

Figure 6.1 Loads and associated deformations.

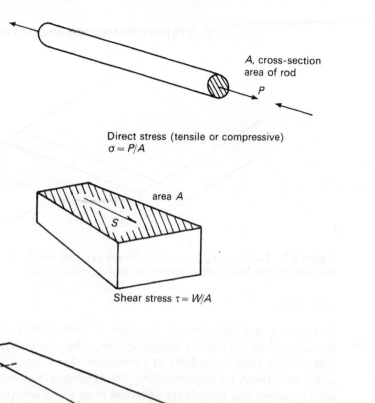

Direct stress (tensile or compressive)
$$\sigma = P/A$$

area A

Shear stress $\tau = W/A$

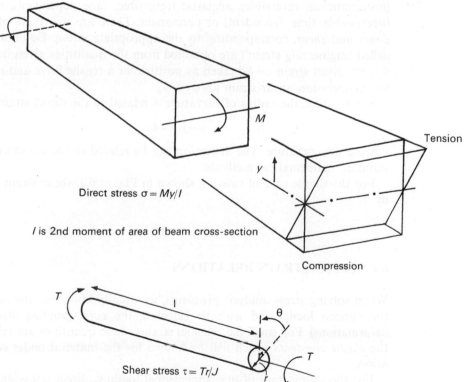

Direct stress $\sigma = My/I$

I is 2nd moment of area of beam cross-section

Shear stress $\tau = Tr/J$

J is polar 2nd moment of area of shaft cross-section

Figure 6.2 Loads and associated stresses.

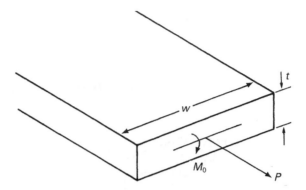

Figure 6.3 Loads acting on a plate: Force per unit width, $N = P/w$; moment per unit width, $M = M_0/w$; direct stress due to P, $\sigma = N/t = P/wt$.

6.3.2 Strain

Strain is a non-dimensional measure of deformation; it can be elastic (instantaneous, reversible), anelastic (reversible, time dependent), inelastic (irreversible, time dependent) or permanent. There are two types of strain, *direct* and *shear*, corresponding to the appropriate stress. Definitions (so-called 'engineering strain') are obtained from the quantities given in Figure 6.1. So, direct strain $\varepsilon = \delta/l$ taken as positive for a tensile force and negative for compression; shear strain $\gamma = \phi_1 + \phi_2$.

For bending, the radius of curvature is related to the direct strain by

$$\varepsilon = y/R = ky,$$

k being the curvature. The deflection can be related to the curvature using standard mathematical methods.

For the simple case of twisting shown in Figure 6.2, shear strain is given by

$$\gamma = r\theta/l$$

6.4 STRESS–STRAIN RELATIONS

When solving stress analysis problems, we usually know the stresses (from the applied loads) and wish to calculate the corresponding strains (or deformations). For small (i.e., elastic) strains, these quantities are related by the *elastic constants* which will be known for the material under consideration.

For the simple case of one-dimensional loading, direct stress and strain are related by Young's modulus, E, i.e.

$$\sigma = E\varepsilon \quad \text{or} \quad \varepsilon = \sigma/E, \tag{6.1}$$

and shear stress and strain are related by the shear modulus, G, i.e.

$$\tau = G\gamma \quad \text{or} \quad \gamma = \tau/G. \tag{6.2}$$

Any longitudinal deformation (in the direction of the force) will be accompanied by a lateral deformation in the opposite sense, as illustrated for a plate in Figure 6.4.

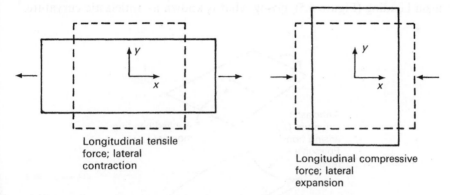

Longitudinal tensile force; lateral contraction

Longitudinal compressive force; lateral expansion

Figure 6.4   Effect of Poisson's ratio.

The strains in the two orthogonal directions are related by a quantity known as Poisson's ratio, v, i.e.

$$v = -\varepsilon_y/\varepsilon_x$$

It follows, then, that if direct stresses are acting in two orthogonal directions, the resultant strain in each direction will be dependent on both stresses, i.e.

$$\varepsilon_x = \frac{\sigma_x}{E} - \frac{v\sigma_y}{E}, \text{ and } \varepsilon_y = -\frac{v\sigma_x}{E} + \frac{\sigma_y}{E}. \tag{6.3}$$

The above equations refer to a material for which stress is proportional to strain, i.e., the material is *linear elastic*. Such materials are said to obey Hooke's Law. Linear behaviour is not always seen, especially for shear loading.

Usually we also need stresses in terms of strain. These are easily found by solving equations 6.3, i.e.

$$\sigma_x = \frac{E}{1 - v^2}(\varepsilon_x + v\varepsilon_y),$$

$$\sigma_y = \frac{E}{1 - v^2}(v\varepsilon_x + \varepsilon_y). \tag{6.4}$$

6.5 BENDING OF PLATES

The influence of Poisson's ratio is seen, also, in the bending of plates. As already stated, bending will result in a linear distribution of direct stress varying through the thickness, from tension on one surface to compression on the other. Associated with the tensile stress we see now that there will be a lateral contraction, and associated with the compressive stress a lateral expansion. The result of this is to bend the plate in the opposite sense to the main bending (Figure 6.5), giving what is known as anticlastic curvature.

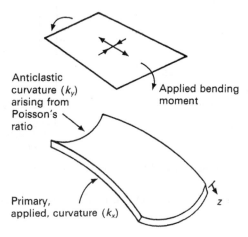

Figure 6.5 Bending deformation of a plate. Tensile stress on upper surface produces lateral contraction (opposite on the lower surface).

Using a generalization of beam bending (section 6.3) we can relate the direct strains and curvatures arising from applied bending moments, as

$$\varepsilon_x = z/R_x = zk_x, \text{ and}$$
$$\varepsilon_y = z/R_y = zk_y,$$

z being the distance measured through the thickness. For an applied twisting moment there will be a corresponding shear strain given by

$$\gamma_{xy} = zk_{xy}.$$

Note the use of a double subscript when dealing with shear. The first subscript relates to the normal of the plane on which the stress acts and the second subscript to the direction of the stress (see also Figure 6.7).

If we substitute the above expressions for direct strain into equations 6.3 we can show that the corresponding direct stresses are

$$\sigma_x = (Ez/(1-v^2))(k_x + vk_y),$$
$$\sigma_y = (Ez/(1-v^2))(vk_x + k_y). \tag{6.5}$$

Methods for finding deflections are given in advanced texts on applied mechanics.

6.6 ISOTROPIC MATERIALS

The equations used above relate to *isotropic* materials, such as unreinforced metals or polymers in certain conditions. In these materials the elastic constrants, E, G and v, do not vary with direction. In other words, if we were to take a sheet of metal with random orientation of the grains, say, and cut from it a series of test specimens, as illustrated in Figure 6.6, the value of E determined from a tensile test would be the same for each specimen.

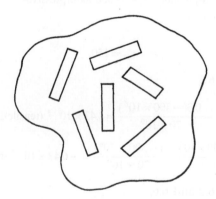

Figure 6.6 Test specimens taken from an isotropic sheet.

A further property of isotropic materials is that only *two* elastic constants are needed to describe (or characterize) the stress–strain behaviour. Usually we use E and v, which are related to G by the equation

$$G = \frac{E}{2(1+v)}.$$ (6.6)

Example 6.1
A plate is subjected to the set of stresses shown in the figure below. Note that the shear stress is shown here as positive. Determine the associated strains if the material is isotropic with $E = 70$ GPa and $v = 0.3$.

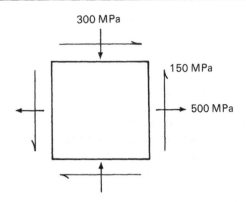

Let

$$\sigma_x = 500, \quad \sigma_y = -300 \text{ (compressive)},$$

and

$$\tau_{xy} = 150.$$

Using equations 6.3

$$\varepsilon_x = \frac{500 \times 10^6}{70 \times 10^9} - \frac{0.3(-300 \times 10^6)}{70 \times 10^9} = 8.43 \times 10^{-3} \text{ (tensile)},$$

$$\varepsilon_y = \frac{-0.3 \times 500 \times 10^6}{70 \times 10^9} + \frac{(-300 \times 10^6)}{70 \times 10^9} = -6.43 \times 10^{-3} \text{ (compressive)}.$$

Using equations 6.2 and 6.6,

$$\gamma = \tau_{xy}/G = 2\tau_{xy}(1+v)/E = 5.57 \times 10^{-3}.$$

6.7 PRINCIPAL STRESS AND STRAIN

When determining the strength of an isotropic material it is conventional to use principal stresses or strains in conjunction with a failure criterion. Principal values are the maxima at a point in a component and are found from the stresses, or strains, at that point, expressed in a convenient set of orthogonal axes. When the stresses vary throughout the component, as is usually the case, it is necessary to calculate the principal values at several points to find the absolute maximum.

In order to find the principal stresses at a point, it is first necessary to obtain an expression for the direct and shear stresses acting on a face at some angle θ to the basic axes as shown in Figure 6.7. Note that all the

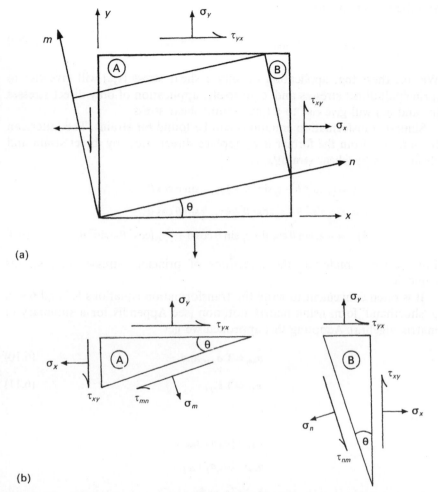

Figure 6.7 (a) Element in n–m axes oriented at θ to element in x–y axes; (b) stresses on faces of triangles A and B used to related stresses in the two axes systems.

stresses are shown as acting in a positive sense. The relevant set of transformation equations is

$$\sigma_n = \sigma_x \cos^2\theta + \sigma_y \sin^2\theta + \tau_{xy}\, 2\sin\theta\cos\theta,$$
$$\sigma_m = \sigma_x \sin^2\theta + \sigma_y \cos^2\theta - \tau_{xy}\, 2\sin\theta\cos\theta,$$
$$\tau_{nm} = -\sigma_x \sin\theta\cos\theta + \sigma_y \sin\theta\cos\theta + \tau_{xy}(\cos^2\theta - \sin^2\theta). \qquad (6.7)$$

By differentiating equation 6.7 with respect to θ the maximum, i.e., principal, values of the direct stresses can be shown to be

$$\sigma_{1,2} = \frac{\sigma_x + \sigma_y}{2} \pm \frac{1}{2}[(\sigma_x - \sigma_y)^2 + 4\tau_{xy}^2]^{1/2} \qquad (6.8a)$$

and the maximum shear stress

$$\tau_{xy_{max}} = \frac{1}{2}[(\sigma_x - \sigma_y)^2 + 4\tau_{xy}^2]^{1/2} \tag{6.8b}$$

We see, then, that application of only a shear stress (τ_{xy}) will give rise to (principal) direct stresses and, conversely, application of only direct stresses (σ_x and σ_y) will give rise to a (maximum) shear stress.

Similar transformation equations can be found for strains. The latter can be obtained from the former if we replace direct stress by direct strain and shear stress by (shear strain)/2, i.e.

$$\varepsilon_x = \varepsilon_x \cos^2\theta + \varepsilon_y \sin^2\theta + \tfrac{1}{2}\gamma_{xy} 2\sin\theta\cos\theta$$

$$\varepsilon_y = \varepsilon_x \sin^2\theta + \varepsilon_y \cos^2\theta - \tfrac{1}{2}\gamma_{xy} 2\sin\theta\cos\theta$$

$$\tfrac{1}{2}\gamma_{xy} = -\varepsilon_x \sin\theta\cos\theta + \varepsilon_y \sin\theta\cos\theta + \tfrac{1}{2}\gamma_{xy}(\cos^2\theta - \sin^2\theta) \tag{6.9}$$

For isotropic materials the directions of principal stresses and strains coincide.

It is often convenient to write the transformation equations 6.7 and 6.9 in a 'shorthand' form using matrix notation (see Appendix for a summary of matrix algebra). Adopting this approach we get

$$\sigma_{nm} = \mathbf{T}\,\sigma_{xy}, \text{ and} \tag{6.10}$$

$$\bar{\varepsilon}_{nm} = \mathbf{T}\,\bar{\varepsilon}_{xy}, \tag{6.11}$$

where

$$\sigma_{nm} = \{\sigma_n\ \sigma_m\ \tau_{nm}\},$$

$$\sigma_{xy} = \{\sigma_x\ \sigma_y\ \tau_{xy}\},$$

$$\bar{\varepsilon}_{nm} = \{\varepsilon_n\ \varepsilon_m\ \tfrac{1}{2}\gamma_{nm}\},$$

$$\bar{\varepsilon}_{xy} = \{\varepsilon_x\ \varepsilon_y\ \tfrac{1}{2}\gamma_{xy}\},$$

and the transformation matrix

$$\mathbf{T} = \begin{bmatrix} m^2 & n^2 & 2mn \\ n^2 & m^2 & -2mn \\ -mn & mn & (m^2 - n^2) \end{bmatrix}, \tag{6.12}$$

where $m = \cos\theta$, $n = \sin\theta$.

We see that the same transformation can be employed for both stress and strain provided we use $\gamma/2$ in the strain matrix, i.e. we use $\bar{\varepsilon}$ instead of ε.

6.8 THIN-WALLED CYLINDERS AND SPHERES

A frequently-encountered situation, in which a two-dimensional set of stresses exists, is that of a thin-walled cylinder or sphere subjected to internal pressure, p.

6.8.1 Cylinder with closed ends

Figure 6.8 shows the stresses in the axial direction (σ_a) and hoop, or circumferential, direction (σ_h) which are given by

$$\sigma_a = pR/2t, \quad \text{and}$$
$$\sigma_h = pR/t, \tag{6.13}$$

t, being the wall thickness and R the tube radius measured to the mid-line of the wall. Following equation 6.3 we can obtain the corresponding strains

$$\varepsilon_a = \frac{\sigma_a}{E} - v\frac{\sigma_h}{E},$$

$$\varepsilon_h = -v\frac{\sigma_a}{E} + \frac{\sigma_h}{E}. \tag{6.14}$$

If we substitute from equation 6.13 into equation 6.14 we get

$$\varepsilon_a = \frac{pR}{Et}\left(\frac{1}{2} - v\right),$$

$$\varepsilon_h = \frac{pR}{Et}\left(1 - \frac{v}{2}\right). \tag{6.15}$$

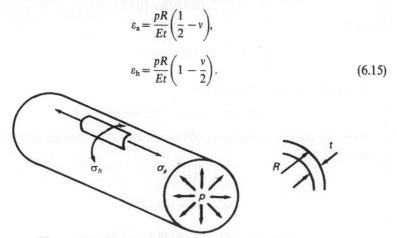

Figure 6.8 Stresses in wall of pressurized cylinder.

6.8.2 Sphere

From the essential symmetry of the sphere seen in Figure 6.9 we see that orthogonal stresses, in any set of axes, must be the same. It can be shown that the stress for a sphere of radius R and wall thickness t is

$$\sigma_s = pR/2t. \tag{6.16}$$

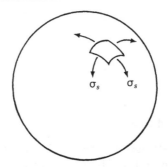

Figure 6.9 Stress in wall of pressurized sphere.

Example 6.2
A closed pipe 2 m long with 500 mm diameter and 20 mm wall thickness is subjected to an internal pressure of 0.5 MPa. Calculate the change of length and diameter caused by the pressure. Take Young's modulus to be 1 GPa and Poisson's ratio as 0.45.

From equation 6.15

$$\varepsilon_a = \frac{pR}{Et}(\tfrac{1}{2} - v) = \frac{0.5 \times 10^6 \times 250}{1 \times 10^9 \times 20}(0.5 - 0.45)$$
$$= 0.3125 \times 10^{-3},$$

and

$$\varepsilon_h = \frac{pR}{Et}\left(1 - \frac{v}{2}\right) = 4.838 \times 10^{-3}.$$

Hence the change in length (L)

$$= \varepsilon_a L = 0.3125 \times 10^{-3} \times 2\,\text{m} = 0.615\,\text{mm}.$$

Now, the change in diameter (D) is related to the change in circumference (C), i.e.

$$C = \pi D$$

$$(C + \delta C) = \pi(D + \delta D)$$

$$C(1 + \varepsilon_c) = \pi D\left(1 + \frac{\delta D}{D}\right).$$

That is,

$$1 + \varepsilon_c = 1 + \frac{\delta D}{D}, \text{ or } \varepsilon_c = \frac{\delta D}{D}$$

but $\varepsilon_c \equiv \varepsilon_h$, so $\delta D = \varepsilon_h D$.

Hence the change in diameter $= 4.838 \times 10^{-3} \times 500 = 2.419\,\text{mm}.$

6.9 FAILURE CRITERIA

Having determined the principal stress or strain we would use these maxima in an appropriate failure criterion to predict whether, or not, our material will fail, failure being taken as yielding, i.e., onset of permanent deformation, or fracture.

For isotropic materials there are a large number of criteria, the appropriate choice depending on the material (is it brittle or ductile?) and the local stress field (one-, two- or three-dimensional). There is no single criterion that is universally applicable.

A simple criterion is the *maximum principal stress theory* which states that failure will occur, in a material under a multi-axial stress field, when the maximum principal stress attains a value equal to the yield stress (σ_y) obtained from a uniaxial tensile test, i.e.

$$\sigma_1 = \sigma_y. \tag{6.17}$$

A more complicated criterion, often known as the *von Mises criterion* is derived from considerations of shear strain energy. The equation defining failure for a three-dimensional stress state is

$$(\sigma_1 - \sigma_2)^2 + (\sigma_2 - \sigma_3)^2 + (\sigma_3 - \sigma_1)^2 = 2\sigma_y^2 \tag{6.18}$$

where σ_1, σ_2 and σ_3 are the principal direct stresses.

Under plane stress conditions, such as found in a thin sheet subjected to biaxial loading, equation 6.18 reduces to

$$\sigma_1^2 + \sigma_2^2 - \sigma_1\sigma_2 = \sigma_y^2. \tag{6.19}$$

6.10 SUMMARY

In this chapter we have reviewed the fundamental parameters and relationships associated with the stress analysis of isotropic materials. The solution of the relevant equations enables us to find stresses and strains (and therefore deformations) and hence assess whether a material will fail under a defined set of loads.

We shall see in later chapters that similar equations appear in the analysis of composites, differing only in detail from those used for isotropic materials. In other words, no new concepts are needed.

FURTHER READING

General

Benham, P. P. and Warnock, F. V. (1982) *Mechanics of Solids and Structures*, Pitman.
Crandall, S. H., Dahl, N. C. and Lardner, T. J. (1978) *An Introduction to the Mechanics of Solids*, McGraw Hill.

PROBLEMS

6.1   A sphere has a diameter of 250 mm and a wall thickness of 10 mm. The material from which the sphere is fabricated is isotropic with Young's modulus of 70 GPa and Poisson's ratio of 0.33.

   Calculate the change in the diameter of the sphere when it is subjected to an internal pressure of 0.75 MPa.

6.2   A thin sheet of plastic is loaded by a two-dimensional set of stresses: $\sigma_x = 40$, $\sigma_y = 20$, $\tau_{xy} = 25$ MPa. The elastic properties of the plastic are $E = 3$ GPa and $v = 0.35$. The yield stress obtained from a simple tensile test is 56 MPa.

   Determine whether the sheet has failed according to (a) the maximum principal stress theory and (b) the von Mises criterion.

SELF-ASSESSMENT QUESTIONS

Indicate whether statements 1 to 8 are true or false.

1. A structural member whose length is much greater than its cross-section dimensions is always referred to as a rod.

   (A) true
   (B) false

2. The thickness of a plate is of similar magnitude to its length and breadth.

   (A) true
   (B) false

3. Cylindrical components can be treated as membranes provided the diameter is much larger than the wall thickness.

   (A) true
   (B) false

4. There are three types of stress; direct, shear and bending stress.

   (A) true
   (B) false

5. A plate subjected to a unidirectional direct stress experiences strain only in the direction of the stress.

   (A) true
   (B) false

6. The elastic stress-strain behaviour of an isotropic material can be characterized by only two elastic constants.

    (A) true
    (B) false

7. Principal stresses are the maximum and minimum shear stresses at a point in a body.

    (A) true
    (B) false

8. The failure of a stressed component can always be predicted by the von Mises criterion.

    (A) true
    (B) false

For each of the statements of questions 9 to 11, one or more of the completions given are correct. Mark the correct completions.

9. Stress

    (A) is the total load on a component,
    (B) is a normalized measure of load,
    (C) arises only from the application of a couple,
    (D) always exists as two components at right angles,
    (E) has units of force/length.

10. Strain

    (A) is an alternative term for displacement,
    (B) has units of 1/length,
    (C) is a non-dimensional measure of deformation,
    (D) is caused by the application of a stress,
    (E) can be related to curvature of a bent beam.

11. Poisson's ratio

    (A) defines the curvature of a plate,
    (B) is an alternative property to shear modulus,
    (C) has the same units as stress,
    (D) relates a longitudinal direct strain to the corresponding transverse strain,
    (E) is the same as shear strain.

Each of the sentences in questions 12 to 16 consists of an assertion followed by a reason. Answer:

    (A) if both assertion and reason are true statements and the reason is a correct explanation of the assertion,

(B) if both assertion and reason are true statements but the reason is not a correct explanation of the assertion,

(C) if the assertion is true but the reason is a false statement,

(D) if the assertion is false but the reason is a true statement,

(E) if both the assertion and reason are false statement.

12. The loads acting on a component are sub-divided into forces and couples *because* forces can be sub-divided into tension and compression.

13. In a plate loaded in its plane the orthogonal direct strains are determined by both direct stresses *because* the shear strain is determined by the shear stress.

14. An isotropic material always behaves in a linear elastic fashion *because* shear modulus is related to Young's modulus and Poisson's ratio.

15. A plate loaded by couples will exhibit anticlastic curvature *because* the direct stresses vary linearly through the thickness.

16. In a pressurized cylinder the axial and hoop stresses have the same magnitude *because* the axial strain is greater than the circumferential strain.

ANSWERS

Problems

6.1   0.011 mm.
6.2 (a) Yes, (b) No.

Self-assessment

1. B; 2. B; 3. A; 4. B; 5. B; 6. A; 7. B; 8. B; 9. B; 10. C, D, E; 11. D; 12. B; 13. B; 14. D; 15. B; 16. E.

Stiffness of unidirectional composites and laminates

7

7.1 INTRODUCTION

When calculating the mechanical properties of composites it is convenient to start by considering a composite in which all the fibres are aligned in one direction (i.e., a unidirectional composite). This basic 'building block' can then be used to predict the behaviour of continuous fibre multi-directional laminates, as well as short fibre, non-aligned systems.

The essential point about a unidirectional fibre composite is that its stiffness (and strength) are different in different directions. This behaviour was mentioned in the chapter on metal matrix composites (see for example Figures 3.11 and 3.14) and contrasts with a metal with a random orientation of grains, or other isotropic material, which has the same elastic properties in all directions.

In a unidirectional composite the fibre distribution implies that the behaviour is essentially isotropic in a cross-section perpendicular to the fibres (Figure 7.1). In other words, if we were to conduct a mechanical test

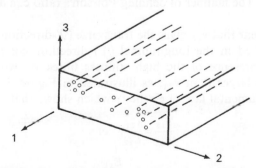

Figure 7.1 Orientation of principal material axes.

by applying a stress in the '2' direction or in the '3' direction (both normal to the fibre axes), we would obtain the same elastic properties from each test. We say the material is *transversely isotropic*. Clearly the properties in the longitudinal ('1') direction are very different from those in the other two directions. We call such a material *orthotropic*. The elastic properties are symmetric with respect to the chosen (1–2–3) axes, which are often called the principal material axes.

7.2 BASIC STRESS–STRAIN RELATIONS

The stress–strain relations (also known as the constitutive equations) can be found directly from those for an isotropic material (equations 6.2 and 6.3, Chapter 6), provided we take account of the fact that the properties are direction-dependent. Considering the composite illustrated in Figure 7.1, we see that when the directions of the applied stresses coincide with the principal (1–2) material axes, the strains in terms of stresses are given by

$$\varepsilon_1 = \frac{\sigma_1}{E_{11}} - v_{21}\frac{\sigma_2}{E_{22}},$$

$$\varepsilon_2 = -v_{12}\frac{\sigma_1}{E_{11}} + \frac{\sigma_2}{E_{22}}, \qquad (7.1)$$

$$\gamma_{12} = \frac{\tau_{12}}{G_{12}},$$

where

E_{11} is the elastic modulus in the '1', or longitudinal direction,
E_{22} is the elastic modulus in the '2', or transverse direction,
G_{12} is the shear modulus in the 1–2 axes,
v_{12} is the 'major' Poisson's ratio, and
v_{21} is the 'minor' Poisson's ratio.

It should be emphasized that the convention for suffixes used here is by no means universal. The manner of defining Poisson's ratio can also differ from text to text.

It should be clear that v_{12} gives the transverse ('2'-direction) strain caused by a strain applied in the longitudinal ('1') direction; conversely for v_{21}. Because of the presence of the high stiffness fibres we would intuitively expect v_{12} to be larger than v_{21}, as illustrated in Figure 7.2. This is confirmed by a fundamental law of elasticity which shows that

$$\frac{v_{12}}{E_{11}} = \frac{v_{21}}{E_{22}}, \quad \text{or}$$

$$v_{12} = v_{21}\frac{E_{11}}{E_{22}}, \qquad (7.2)$$

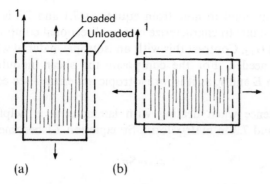

Figure 7.2 Illustration of orthogonal contractions related to (a) v_{12}; (b) v_{21}.

and as

$$\frac{E_{11}}{E_{22}} > 1, \quad v_{12} > v_{21}.$$

To get stresses in terms of strains we can rearrange equations 7.1 to give

$$\sigma_1 = \frac{E_{11}\varepsilon_1}{1 - v_{12}v_{21}} + \frac{v_{21}E_{11}\varepsilon_2}{1 - v_{12}v_{21}},$$

$$\sigma_2 = \frac{v_{12}E_{22}\varepsilon_1}{1 - v_{12}v_{21}} + \frac{E_{22}\varepsilon_2}{1 - v_{12}v_{21}}, \tag{7.3}$$

$$\tau_{12} = G_{12}\gamma_{12}.$$

Note the similarity to equations 6.4 of Chapter 6.

Representative experimentally-determined values of elastic properties are given in Table 7.1 for carbon, glass and 'Kevlar' fibre-reinforced epoxy resins (CFRP, GFRP, KFRP respectively). As we shall see later, it is sometimes possible to calculate elastic constants from the properties of the fibre and matrix.

Table 7.1 Representative elastic properties of unidirectional fibre-reinforced epoxy resins

Material	Fibre volume fraction v_f	$E_{11}(GPa)$	$E_{22}(GPa)$	v_{12}	$G_{12}(GPa)$
CFRP (AS fibre)	0.66	140.0	8.96	0.30	7.10
CFRP (IM6 fibre)	0.65	200.0	11.10	0.32	8.35
GFRP (E-glass fibre)	0.46	35.0	8.22	0.26	4.10
KFRP (Kevlar-49 fibre)	0.60	76.0	5.50	0.33	2.35

The important point to note from equations 7.1 and 7.3 is that we need *four* elastic constants to characterize our unidirectional composite; E_{11}, E_{22}, v_{12} (or v_{21}) and G_{12}. Contrast this with an isotropic material where only two quantities are needed. For the composite the shear modulus cannot be calculated from E and v, as it is for isotropic materials – see equation 6.6 in Chapter 6.

For convenience when dealing with laminates it is helpful to rewrite equations 7.1 and 7.3 in matrix form. So equation 7.1 becomes

$$\varepsilon_{12} = S\sigma_{12} \tag{7.4}$$

where

$$\varepsilon_{12} = \{\varepsilon_1 \varepsilon_2 \gamma_{12}\}, \text{ and}$$

$$\sigma_{12} = \{\sigma_1 \sigma_2 \tau_{12}\}$$

and we define the compliance matrix

$$S = \begin{bmatrix} \dfrac{1}{E_{11}} & -\dfrac{v_{21}}{E_{22}} & 0 \\[2ex] -\dfrac{v_{12}}{E_{11}} & \dfrac{1}{E_{22}} & 0 \\[2ex] 0 & 0 & \dfrac{1}{G_{12}} \end{bmatrix};$$

note that $S_{12} = S_{21}$, and equation 7.3 becomes

$$\sigma_{12} = Q\varepsilon_{12}, \tag{7.5}$$

where the stiffness matrix is defined by

$$Q = \begin{bmatrix} \dfrac{E_{11}}{1-v_{12}v_{21}} & \dfrac{v_{21}E_{11}}{1-v_{12}v_{21}} & 0 \\[2ex] \dfrac{v_{12}E_{22}}{1-v_{12}v_{21}} & \dfrac{E_{22}}{1-v_{12}v_{21}} & 0 \\[2ex] 0 & 0 & G_{12} \end{bmatrix};$$

note that $Q_{12} = Q_{21}$.

We see that the first column of S gives the strains caused by a unit value of σ_1, the second column of Q gives the stresses needed to cause a unit value of ε_2, and so on. It should be clear from the properties of matrices (see Appendix) that $Q = S^{-1}$.

Example 7.1

Suppose we take a composite with properties representative of CFRP from Table 7.1, i.e.

$$E_{11} = 138.0, E_{22} = 8.96, G_{12} = 7.10 \text{ GPa}; v_{12} = 0.30.$$

Using equation 7.2 we find that $v_{21} = 0.0195$ and substituting into equations 7.4 and 7.5 we get the compliance and stiffness matrices as

$$\mathbf{S} = 10^{-3} \begin{bmatrix} 7.25 & -2.17 & 0 \\ -2.17 & 111.61 & 0 \\ 0 & 0 & 140.85 \end{bmatrix} \text{GPa}^{-1},$$

$$\mathbf{Q} = \begin{bmatrix} 138.81 & 2.70 & 0 \\ 2.70 & 9.01 & 0 \\ 0 & 0 & 7.10 \end{bmatrix} \text{GPa}.$$

7.3 OFF-AXIS LOADING OF A UNIDIRECTIONAL COMPOSITE

In later sections and chapters we shall refer frequently to *laminates*. These plate-like entities are constructed by assembling unidirectional layers (or laminae, or plies) one on top of another, the direction of the fibres often being changed from layer to layer. Consequently there will be layers for which the fibres are no longer aligned with the applied stresses (the situation considered in the previous section). We term these *rotated layers* and say that they are subjected to *off-axis loading*.

To prepare ourselves for the analysis of laminates, it is useful at this stage to consider in isolation one of these rotated laminae. The situation, which is illustrated in Figure 7.3, is seen to correspond to that discussed in section 6.7 and illustrated in Figure 6.7 of Chapter 6. The applied stresses, as before, are parallel to the plate edges (i.e. the x–y axes), and our principal material directions (1–2 axes), which correspond to the $n - m$ axes used previously (Figure 6.7), are at an angle θ to the x–y axes.

It follows, then, that the equations given before (equations 6.7 and 6.9, Chapter 6) can be used directly to obtain the stresses and strains in our unidirectional composite, referred to axes along and transverse to the fibres; we simply replace σ_n by σ_1, σ_m by σ_2 and τ_{nm} by τ_{12} (similarly for the strains). Obviously the transformation used before will also apply here.

We can write, then

$$\sigma_{12} = \mathbf{T}\sigma_{xy}, \text{ and} \tag{7.6}$$

$$\bar{\varepsilon}_{12} = \mathbf{T}\bar{\varepsilon}_{xy}, \tag{7.7}$$

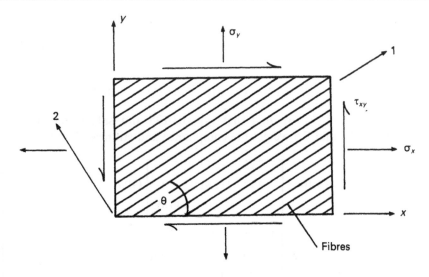

Figure 7.3   Unidirectional lamina with principal axes rotated by θ relative to the x–y axes.

where

$$\sigma_{12} = \{\sigma_1 \quad \sigma_2 \quad \tau_{12}\},$$
$$\sigma_{xy} = \{\sigma_x \quad \sigma_y \quad \tau_{xy}\},$$
$$\bar{\varepsilon}_{12} = \{\varepsilon_1 \quad \varepsilon_2 \quad \tfrac{1}{2}\gamma_{12}\},$$
$$\bar{\varepsilon}_{xy} = \{\varepsilon_x \quad \varepsilon_y \quad \tfrac{1}{2}\gamma_{xy}\},$$

and

$$\mathbf{T} = \begin{bmatrix} m^2 & n^2 & 2mn \\ n^2 & m^2 & -2mn \\ -mn & mn & (m^2-n^2) \end{bmatrix},$$

with $m = \cos\theta$, and $n = \sin\theta$.

If we wish to transform in the opposite direction then from equations 7.6 and 7.7 we get

$$\sigma_{xy} = \mathbf{T}^{-1}\sigma_{12}, \text{ and} \tag{7.6a}$$

$$\bar{\varepsilon}_{xy} = \mathbf{T}^{-1}\bar{\varepsilon}_{12}. \tag{7.7a}$$

where the inverse transformation matrix is

$$\mathbf{T}^{-1} = \begin{bmatrix} m^2 & n^2 & -2mn \\ n^2 & m^2 & 2mn \\ mn & -mn & (m^2-n^2) \end{bmatrix}.$$

We see that we can re-write equation 7.5 as

$$\sigma_{12} = \mathbf{QR}\bar{\varepsilon}_{12}, \tag{7.8}$$

where

$$\varepsilon_{12} = \mathbf{R}\bar{\varepsilon}_{12},$$

$$\mathbf{R} = \begin{bmatrix} 1 & 0 & 0 \\ 0 & 1 & 0 \\ 0 & 0 & 2 \end{bmatrix}.$$

As with any elasticity analysis we wish to determine the strains for a known set of applied stresses (or vice versa). We can do this provided we know the elastic properties of the material. The situation we are faced with here is that whilst we know the properties referred to the 1–2 axes, we do not know them with reference to the x–y axes. So, before we can solve the problem, we need to do some mathematical manipulation.

Now, from equation 7.6a we have

$$\sigma_{xy} = \mathbf{T}^{-1}\sigma_{12},$$

and using equation 7.8 we get

$$\sigma_{xy} = \mathbf{T}^{-1}\mathbf{QR}\bar{\varepsilon}_{12},$$

which combined with equation 7.7 gives

$$\sigma_{xy} = \mathbf{T}^{-1}\mathbf{QRT}\bar{\varepsilon}_{xy} = \mathbf{T}^{-1}\mathbf{QRTR}^{-1}\varepsilon_{xy}.$$

We write finally

$$\sigma_{xy} = \bar{\mathbf{Q}}\varepsilon_{xy}. \tag{7.9}$$

Note that we have returned to $\varepsilon_{xy}(=\{\varepsilon_x\ \varepsilon_y\ \gamma_{xy}\})$ rather than retaining $\bar{\varepsilon}_{xy}(=\{\varepsilon_x\ \varepsilon_y\ \frac{1}{2}\gamma_{xy}\})$.

This makes equation 7.9 consistent with equation 7.5.

The transformed stiffness matrix

$$\bar{\mathbf{Q}} = \mathbf{T}^{-1}\mathbf{QRTR}^{-1},$$

the elements of which are:

$$\bar{Q}_{11} = Q_{11}m^4 + 2(Q_{12} + 2Q_{33})n^2m^2 + Q_{22}n^4,$$

$$\bar{Q}_{22} = Q_{11}n^4 + 2(Q_{12} + 2Q_{33})n^2m^2 + Q_{22}m^4,$$

$$\bar{Q}_{12} = (Q_{11} + Q_{22} - 4Q_{33})n^2m^2 + Q_{12}(m^4 + n^4), \tag{7.10}$$

$$\bar{Q}_{33} = (Q_{11} + Q_{22} - 2Q_{12} - 2Q_{33})n^2m^2 + Q_{33}(m^4 + n^4),$$

$$\bar{Q}_{13} = (Q_{11} - Q_{12} - 2Q_{33})nm^3 + (Q_{12} - Q_{22} + 2Q_{33})n^3m,$$

$$\bar{Q}_{23} = (Q_{11} - Q_{12} - 2Q_{33})n^3m + (Q_{12} - Q_{22} + 2Q_{33})nm^3.$$

The terms Q_{11}, etc., are found from equation 7.5.

We see that knowledge of the orientation (θ) and unidirectional properties (Q) in the principal directions enables us to calculate the stiffness of the rotated lamina. In the general form derived we would call this a 'generally orthotropic lamina'. If conditions are such that $\bar{Q}_{13} = \bar{Q}_{23} = 0$, we have what is called a 'specially orthotropic lamina'.

Example 7.2
For the same material as in Example 7.1, calculate the transformed stiffness matrix for a unidirectional lamina with the fibres oriented at 45° to the stress axes.

In this case $\theta = 45°$ and since $\cos 45° = \sin 45°$, $m = n = 0.7071$. From Example 7.1 we have

$$Q_{11} = 138.81, \quad Q_{12} = 2.70, \quad Q_{13} = 0,$$
$$Q_{21} = 2.70, \quad Q_{22} = 9.01, \quad Q_{23} = 0,$$
$$Q_{31} = 0, \quad Q_{32} = 0, \quad Q_{33} = 7.10 \text{ (units GPa)}.$$

Using equation 7.10 we can get the elements of $\bar{Q}$ so,

$$\bar{Q}_{11} = 138.81 \times 0.25 + 2(2.70 + 2 \times 7.10)0.5 \times 0.5 + 9.01 \times 0.25$$
$$= 45.41 \text{ GPa}.$$

Repeating for the other terms gives

$$\bar{Q} = \begin{bmatrix} 45.41 & 31.21 & 32.45 \\ 31.21 & 45.41 & 32.45 \\ 32.45 & 32.45 & 35.60 \end{bmatrix} \text{GPa}.$$

The value of $\bar{Q}_{11}$ in example 7.2 is equivalent to the composite's modulus in the stress direction. Clearly there is a large reduction compared to the original aligned state (Q_{11}) (see also Chapter 10, Figure 10.13).

If we require strains in terms of stresses then we invert equation 7.9 to give

$$\varepsilon_{xy} = \bar{Q}^{-1}\sigma_{xy}, \text{ or}$$
$$\varepsilon_{xy} = \bar{S}\sigma_{xy}, \tag{7.11}$$

where $\bar{S}$ is the transformed compliance matrix, the elements of which can be obtained by a similar process to that used for finding the elements of $\bar{Q}$, i.e.

$$\bar{S}_{11} = S_{11}m^4 + (2S_{12} + S_{33})n^2m^2 + S_{22}n^4,$$
$$\bar{S}_{22} = S_{11}n^4 + (2S_{12} + S_{33})n^2m^2 + S_{22}m^4,$$
$$\bar{S}_{12} = (S_{11} + S_{22} - S_{33})n^2m^2 + S_{12}(m^4 + n^4),$$
$$\bar{S}_{33} = 2(2S_{11} + 2S_{22} - 4S_{12} - S_{33})n^2m^2 + S_{33}(m^4 + n^4), \tag{7.12}$$
$$\bar{S}_{13} = (2S_{11} - 2S_{12} - S_{33})m^3n + (2S_{12} - 2S_{22} + S_{33})mn^3,$$
$$\bar{S}_{23} = (2S_{11} - 2S_{12} - S_{33})mn^3 + (2S_{12} - 2S_{22} + S_{33})m^3n.$$

Example 7.3
Calculate, for the lamina of Example 7.2, the transformed compliance matrix.

From Example 7.1

$$S = 10^{-3} \begin{bmatrix} 7.25 & -2.17 & 0 \\ -2.17 & 111.61 & 0 \\ 0 & 0 & 140.85 \end{bmatrix} \text{GPa}^{-1}.$$

So from equation 7.12

$$\bar{S}_{11} = \{7.25 \times 0.25 + (-2 \times 2.17 + 140.85) \times 0.5 \times 0.5 + 111.61 \times 0.25\}$$
$$= 63.84 \times 10^{-3} \text{GPa}^{-1}.$$

Repeating for the other terms gives

$$\bar{S} = \begin{bmatrix} 63.84 & -6.58 & -52.18 \\ & 63.84 & -52.18 \\ \text{symmetric} & & 123.20 \end{bmatrix} 10^{-3} \text{GPa}^{-1}.$$

Example 7.4
The rotated lamina considered in Example 7.2 is subjected to the uniaxial stress system shown below. Calculate the corresponding strains.

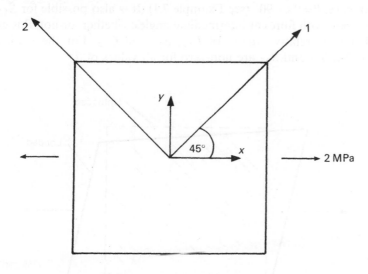

We have

$$\sigma_x = 2 \text{ MPa}, \quad \sigma_y = \tau_{xy} = 0.$$

Using equation 7.11, and the $\bar{S}$ found in Example 7.3 we have

$$
\begin{bmatrix} \varepsilon_x \\ \varepsilon_y \\ \gamma_{xy} \end{bmatrix} = 10^{-3} \begin{bmatrix} 63.84 & -6.58 & -52.18 \\ -6.58 & 63.84 & -52.18 \\ -52.18 & -52.18 & 123.20 \end{bmatrix} \begin{bmatrix} 2 \\ 0 \\ 0 \end{bmatrix},
$$

giving

$$
\varepsilon_x = 63.84 \times 10^{-3} \frac{1}{\text{GPa}} \times 2\,\text{MPa}
$$

$$
= 127.7 \times 10^{-6}
$$

$$
\varepsilon_y = -13.2 \times 10^{-6}
$$

$$
\gamma_{xy} = -104.4 \times 10^{-6}
$$

So, we see from Example 7.4 that in addition to an extension in the direction of the applied stress and a lateral contraction, as we would expect, there is also a negative shear strain. The deformed lamina is shown in Figure 7.4. This phenomenon, which is known as extension–shear coupling, would not be observed with an isotropic material. We see that the shear strain is determined by the $\bar{S}_{13}$ term, when only σ_x is acting. Had we applied only σ_y, there would have been a shear strain arising from the $\bar{S}_{23}$ term. In other words there will be no extension–shear coupling if $\bar{S}_{13} = \bar{S}_{23} = 0$. This is clearly the case when the fibres of a unidirectional layer are parallel to the stress axes, i.e. $\theta = 0$ or $90°$ (see Example 7.1). It is also possible for $\bar{S}_{13}$ and $\bar{S}_{23}$ to be zero with fibres at intermediate angles; whether, or not, this occurs depends on the relative values of E_{11}, E_{22} and G_{12}. For the material of Example 7.1, $\bar{S}_{13}$ and $\bar{S}_{23}$ are non-zero for all values of θ.

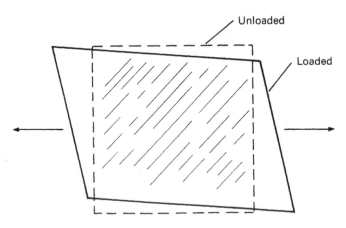

Figure 7.4   Deformation of lamina discussed in Example 7.4.

Example 7.5
The rotated lamina considered in Example 7.2 is subjected to the stresses shown in the figure below. Note that both the vertical direct stress and shear stress are acting in a negative sense (according to the convention shown in Figure 6.7). Calculate the stresses in the 1–2 directions and the strains in the *x–y* directions.

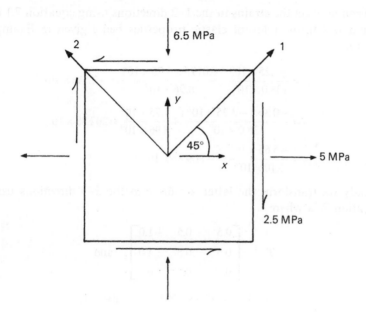

We have

$$\sigma_x = 5, \quad \sigma_y = -6.5, \quad \tau_{xy} = -2.5 \text{ MPa}$$

Using equation 7.6 we can get the stresses in the 1–2 axes, i.e.

$$\sigma_{12} = \mathbf{T}\sigma_{xy}.$$

Here

$$\mathbf{T} = \begin{bmatrix} 0.5 & 0.5 & 1.0 \\ 0.5 & 0.5 & -1.0 \\ -0.5 & 0.5 & 0 \end{bmatrix}, \quad \text{and}$$

$$\sigma_{xy} = \{5 \quad -6.5 \quad -2.5\} \text{ MPa}.$$

Hence

$$\sigma_1 = 0.5 \times 5 - 0.5 \times 6.5 - 1.0 \times 2.5 = -3.25 \text{ MPa},$$

$$\sigma_2 = 0.5 \times 5 - 0.5 \times 6.5 + 1.0 \times 2.5 = 1.75 \text{ MPa},$$

$$\tau_{12} = -0.5 \times 5 - 0.5 \times 6.5 + 0 \times 2.5 = -5.75 \text{ MPa},$$

giving

$$\sigma_{12} = \{-3.25 \quad 1.75 \quad -5.75\} \text{ MPa}.$$

We can now get the strains in the 1–2 directions using equation 7.1 (or equation 7.4), the relevant elastic properties being given in Example 7.1, i.e.

$$\varepsilon_1 = \frac{-3.25 \times 10^6}{138.0 \times 10^9} - \frac{0.0195 \times 1.75 \times 10^6}{8.96 \times 10^9} = -27.36 \times 10^{-6},$$

$$\varepsilon_2 = \frac{-0.3 \times (-3.25 \times 10^6)}{138.0 \times 10^9} + \frac{1.75 \times 10^6}{8.96 \times 10^9} = 202.38 \times 10^{-6},$$

$$\gamma_{12} = \frac{-5.65 \times 10^6}{7.10 \times 10^9} = -809.86 \times 10^{-6}.$$

Finally we transform the latter strains into the x–y directions using equation 7.7a where

$$T^{-1} = \begin{bmatrix} 0.5 & 0.5 & -1.0 \\ 0.5 & 0.5 & 1.0 \\ 0.5 & -0.5 & 0 \end{bmatrix}, \quad \text{and}$$

$$\bar{\varepsilon}_{12} = 10^{-6}\{-27.36 \quad 202.38 \quad -404.93\}.$$

Hence

$$\bar{\varepsilon}_{xy} = 10^{-6}\{492.44 \quad -317.45 \quad -114.87\},$$

giving finally

$$\varepsilon_{xy} = 10^{-6}\{492.44 \quad -317.45 \quad -229.74\}.$$

Alternatively, we could use the $\bar{S}$ matrix, from Example 7.3, to give the strains directly.

7.4 STIFFNESS OF LAMINATES

Thin sheet constructions, known as laminates, are an important class of composites. They are made by stacking together, usually, unidirectional layers (also called plies or laminae) in predetermined directions and thicknesses to give the desired stiffness and strength properties. Such constructions are frequently encountered. The skins of aeroplane wings and tails, the hull sides and decking of ships, the sides and bottom of water tanks are typical examples. Even cylindrical components, such as filament wound tanks, can be treated as laminates, provided the radius-to-thickness ratio is sufficiently large (say > 50). Often the layers will be cut from prepreg material and cured in an autoclave or hot press. The alternative method using dry fabric and wet impregnation is however frequently used, particularly with glass/polyester combinations. Laminates will be typically between 4 and 40 layers, each ply being around 0.125 mm thick if it is carbon or glass fibre/epoxy prepreg. Typical lay-ups (the arrangement of fibre orientations) are cross-ply, angle ply and quasi-isotropic.

When making a laminate we must decide on the order in which the plies are placed through the thickness (known as the stacking sequence). As we shall see later, this has an important influence on the flexural performance of the laminate.

There is an established convention for denoting both the lay-up and stacking sequence of a laminate. Thus, a 4-ply cross-ply laminate which has ply fibre orientations in the sequence $0°, 90°, 90°, 0°$ from the upper to the lower surface, would be denoted $(0/90)_s$. The suffix 's' means that the stacking sequence is symmetric about the mid-thickness of the laminate. Laminates denoted by $(0/45/90)_s$ and $(45/90/0)_s$ have the same lay-up but different stacking sequences.

We have already seen in the previous section how we can describe the stiffness of a unidirectional ply, when subjected to in-plane loading, both when the fibres are aligned with the loads, or are at some angle. We shall now use those results to formulate the behaviour of a laminate.

The way in which they are used means that laminated plates may be subjected to both in-plane and transverse (normal to the plate) loading. In other words a plate will stretch and bend, and both these effects must be taken into account when describing the overall behaviour of the plate. The two effects are allowed for by considering the total strain to be the superposition of separate in-plane strains, ε^0, constant across the plate thickness, and strains caused by bending, linear across the thickness, as shown in Figure 7.5. As noted in

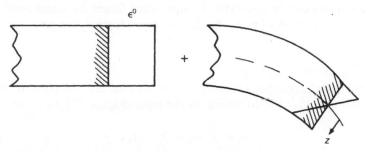

Figure 7.5 The two components of laminate strain: ε^0 in-plane, constant over the thickness; $\varepsilon_x = zk$, bending, linear variation over the thickness.

Chapter 6, the bending strains can be defined in terms of the plate curvatures. For example, $\varepsilon_x = z k_x$, z being the coordinate normal to the plate measured from the laminate mid-plane, and is positive downwards (as shown in Figure 7.5).

So we have

$$
\begin{bmatrix} \varepsilon_x \\ \varepsilon_y \\ \gamma_{xy} \end{bmatrix} = \begin{bmatrix} \varepsilon_x^0 \\ \varepsilon_y^0 \\ \gamma_{xy}^0 \end{bmatrix} + z \begin{bmatrix} k_x \\ k_y \\ k_{xy} \end{bmatrix}
\tag{7.13}
$$

or

$$
\varepsilon_{xy} = \varepsilon^0 + z \mathbf{k}
\tag{7.14}
$$

Because all the plies are 'bonded together' in the manufacturing process, we assume that they each have the same in-plane strains and curvatures. So, for any one layer, say the jth (Figure 7.6) we have, using equation 7.9,

$$
\sigma_{xy_j} = \bar{\mathbf{Q}}_j \varepsilon^0 + z \bar{\mathbf{Q}}_j \mathbf{k}
\tag{7.15}
$$

$\bar{\mathbf{Q}}_j$ being the transformed stiffness matrix for the layer.

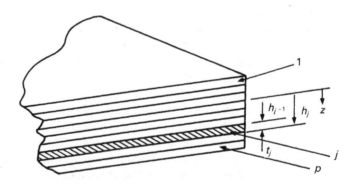

Figure 7.6 Definition of plies within a laminate.

These stresses can be converted to equivalent forces (or stress resultants) acting on a unit width of plate. So, for example, from σ_x we get

$$
N_{x_j} = \sigma_{x_j} t_j,
$$

t_j being the thickness of layer j.

If we add up the resultants for all the plies the total must be equal to the external force (per unit width) acting on the plate (Figure 7.7). In other words

$$
N_x = \sum_{j=1}^{p} N_{x_j} = \sum_{j=1}^{p} \sigma_{x_j} t_j,
\tag{7.16}
$$

where σ_{x_j} can be found from the first row of equations 7.15.

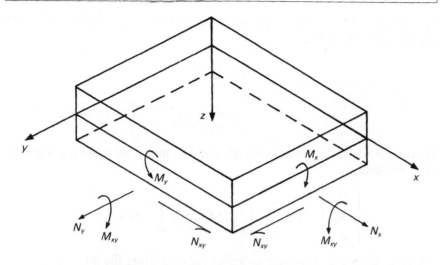

Figure 7.7 Loads acting on a laminate.

We can repeat this process for the other components of force, N_y and N_{xy}, giving us, finally,

$$\mathbf{N} = \begin{bmatrix} N_x \\ N_y \\ N_{xy} \end{bmatrix} = \sum_{j=1}^{p} \int_{h_{j-1}}^{h_j} \bar{\mathbf{Q}} \boldsymbol{\varepsilon}^0 \, dz + \int_{h_{j-1}}^{h_j} \bar{\mathbf{Q}} \mathbf{k} z \, dz \ . \tag{7.17}$$

We note that $h_j - h_{j-1} = t_j$, the thickness of the layer.

Evaluation of these integrals is not difficult since $\boldsymbol{\varepsilon}^0$ and $\mathbf{k}$ are not functions of z and, within a layer, $(h_{j-1} \to h_j)$, $\bar{\mathbf{Q}}_j$ is not a function of z. Hence we get

$$\mathbf{N} = \sum_{j=1}^{p} \ \bar{\mathbf{Q}} \boldsymbol{\varepsilon}^0 \int_{h_{j-1}}^{h_j} dz + \bar{\mathbf{Q}} \mathbf{k} \int_{h_{j-1}}^{h_j} z \, dz \ .$$

Also $\boldsymbol{\varepsilon}^0$ and $\mathbf{k}$ are not functions of j so we finally have

$$\mathbf{N} = \boldsymbol{\varepsilon}^0 \sum_{j=1}^{p} \bar{\mathbf{Q}}_j \int_{h_{j-1}}^{h_j} dz + \mathbf{k} \sum_{j=1}^{p} \bar{\mathbf{Q}}_j \int_{h_{j-1}}^{h_j} z \, dz$$

or

$$\mathbf{N} = \mathbf{A} \boldsymbol{\varepsilon}^0 + \mathbf{B} \mathbf{k}. \tag{7.18}$$

Now, the equivalent force on a layer will have a moment about the mid-plane. Again using σ_x only to illustrate the process, we have $M_{xj} = \sigma_{xj} t_j z_j$, z_j being the distance from the laminate mid-plane to the mid-thickness of the ply. (Note that the definition of M_x is such that it causes direct stress σ_x, i.e. it is not a couple about the x-axis). Adding together these moments for all the plies will give the external moment (per unit width) acting on the plate (Figure 7.7), i.e.

$$M_x = \sum_{j=1}^{p} M_{xj}$$

$$= \sum_{j=1}^{p} \sigma_{xj} t_j z_j.$$

We can get similar expressions for M_y and M_{xy} and, if we again use equation 7.13, we obtain

$$\begin{bmatrix} M_x \\ M_y \\ M_{xy} \end{bmatrix} = \mathbf{M} = \sum_{j=1}^{p} \int_{h_{j-1}}^{h_j} \bar{\mathbf{Q}}_j \varepsilon^0 z_j \, dz + \int_{h_{j-1}}^{h_j} \bar{\mathbf{Q}}_j k z_j^2 \, dz \ .$$

Adopting the same procedure as before for evaluating the integrals gives

$$\mathbf{M} = \varepsilon^0 \sum_{j=1}^{p} \bar{\mathbf{Q}}_j \int_{h_{j-1}}^{h_j} z \, dz + \mathbf{k} \sum_{j=1}^{p} \bar{\mathbf{Q}}_j \int_{h_{j-1}}^{h_j} z^2 \, dz$$

or

$$\mathbf{M} = \mathbf{B}\varepsilon^0 + \mathbf{D}\mathbf{k}. \tag{7.19}$$

Equations 7.18 and 7.19 are known collectively as the *plate constitutive equations* written as

$$\begin{bmatrix} \mathbf{N} \\ \mathbf{M} \end{bmatrix} = \begin{bmatrix} \mathbf{A} & \mathbf{B} \\ \mathbf{B} & \mathbf{D} \end{bmatrix} \begin{bmatrix} \varepsilon^0 \\ \mathbf{k} \end{bmatrix} \tag{7.20}$$

The elements of the **A**, **B** and **D** matrices are

$$A_{rs} = \sum_{j=1}^{p} \bar{Q}_{rs_j}[h_j - h_{j-1}] = \sum_{j=1}^{p} \bar{Q}_{rs_j} t_j,$$

$$B_{rs} = \tfrac{1}{2} \sum_{j=1}^{p} \bar{Q}_{rs_j}[h_j^2 - h_{j-1}^2], \tag{7.21}$$

$$D_{rs} = \tfrac{1}{3} \sum_{j=1}^{p} \bar{Q}_{rs_j}[h_j^3 - h_{j-1}^3].$$

Example 7.6
To illustrate the form of the **A**, **B** and **D** matrices, and the calculations involved, we consider a very simple example.

Suppose we take a laminate composed of two unidirectional layers, each 1 mm thick, the fibres being at 45° to the plate edges. The material is the same as that used in Examples 7.1 and 7.2.

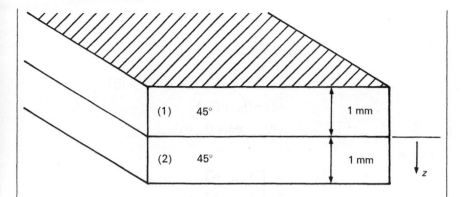

Clearly for ply (1),

$$h_j = 0, h_{j-1} = -1,$$

and for ply (2),

$$h_j = 1, h_{j-1} = 0.$$

From Example 7.2

$$\bar{Q}_1 = \bar{Q}_2 = \begin{bmatrix} 45.41 & 31.21 & 32.45 \\ & 45.41 & 32.45 \\ & & 35.60 \end{bmatrix} \text{GPa.}$$

Using equations 7.21

$$A_{rs} = \sum_{j=1}^{p} \bar{Q}_{rs_j} [h_j - h_{j-1}]$$

where, here, $p = 2$, so

$$A_{rs} = \bar{Q}_{rs_1}[0 - (-1)] + \bar{Q}_{rs_2}[1 - 0] = 2\bar{Q}_{rs_1} \text{ (or } 2\bar{Q}_{rs_2}).$$

That is

$$A_{11} = 2 \text{ mm} \times 45.41 \text{ GPa} = 90.82 \text{ kN/mm}$$

(note that we conveniently use GPa = kMPa and MPa = MN/m² = N/mm²),

$$A_{12} = 2 \times 31.21 = 62.42, \text{ etc.,}$$

giving

$$A = \begin{bmatrix} 90.82 & 62.42 & 64.90 \\ & 90.82 & 64.90 \\ \text{symmetric} & & 71.20 \end{bmatrix} \text{kN/mm.}$$

Similarly

$$B_{rs} = \tfrac{1}{2} \sum_{1}^{2} \bar{Q}_{rs_j}[h_j^2 - h_{j-1}^2]$$
$$= \tfrac{1}{2}\{\bar{Q}_{rs_1}[0-(-1)^2] + \bar{Q}_{rs_2}[1^2 - 0]\}$$
$$= \tfrac{1}{2}\{-\bar{Q}_{rs_1} + \bar{Q}_{rs_2}\}$$
$$= 0, \quad \text{since } \bar{Q}_{rs_1} = \bar{Q}_{rs_2},$$

that is

$$\mathbf{B} = \begin{bmatrix} 0 & 0 & 0 \\ 0 & 0 & 0 \\ 0 & 0 & 0 \end{bmatrix} \text{kN.}$$

And finally,

$$D_{rs} = \tfrac{1}{3} \sum_{1}^{2} \bar{Q}_{rs_j}[h_j^3 - h_{j-1}^3]$$
$$= \tfrac{1}{3}\{\bar{Q}_{rs_1}[0-(-1)^3] + \bar{Q}_{rs_2}[1^3 - 0]\}$$
$$= \tfrac{1}{3}\{\bar{Q}_{rs_1} + \bar{Q}_{rs_2}\} = \tfrac{2}{3}\bar{Q}_{rs_1}$$

that is

$$\mathbf{D} = \begin{bmatrix} 30.27 & 20.81 & 21.63 \\ & 30.27 & 21.63 \\ \text{symmetric} & & 23.73 \end{bmatrix} \text{kN mm.}$$

Example 7.7
A laminate, which is 6 mm thick, is composed of an upper unidirectional layer, 2 mm thick, with the fibres aligned at 45° to the plate edges, and a lower unidirectional layer, 4 mm thick, with the fibres parallel to the *x*-axis. The material is the same as in the last example. Find the **A**, **B** and **D** matrices.

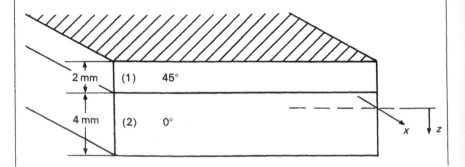

We determine the positions, relative to the laminate mid-plane, of the upper and lower ply surfaces.

So for ply (1),

$$h_j = -1, \ h_{j-1} = -3,$$

and for ply (2),

$$h_j = 3, \ h_{j-1} = -1.$$

From the earlier examples we have

$$\bar{Q}_1 = \begin{bmatrix} 45.41 & 31.21 & 32.45 \\ & 45.41 & 32.45 \\ \text{symmetric} & & 35.60 \end{bmatrix} \text{GPa}, \quad \text{(Example 7.2)}$$

and

$$\bar{Q}_2 = \begin{bmatrix} 138.81 & 2.70 & 0 \\ & 9.01 & 0 \\ \text{symmetric} & & 7.10 \end{bmatrix} \text{GPa}. \quad \text{(Example 7.1)}$$

Using equations 7.21 we have

$$A_{rs} = \sum_{j=1}^{p} \bar{Q}_{rs_j}[h_j - h_{j-1}] \quad \text{where, here, } p = 2, \text{ so}$$

$$A_{rs} = \bar{Q}_{rs_1}[-1 - (-3)] + \bar{Q}_{rs_2}[3 - (-1)]$$

$$= 2\bar{Q}_{rs_1} + 4\bar{Q}_{rs_2}.$$

That is

$$A_{11} = 2\bar{Q}_{11_1} + 4\bar{Q}_{11_2}$$

$$= 2 \times 45.41 + 4 \times 138.81$$

$$= 646.1 \text{ kN/mm},$$

$$A_{12} = 2\bar{Q}_{12_1} + 4\bar{Q}_{12_2}$$

$$= 2 \times 31.21 + 4 \times 2.70$$

$$= 73.23 \text{ kN/mm},$$

and so on for the other terms giving

$$A = \begin{bmatrix} 646.10 & 73.23 & 64.90 \\ & 126.90 & 64.90 \\ \text{symmetric} & & 99.61 \end{bmatrix} \text{kN/mm}.$$

Similarly

$$B_{rs} = \tfrac{1}{2} \sum_{1}^{2} \bar{Q}_{rs_j} [h_j^2 - h_{j-1}^2]$$

$$= \tfrac{1}{2} \{ \bar{Q}_{rs_1} [(-1)^2 - (-3)^2] + \bar{Q}_{rs_2} [3^2 - (-1)^2] \}$$

$$= 4\{ -\bar{Q}_{rs_1} + \bar{Q}_{rs_2} \}.$$

That is

$$B_{11} = -4 \times 45.41 + 4 \times 138.81$$

$$= 373.6 \text{ kN},$$

and so on for the other terms giving

$$\mathbf{B} = \begin{bmatrix} 373.6 & -114.0 & -129.8 \\ & -145.6 & -129.8 \\ \text{symmetric} & & -114.0 \end{bmatrix} \text{kN},$$

and finally

$$D_{rs} = \tfrac{1}{3} \sum_{1}^{2} \bar{Q}_{rs_j} [h_j^3 - h_{j-1}^3]$$

$$= \tfrac{1}{3} \{ \bar{Q}_{rs_1} [(-1)^3 - (-3)^3] + \bar{Q}_{rs_2} [3^3 - (-1)^3] \}$$

$$= \tfrac{1}{3} \{ 26\bar{Q}_{rs_1} + 28\bar{Q}_{rs_2} \}.$$

That is

$$D_{11} = \tfrac{26}{3} \times 45.41 + \tfrac{28}{3} \times 138.81$$

$$= 1689.1 \text{ kN mm, etc.},$$

giving

$$\mathbf{D} = \begin{bmatrix} 1689.1 & 295.7 & 281.2 \\ & 477.7 & 281.2 \\ \text{symmetric} & & 374.8 \end{bmatrix} \text{kN mm.}$$

7.5 THE A, B AND D MATRICES

We have already seen when considering a unidirectional composite that certain terms in the compliance, or stiffness, matrix are associated with coupling between particular deformations and loads (see Example 7.4 and the discussion following). Examination of the plate constitutive equations 7.18 and 7.19 allows us to identify similar couplings for laminates.

Suppose we write equations 7.18 and 7.19 in expanded form:

$$
\begin{bmatrix} N_x \\ N_y \\ N_{xy} \end{bmatrix} = \begin{bmatrix} A_{11} & A_{12} & A_{13} \\ A_{21} & A_{22} & A_{23} \\ A_{31} & A_{32} & A_{33} \end{bmatrix} \begin{bmatrix} \varepsilon_x^0 \\ \varepsilon_y^0 \\ \varepsilon_{xy}^0 \end{bmatrix} + \begin{bmatrix} B_{11} & B_{12} & B_{13} \\ B_{21} & B_{22} & B_{23} \\ B_{31} & B_{32} & B_{33} \end{bmatrix} \begin{bmatrix} k_x \\ k_y \\ k_{xy} \end{bmatrix},
$$

$$
\begin{bmatrix} M_x \\ M_y \\ M_{xy} \end{bmatrix} = \begin{bmatrix} B_{11} & B_{12} & B_{13} \\ B_{21} & B_{22} & B_{23} \\ B_{31} & B_{32} & B_{33} \end{bmatrix} \begin{bmatrix} \varepsilon_x^0 \\ \varepsilon_y^0 \\ \varepsilon_{xy}^0 \end{bmatrix} + \begin{bmatrix} D_{11} & D_{12} & D_{13} \\ D_{21} & D_{22} & D_{23} \\ D_{31} & D_{32} & D_{33} \end{bmatrix} \begin{bmatrix} k_x \\ k_y \\ k_{xy} \end{bmatrix},
$$

noting, of course that $A_{21} = A_{12}$, etc.

We can then make the following associations:

A_{13} and A_{23} relate in-plane direct forces to in-plane shear strain, or in-plane shear force to in-plane direct strains.

B_{11}, B_{12} and B_{22} relate in-plane direct forces to plate curvatures, or bending moments to in-plane direct strains.

B_{13} and B_{23} relate in-plane direct forces to plate twisting, or torque to in-plane direct strains.

B_{33} relates in-plane shear force to plate twisting, or torque to in-plane shear strain.

D_{13} and D_{23} relate bending moments to plate twisting, or torque to plate curvatures.

In certain circumstances some of the couplings listed above can be undesirable. They can sometimes be eliminated by appropriate construction of the laminate. We can see this more clearly by examination of equations 7.21.

If $A_{13} = A_{23} = 0$ there will be no coupling between direct stresses and shear strains (or shear stresses and direct strains) – see Example 7.4. This can be achieved if we have a laminate in which all plies have $0°$ and/or $90°$ fibre orientations (a unidirectional or cross-ply laminate), or if the lay-up is *balanced*, i.e. for every layer with a $+\theta$ orientation there is an identical lamina with a $-\theta$ orientation. The laminate need not be symmetric. Examples of balanced lay-ups are: $(+30/-30°)$ – an *angle-ply*; $(0/+45/-45°)$; $(90/+25/-25°)_s$.

Bending–membrane coupling can be avoided if the **B** matrix is zero. This is very easily achieved by making the laminate *symmetric* about its mid-plane – see Example 7.6. In practice laminates usually have a symmetric stacking sequence.

The phenomenon of bending–twisting coupling is eliminated if $D_{13} = D_{23} = 0$. This is achieved with unidirectional or cross-ply laminates, or with *balanced anti-symmetric* lay-ups, i.e. for every layer at $+\theta$ orientation and a given distance above the mid-plane there is a layer with *identical* thickness and properties oriented at $-\theta$ and the same distance below the mid-plane. Such a lay-up is *not* symmetric (i.e. $\mathbf{B} \neq 0$). Examples are: $(+30/-30\,°)$ – an angle-ply; $(+45/-45/0/90/0/+45/-45\,°)$.

The preference, in practice, for symmetrical laminates means that D_{13} and $D_{23} \neq 0$. However, these terms tend to zero for thick multi-layer symmetric laminates (see Example 7.8). Also, there are occasions when this coupling can be put to advantage, as in the aeroelastic tailoring of the skins of aircraft with swept forward wings. If the lay-up is correctly chosen the wings can be made to twist (nose down) in a stable fashion, as they bend (upwards) under the aerodynamic loads.

Example 7.8

A 4 mm thick symmetric laminate is composed of layers of unidirectional material having the elastic properties given in Example 7.1. The fibre angles are alternately at $+45\,°$ or $-45\,°$, but the ply thickness are changed as shown in figures (a)–(d) below. Determine D_{13} and D_{23} for the configurations illustrated.

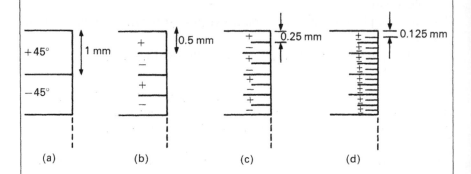

From equation 7.21

$$D_{13} = \tfrac{1}{3} \sum_{j=1}^{p} \bar{Q}_{13_j}[h_j^3 - h_{j-1}^3],$$

$$D_{23} = \tfrac{1}{3} \sum_{j=1}^{p} \bar{Q}_{23_j}[h_j^3 - h_{j-1}^3].$$

Now $\bar{Q}_{13} = \bar{Q}_{23} = 32.45\,\text{GPa}$ (from Example 7.2) for the $+45\,°$ layers. From equations 7.10 we see that the value is the same, but the sign changed for the $-45\,°$ layers. Clearly, here, $D_{13} = D_{23}$.

For (a):

$$D_{13} = \tfrac{1}{3}\{32.45((-1)^3 - (-2)^3) - 32.45(0^3 - (-1)^3) - 32.45(1^3 - 0^3) + 32.45(2^3 - 1^3)\}$$
$$= 129.80\,\text{kN mm}.$$

For (b):

$$D_{13} = \tfrac{1}{3}\{32.45((-1.5)^3 - (-2)^3) - 32.45((-1)^3 - (-1.5)^3)$$
$$+ 32.45((-0.5)^3 - (-1)^3) - 32.45(0^3 - (-0.5)^3)$$
$$- 32.45(0.5^3 - 0^3) + 32.45(1^3 - 0.5^3)$$
$$- 32.45(1.5^3 - 1^3) + 32.45(2^3 - 1.5^3)\}$$
$$= 64.9 \text{ kN mm.}$$

Similarly for (c):

$$D_{13} = 32.45 \text{ kN mm,}$$

and for (d)

$$D_{13} = 16.23 \text{ kN mm.}$$

For some lay-ups *in-plane* behaviour of the laminate is such that it appears to be isotropic (it is then called *quasi-isotropic*). Examples are $(0/90/\pm 45°)$ and $(0/\pm 60°)$ lay-ups. For such laminates it can be shown that

$$A_{11} = A_{22},$$
$$A_{11} - A_{12} = 2A_{33},$$
$$A_{13} = A_{23} = 0.$$

7.6 USING THE LAMINATE CONSTITUTIVE EQUATIONS

It is clear from the examples given above that evaluation of the **A**, **B** and **D** matrices whilst not being particularly difficult is certainly tedious. Fortunately the calculations involved are easily implemented on a computer and several software packages are available for use on PCs.

The use of the laminate constitutive equations, whilst obviously applicable to flat panels, is by no means restricted to such components. Cylindrical items can be analysed provided the radius-to-thickness ratio is reasonably large (say > 50). Also the equations can be used for laminates constructed from other than unidirectional prepregs. Laminates made from layers of woven or isotropic materials can be treated merely by inserting the relevant values into the layer stiffness matrix. Equally, sandwich panels can be analysed by using for the central layer a thickness and stiffnesses appropriate to the chosen core material.

7.7 SUMMARY

In this chapter we started by presenting the stress–strain relations for a unidirectional laminate. Two situations were considered; first, when the stresses and material principal directions are aligned and second, when these two sets of axes are not aligned. The second situation is important because it was required later on when we considered laminates.

The important point to remember is that no new concepts have been used; the equations are essentially the same as those encountered in Chapter 6 for isotropic materials. Clearly the equations become more complicated because of the need to account for the different stiffnesses parallel and transverse to the fibres.

The properties of a unidirectional ply were then used to derive the laminate constitutive equations, i.e. the equations that related applied loads to resultant strains and curvatures. It was shown that some of the coupling phenomena can be eliminated by the correct choice of lay-up and/or stacking sequence, e.g. symmetric, balanced or anti-symmetric laminates.

FURTHER READING

General

Agarwal, B. D. and Broutman, L. J. (1990), *Analysis and Performance of Fiber Composites*, 2nd edn., Wiley.

PROBLEMS

7.1   A unidirectional GFRP lamina, 1 mm thick, has the following elastic properties: $E_{11} = 38.0$, $E_{22} = 8.0$, $G_{12} = 4.2$ GPa and $v_{12} = 0.25$. Calculate the compliance and stiffness matrices.

7.2   The lamina of Problem 7.1 has its fibres oriented at 30° to the x-axis as shown in the figure below. Calculate the strains in the 1–2 and x–y axes

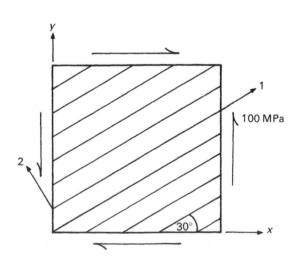

when a shear stress of 100 MPa is applied to the lamina: (a) by using transformation equations, and (b) directly from the compliance matrix.

7.3 Derive the terms in the transformed compliance matrix for an off-axis unidirectional composite.

7.4 A sandwich plate has the configuration shown below. Each 0.125 mm thick ply has the properties given in Problem 7.1, with the fibres oriented as indicated. The isotropic foam core has the following elastic properties: $E = 1$ GPa, $v = 0.3$. Calculate the **A**, **B** and **D** matrices.

SELF-ASSESSMENT QUESTIONS

Indicate whether statements 1 to 7 are true or false.

1. For a transversely isotropic material the elastic properties in the longitudinal direction are the same as those transverse to the fibres.

   (A) true
   (B) false

2. The principal material axes are at $45°$ to the fibre direction.

   (A) true
   (B) false

3. A unidirectional lamina can be characterized by only four elastic constants.

   (A) true
   (B) false

4. The major and minor Poisson's ratios are independent elastic properties.

   (A) true
   (B) false

5. The shear modulus of a unidirectional lamina can be determined from the longitudinal modulus and the major Poisson's ratio.

(A) true
(B) false

6. The **B** matrix is zero for a laminate with an antisymmetric lay-up.

(A) true
(B) false

7. The D_{13} and D_{23} terms are zero for a laminate with a balanced, symmetric lay-up.

(A) true
(B) false

For each of the statements of questions 8 and 9, one or more of the completions given are correct. Mark the correct completions.

8. Compliance

(A) is the inverse of stiffness,
(B) is a measure of the flexibility of a material,
(C) is directly proportional to Young's modulus,
(D) has units of force/area,
(E) gives stress for a unit strain.

9. Transformed stiffness matrix

(A) is the inverse of the stiffness matrix,
(B) is obtained by changing the x- for the y-axes,
(C) is the stiffness in the principal material directions of a lamina oriented at an angle to the x- and y-axes,
(D) is the stiffness of a rotated lamina in the x- and y-axes,
(E) is determined from the stiffness coefficients and functions of the angle of rotation.

Each of the sentences in questions 10 to 15 consists of an assertion followed by a reason. Answer:

(A) if both assertion and reason are true statements and the reason is a correct explanation of the assertion,
(B) if both assertion and reason are true statements but the reason is not a true explanation of the assertion,
(C) if the assertion is true but the reason is a false statement,
(D) if the assertion is false but the reason is a true statement,
(E) if both the assertion and reason are false statements.

10. For a unidirectional lamina the major Poisson's ratio is always larger than the minor Poisson's ratio *because* the transverse modulus is always smaller than the longitudinal modulus.

11. For a unidirectional lamina the S_{11} term of the compliance matrix is always greater than S_{22} *because* the stiffness term Q_{11} is always greater than Q_{22}.

12. For a unidirectional lamina the terms Q_{13} and Q_{23} of the stiffness matrix are always zero *because* the shear modulus is much smaller than the longitudinal modulus.

13. A unidirectional lamina subjected to off-axis loading will exhibit extension–shear coupling *because* the terms $\bar{Q}_{13}$ and $\bar{Q}_{23}$ are not zero.

14. Laminates with the same lay-up but different stacking sequences have the same **A** matrix *because* the **A** matrix depends only on ply thickness.

15. The laminate denoted by $(+\theta/-\theta/+\theta/90/-\theta/+\theta/-\theta)$ has D_{13} and $D_{23} \neq 0$ *because* a layer at $90°$ does not influence these terms.

ANSWERS

Problems

7.1

$$\text{compliance: } 10^{-3} \begin{bmatrix} 26.32 & -6.58 & 0 \\ & 125.0 & 0 \\ & & 288.1 \end{bmatrix} \text{GPa}^{-1},$$

$$\text{stiffness: } \begin{bmatrix} 38.51 & 2.03 & 0 \\ & 8.11 & 0 \\ & & 4.2 \end{bmatrix} \text{GPa}.$$

7.2

$$\varepsilon_{xy} = \{-0.490 \quad -0.414 \quad 1.276\}\%,$$
$$\varepsilon_{12} = \{\ 0.082 \quad -0.985 \quad 0.704\}\%.$$

7.4

$$\mathbf{A} = \begin{bmatrix} 28.15 & 7.40 & 0 \\ & 18.38 & 0 \\ & & 8.49 \end{bmatrix} \text{kN/mm}, \quad \mathbf{B} = 0,$$

$$\mathbf{D} = \begin{bmatrix} 711.6 & 162.9 & 3.45 \\ & 345.0 & 1.02 \\ & & 203.7 \end{bmatrix} \text{kN mm.}$$

Self-assessment

1. B; 2. B; 3. A; 4. B; 5. B; 6. B; 7. B; 8. A, B; 9. D, E; 10. A; 11. D; 12. B; 13. A; 14. C; 15. D.

Micromechanics of unidirectional composites

8

8.1 MACROMECHANICS AND MICROMECHANICS

In the previous chapter we developed equations that describe the stress–strain behaviour of a lamina and the load-deformation behaviour of a laminate. These equations were based on the elastic properties of the lamina, and, as such, ignored the microscopic nature of the material. In other words we took no direct account of the fact that we are dealing with fibre-reinforced materials, we merely acknowledged that they were non-isotropic. We refer to this as *macromechanics* analysis.

Because the starting point of a significant proportion of composites' manufacture is the combination of fibres and matrix, it would be very helpful if we could predict the behaviour of the composite (laminate) from a knowledge of the properties of the constituents alone. As we shall see in this chapter there are many limitations to such *micromechanics* analyses. However, studying performance on a micro scale is essential if we are to understand fully what controls the strength, toughness, etc., of composites.

8.2 MICROMECHANICS MODELS FOR STIFFNESS

8.2.1 Longitudinal stiffness

We stated in Chapter 7 that the unidirectional ply forms a useful building block for many studies of composites. Much micromechanics analysis has, therefore, been devoted to this simple system. For example, predictions may be made using the Halpin–Tsai equations (Chapter 10). The most successful application is the prediction of the stiffness parallel to the fibres, i.e. the longitudinal stiffness. We denoted this as E_{11} in Chapter 7. To obtain this stiffness consider an array of uniform parallel continuous fibres, perfectly bonded to the matrix, and loaded parallel to the fibres as shown in Figure 8.1. The assumption of perfect bonding means that there is no slipping at the fibre/matrix interface and, hence, the imposed load along the fibre direction will produce strains (ε) that are equal in all components (composite, fibres, matrix), i.e.

$$\varepsilon_c = \varepsilon_f = \varepsilon_m, \tag{8.1}$$

where the subscripts c, f and m refer to the composite, fibre and matrix respectively.

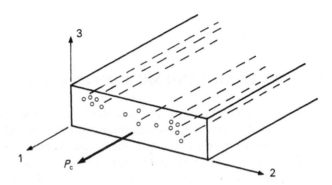

Figure 8.1 Unidirectional composite loaded by a force P_c parallel to the fibres; definition of axes.

Now, the load on the fibres is

$$P_f = \sigma_f A_f,$$

on the matrix is

$$P_m = \sigma_m A_m, \tag{8.2}$$

and on the composite is

$$P_c = \sigma_c A_c,$$

where A is the cross-section area and σ represents stress.

But, since, for equilibrium $P_c = P_f + P_m$ we get, on substituting from equation 8.2

$$\sigma_c A_c = \sigma_f A_f + \sigma_m A_m$$

or

$$\sigma_c = \frac{\sigma_f A_f}{A_c} + \frac{\sigma_m A_m}{A_c}. \tag{8.3}$$

But, for parallel fibres the area fraction is the same as the volume fraction (see section 1.5 for definition of volume fraction), i.e.

$$v_f = \frac{A_f}{A_c} \text{ and } v_m = \frac{A_m}{A_c}$$

Hence, equation 8.3 becomes

$$\sigma_c = \sigma_f v_f + \sigma_m v_m. \tag{8.4}$$

We further assume *linear elastic* behaviour, so modulus (E) and stress are related by

$$\sigma_c = E_c \varepsilon_c, \quad \sigma_f = E_f \varepsilon_f, \quad \sigma_m = E_m \varepsilon_m,$$

and if we substitute these expressions in equation 8.4 we get

$$E_c \varepsilon_c = E_f \varepsilon_f v_f + E_m \varepsilon_m v_m,$$

and using equation 8.1 gives for the longitudinal modulus of the composite

$$E_c = E_{11} = E_f v_f + E_m v_m \tag{8.5}$$

A relationship of this form is known as the *Rule* or *Law of Mixtures*. If there is more than one type of fibre, equation 8.5 becomes

$$E_{11} = E_m v_m + E_{f_1} v_{f_1} + E_{f_2} v_{f_2} + \cdots \tag{8.5a}$$

Example 8.1
A unidirectional composite is composed of 65% by volume of carbon fibres (modulus 240 GPa) in an epoxy resin matrix (modulus 4 GPa). Calculate the longitudinal modulus of the composite

Using the Rule of Mixtures, (equation 8.5)

$$E_{11} = E_f v_f + E_m v_m.$$

Here, $v_f = 0.65$, and hence $v_m = 0.35$, so

$$E_{11} = 0.65 \times 240 + 0.35 \times 4$$
$$= 156 + 1.4$$
$$= 157.4 \text{ GPa}.$$

The predictions from equation 8.5 agree well (within 5%) with data from carefully controlled experiments for tensile loading. Predictions are not so good for compressive loading because the experimental results are very sensitive to the design of the equipment and the alignment of the fibres in the specimen.

We see from Example 8.1 that we can, without serious error, neglect the matrix provided that $E_f \gg E_m$. Such an approximation would not be justified for MMCs and CMCs.

8.2.2 Transverse stiffness

To obtain an expression for the transverse stiffness, denoted as E_{22} in Chapter 7, we consider a load applied at right angles to the fibre direction, as shown in Figure 8.2(a). We replace the real composite by the simple model shown in Figure 8.2(b). It is assumed that the lengths are proportional to the volume fractions and that the force (or stress) is the same in each constituent. Recall that in section 8.2.1 we assumed that the *strains* were the same.

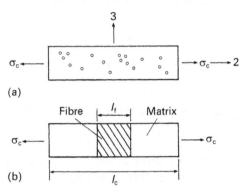

(a)

(b)

Figure 8.2 (a) Unidirectional composite loaded transversely to the fibres; (b) simplified model.

Now, the (transverse) extension of the composite (δ_c) is the sum of the extension in the matrix (δ_m) and fibres (δ_f), i.e.

$$\delta_c = \delta_f + \delta_m \qquad (8.6)$$

or

$$\varepsilon_c l_c = \varepsilon_f l_f + \varepsilon_m l_m,$$

that is

$$\varepsilon_c = \varepsilon_f \frac{l_f}{l_c} + \varepsilon_m \frac{l_m}{l_c},$$

but

$$\frac{l_f}{l_c} = v_f \text{ and } \frac{l_m}{l_c} = v_m,$$

so

$$\varepsilon_c = \varepsilon_f v_f + \varepsilon_m v_m.$$

As before we assume elastic behaviour, i.e. $E = \sigma/\varepsilon$, so

$$\frac{\sigma_c}{E_c} = \frac{\sigma_f}{E_f} v_f + \frac{\sigma_m}{E_m} v_m,$$

but

$$\sigma_c = \sigma_f = \sigma_m,$$

giving the transverse modulus of the composite as

$$\frac{1}{E_c} = \frac{1}{E_{22}} = \frac{v_f}{E_f} + \frac{v_m}{E_m}. \tag{8.7}$$

It is usual to take E_f as the longitudinal value, if only because the transverse fibre modulus is extremely difficult to determine. This assumption will only be correct for isotropic fibres.

As might be expected, because of the enormous simplifications made in formulating the above model, equation 8.7 does not give very good predictions for E_{22}. Extremely complicated elasticity or finite element analyses are required to produce a more accurate model. However, the simple model is useful in allowing us to assess the effectiveness of different fibres and matrices. For PMCs E_{22} is, unlike E_{11}, strongly influenced by the magnitude of E_m.

Example 8.2
Compare the longitudinal and transverse stiffnesses of two composites with the same matrix but different fibres. In the first case $E_f/E_m = 50$ and in the second case $E_f/E_m = 25$. Take $v_f = 0.5$.

Now

$$E_{11} = E_f v_f + E_m v_m,$$

or

$$\frac{E_{11}}{E_m} = \frac{E_f}{E_m} v_f + v_m.$$

Also

$$\frac{1}{E_{22}} = \frac{v_f}{E_f} + \frac{v_m}{E_m},$$

or

$$\frac{E_m}{E_{22}} = \frac{v_f}{E_f/E_m} + v_m.$$

For

$$E_f/E_m = 50$$

$$\frac{E_{11}}{E_m} = 50 \times 0.5 + 0.5 = 25.5,$$

$$\frac{E_m}{E_{22}} = \frac{0.5}{50} + 0.5 = \frac{25.5}{50},$$

or

$$\frac{E_{22}}{E_m} = 1.96.$$

Also

$$\frac{E_{11}}{E_{22}} = \frac{25.5}{1.96} = 12.8.$$

For $E_f/E_m = 25$

$$\frac{E_{11}}{E_m} = 25 \times 0.5 + 0.5 = 13.0,$$

$$\frac{E_m}{E_{22}} = \frac{0.5}{25} + 0.5 = \frac{13}{25},$$

or

$$\frac{E_{22}}{E_m} = 1.92.$$

Also

$$\frac{E_{11}}{E_{22}} = \frac{13.0}{1.92} = 6.77.$$

So we see that, for this volume fraction, changing the fibre to one with double the modulus results in (almost) a doubling of the composite's longitudinal modulus, but in only about a 2% increase in the composite's transverse modulus.

8.2.3 Shear modulus

The in-plane shear modulus, denoted as G_{12} in Chapter 7, of a unidirectional composite can be obtained from a similar model to that used for obtaining transverse modulus. In this case it is the shear stress that is the same on both constituents.

The analysis follows that derived in section 8.2.2, giving the result

$$\frac{1}{G_{12}} = \frac{v_f}{G_f} + \frac{v_m}{G_m} \tag{8.8}$$

As with E_{22}, G_{12} is strongly influenced by the magnitude of the modulus of the matrix, G_m. In practice it is often difficult to obtain a meaningful value for G_f; one possibility is to calculate a result from E_f and Poisson's ratio, v_f. However, it may also be necessary to assume v_f as this is, also, extremely difficult to measure.

The model suffers from the same limitations as that used for transverse modulus, very complicated analyses being needed to give a better prediction.

8.2.4 Poisson's ratio

The transverse modulus model also serves to provide a prediction for the major Poisson's ratio, v_{12}. However, in contrast to the derivation for E_{22}, the load is now applied parallel to the fibres.

The transverse deformation of the model is illustrated in Figure 8.3, the various components of the deformation being obtained from the axial strain and corresponding Poisson's ratio, i.e.

$$\delta_f = t_f(\varepsilon_{22})_f = -t_f v_f \varepsilon_f,$$

$$\delta_m = t_m(\varepsilon_{22})_m = -t_m v_m \varepsilon_m,$$

$$\delta_c = t_c(\varepsilon_{22})_c = -t_c v_{12} \varepsilon_c.$$

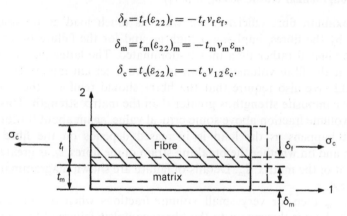

Figure 8.3 Simplified model for derivation of Poisson's ratio.

Now $\delta_c = \delta_f + \delta_m$ (see derivation of E_{22}) and the longitudinal strains are equal, i.e. $\varepsilon_f = \varepsilon_m = \varepsilon_c$ as assumed in the derivation of E_{11}. Hence we get

$$t_c v_{12} = t_f v_f + t_m v_m, \tag{8.9}$$

or

$$v_{12} = v_f \frac{t_f}{t_c} + v_m \frac{t_m}{t_c},$$

but

$$\frac{t_f}{t_c} = v_f \text{ and } \frac{t_m}{t_c} = v_m.$$

So, finally,

$$v_{12} = v_f v_f + v_m v_m. \tag{8.10}$$

8.3 MICROMECHANICS MODELS FOR STRENGTH

It is far more difficult to obtain a prediction for strength than for stiffness. This is because of several factors: the random nature of failure and hence the need to employ statistical methods; the number of failure modes that can cause composite failure (fibre, matrix or interface failure); the very local nature of failure initiation and the influence of the associated stress field, itself determined by the details of fibre packing. As for stiffness, methods for predicting longitudinal performance are better than those for transverse and shear performance.

8.3.1 Longitudinal tensile strength ($\hat{\sigma}_{1T}$)

For maximum fibre efficiency we want as much load as possible to be carried by the fibres, implying a high v_f, and for the failure process to be fibre-dominated rather than matrix-dominated. The latter requirement implies that the fibre volume fraction is above a certain minimum, typically about 0.1. We also require that the fibres should reinforce the matrix, i.e. that the composite strength is greater than the matrix strength. This implies a fibre volume fraction above some critical value, again about 0.1 for PMCs.

What happens as the load is increased depends on the fibre volume fraction and on whether the breaking strain of the fibres, $\hat{\varepsilon}_f$, is greater or less than that of the matrix, $\hat{\varepsilon}_m$. Details of failure are shown diagrammatically in Figure 8.4.

If $\hat{\varepsilon}_f > \hat{\varepsilon}_m$ then for very small volume fractions when the matrix cracks (first), the load is thrown on to the fibres: complete failure of the lamina will follow as there are insufficient fibres to take the load. Using equation 8.4 we can obtain an expression for the corresponding strength of the composite, i.e.

$$\hat{\sigma}_{1T} = \sigma_f^* v_f + \hat{\sigma}_m (1 - v_f), \tag{8.11}$$

where σ_f^* is the fibre stress at a strain corresponding to matrix failure, $\hat{\sigma}_f$ is the fibre strength and $\hat{\sigma}_m$ is the matrix strength. At large fibre volume fractions, on the other hand, the fibres will take most of the load and, after matrix failure, will determine the composite's strength, i.e.

$$\hat{\sigma}_{1T} = \hat{\sigma}_f v_f. \tag{8.12}$$

The crossover point between matrix-dominated failure and fibre-dominated failure, v_f^1, is found to be

$$v_f^1 = \hat{\sigma}_m / (\hat{\sigma}_f - \hat{\sigma}_f^* + \hat{\sigma}_m). \tag{8.13}$$

For FRP $\hat{\sigma}_f \gg \hat{\sigma}_m$, and v_f^1 is usually small (typically 0.1).

If $\hat{\varepsilon}_f < \hat{\varepsilon}_m$ then at large volume fractions, once the fibres break, the matrix cannot take the extra load and breaks, giving the composite strength from

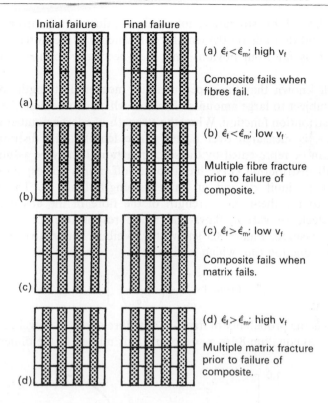

Figure 8.4 Change in failure behaviour with relative values of fibre and matrix failure strains $(\hat{\varepsilon}_f, \hat{\varepsilon}_m)$, and with fibre volume fraction. The fibre is the shaded component.

equation 8.4 as

$$\hat{\sigma}_{1T} = \hat{\sigma}_f v_f + \sigma_m^*(1 - v_f), \tag{8.14}$$

where σ_m^* is the matrix stress at a strain corresponding the fibre failure. For PMCs the second term can be ignored.

At very small fibre volume fractions, the fibres can be ignored and when they break the matrix will carry the extra load, the composite's strength then being given by

$$\hat{\sigma}_{1T} = \hat{\sigma}_m(1 - v_f). \tag{8.15}$$

As for the previous situation it can be shown that the fibre volume fraction for mode change in very small for FRP.

In summary, for most PMCs, equation 8.12 gives a reasonable description of physical load-carrying capability, but it should be recognized that the onset of cracking (of fibres or matrix) may be a more useful design criterion.

The above treatment neglects some significant features, such as the variation of strength from fibre to fibre and the distribution of flaw sizes along each fibre. Fibre strength is not a unique value; it varies from one fibre to another and depends on the length over which it is measured. Thus, the choice of $\hat{\sigma}_f$ is far from obvious and it is necessary to use a statistical approach.

It is well known that the experimentally measured strengths of brittle solids are subject to large amounts of scatter which can be described by the Weibull distribution function. When the strength of a large number of fibres, diameter $2r$, is measured over a test gauge length l the distribution of strengths can be represented by the same function. It may be assumed that the Weibull distribution curve has two cut-off points corresponding to a lower strength limit σ_1 and an upper strength limit σ_u. The latter is equivalent to the theoretical strength of the fibre in the absence of any internal defects or surface flaws which are responsible for the reduced strengths measured. Figure 8.5 shows the Weibull cumulative probability distribution function $G(\sigma)$ which is given by

$$G(\sigma)=1-\left[1-\left(\frac{\sigma-\sigma_1}{\sigma_0}\right)^m\right]^{\omega},\tag{8.16}$$

where $G(\sigma)$ is the probability of fracture of a fibre at a stress level equal to or less than σ. In equation 8.16, $\sigma_0=\sigma_u-\sigma_1$; ω is related to the dimensions of

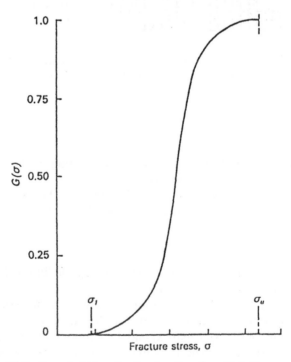

Figure 8.5   Weibull probability distribution function $G(\sigma)$. (Source: Hull, 1981.)

the test sample, which for fibres with a constant diameter is given by $l/2r$; and m, known as the Weibull modulus, is inversely related to the coefficient of variation.

It is often useful to assess fibre strength by measuring the fracture strength σ_b of a bundle of fibres which, as individual fibres, exhibit the distribution described by equation 8.16 and have a mean strength σ. It is found that σ_b is less than σ and that the ratio σ_b/σ depends on the amount of scatter in the strength of the individual fibres. Very brittle fibres, which usually show a large amount of scatter, have values of m between 2 and 5, and the bundle strength is 50–65% of the mean strength. Typically, glass fibres have values of m between 5 and 15, and the bundle strength is 65–80% of the mean strength. When m is infinite, i.e. the coefficient of variation is zero, all the fibres have the same strength and the bundle strength is equal to the individual fibre strength. This assumption is, of course, made in deriving the rule of mixture equations for composite strength (equations 8.11 and 8.14).

If continuous fibres are to be effective as reinforcing agents, flaws must be separated by more than the critical length (see Chapter 10 for further details). In practice this is not usually a problem because flaws, in good quality fibres, are many millimeters apart, whereas the critical length is less than 1 mm.

8.3.2   Longitudinal compressive strength ($\hat{\sigma}_{1C}$)

The complex nature of tensile failure, involving fibre, matrix or interface failure, is also seen in compression, but with the added possibilities of fibre buckling and matrix shear deformation.

The usual starting point for prediction of longitudinal compressive strength in terms of the properties of the fibre and matrix is work by Rosen who envisaged that failure under these loading conditions would be associated with the buckling of fibres as in column buckling of struts. The buckling is restricted by the surrounding matrix so that the buckling stress and hence the longitudinal compressive strength will depend on the elastic properties of the matrix. At low v_f extensional, or out-of-phase, buckling is predicted, the compressive strength being given by

$$\hat{\sigma}_{1C} = 2v_f[v_f E_m E_f/3(1-v_f)]^{1/2}. \tag{8.17}$$

The restriction to low v_f means that this mode is not relevant to commercially important composite materials. At high v_f a shear, or in-phase, mode of buckling is predicted giving

$$\hat{\sigma}_{1C} = G_m/(1-v_f). \tag{8.18}$$

The latter equation indicates that $\hat{\sigma}_{1C}$ increases with v_f but is dominated by the shear modulus of the matrix. The agreement between experimentally determined and predicted strengths is poor, the predicted values being much greater than the experimental results. This suggests that either the micro-buckling model is completely wrong or that the underlying assumptions have been seriously over-simplified.

Any factor which leads to a reduction in the support that the matrix and the surrounding fibres give to a particular fibre to prevent it microbuckling will lead to a reduction in σ_{1C} according to equation 8.18. In real systems this support can be reduced by the following effects: fibre bunching which results in local resin rich regions; the presence of voids; poor alignment of fibres or fibre waviness which results in some fibres being preferentially oriented for easy buckling; fibre debonding which may occur as a result of differences between the Poisson's ratios of the matrix and fibre; and viscoelastic deformation of the matrix which results in an effective reduction of the matrix shear modulus. A recent approach which accounts for initial fibre waviness and the matrix shear yield stress gives a more accurate prediction.

An additional factor which applies to carbon and Kevlar 49 is the non-isotropic elastic properties of these fibres. The derivation of equation 8.18 assumes that both the fibre and the matrix are isotropic. Both carbon and Kevlar 49 fibres have low transverse and shear moduli which means that predictions based on isotropic properties will over-estimate the buck-ling strength of the composite material. In addition, the low compressive strength of Kevlar 49 results in premature failure of the fibres by yielding. The fracture surface usually has many longitudinal splits, and some splaying of the fibres occurs.

Failure strains in compression can be very close to those seen in tension; however, results are very dependent on the test method.

8.3.3 Transverse tensile strength ($\hat{\sigma}_{2T}$)

There is no simple relation for predicting the transverse tensile strength. Unlike the longitudinal tensile strength which is determined almost entirely by a single factor, i.e. the fibre strength, the transverse strength is governed by many factors including the properties of the fibre and matrix, the interface bond strength, the presence and distribution of voids, and the internal stress and strain distribution. The most clear-cut feature of the transverse strength is that for FRP it is usually less than the strength of the parent resin so that, in contrast to their effect on the transverse modulus, we find that the fibres have a negative reinforcing effect.

The transverse tensile strength of a unidirectional composite in which there is little or no interface bonding is determined by the strength of the resin. To a first approximation the fibres can be regarded as cylindrical holes. For a square array of fibres the predicted strength, provided that the resin is not notch sensitive, is given by

$$\hat{\sigma}_{2T} = \hat{\sigma}_{m}[1 - 2(v_{f}/\pi)^{1/2}]. \tag{8.19}$$

At $v_f = 0.785$ the fibres are touching and the strength falls to zero. A similar relation can be obtained for other fibre arrays. We would expect a further reduction in transverse strength if the stress concentrations around the holes are not relieved by plastic flow.

When the fibres are strongly bonded to the matrix, the transverse strength is dependent on interface strength as well as on the strength of the matrix. If the interface bond does not fail there is a stress and strain magnification in the matrix which is a maximum between the fibres. Calculations of the stress fields in the composite show that the maximum tensile stress concentration occurs midway between the fibres with a value of approximately 2.0 for $v_f \approx 0.5$. Although matrix failure will depend on the full triaxial state of stress, it is clear that these stress concentration effects will result in a reduction in strength compared with the matrix by a factor of about two. However, it must be noted that the tensile strength of brittle matrices is dominated by the effect of flaws and so the interpretation of the strength reduction solely in terms of the stress concentration effects of fibres is particularly difficult.

The stress magnification approach is not appropriate for matrices which show a pronounced nonlinear stress–strain response. The reduction in transverse strength should then be accounted for in terms of the strain magnification, which is a maximum between fibres. Very high strain magnification values, perhaps as high as five, are obtained when the fibres are closely spaced.

Microscopic studies of the fracture of unidirectional composites confirm that transverse cracks are nucleated in regions of dense packing and also propagate preferentially through such regions. A main crack will often follow the fibre–matrix interface in regions of dense packing. Cracks can nucleate ahead of the main crack by debonding, or by resin fracture very close to the fibre–matrix interface in regions of maximum radial tensile stress or strain.

Other sources of low transverse tensile strength include debonding at the interface before cohesive failure of the resin or of the fibre. The latter may be particularly relevant to carbon fibres which have layered structures oriented parallel to the surface. In both cases sharp cracks may be formed which can lead to catastrophic failure.

Because of the importance of achieving some transverse strength, considerable effort has been devoted to modifying the matrix structure so as to minimize the stress concentration effects. Two possibilities are of interest: (i) for PMCs the introduction of a very fine dispersion of rubber particles into the brittle resin; these particles are known to increase the fracture toughness of resins without too big a reduction in strength and stiffness. This effect is likely to be of most value in materials with low v_f, (ii) use of an intermediate layer at the fibre–matrix interface which may result in a change in the stress pattern and a reduction of the stress magnification. Such an approach can be envisaged for MMCs.

8.3.4 Transverse compressive strength ($\hat{\sigma}_{2C}$)

As for transverse tension, no satisfactory method of strength prediction exists for compression strength. The transverse compression situation is complicated by the fact that two different failure modes are possible.

A failure involving matrix shear failure and/or fibre–matrix interface failure is most likely as illustrated in Figure 8.6. In this case the strength will be lower than the longitudinal compressive strength, but higher than the transverse tensile strength. If the composite is constrained to prevent failure in the transverse plane, shear failure of the fibres takes place giving a strength approaching the longitudinal value.

Figure 8.6 Shear failure of unidirectional composite subjected to transverse compressive load.

8.3.5 In-plane shear failure ($\hat{\tau}_{12}$)

The in-plane, or intralaminar, shear strength depends on the direction of the shear displacements. The two extremes correspond to shear stress applied parallel to the fibres or perpendicular to the fibres. Shear failure associated with the latter is unlikely as it would imply failure of the fibres; matrix or interface failure is far more likely. The strength will be dominated by matrix properties because crack propagation can occur entirely by shear of the matrix without disturbing or fracturing the fibres. Thus $\hat{\tau}_{12}$ will be dependent on the viscoelastic properties of the resin in a similar way to the transverse tensile properties described above.

For a given matrix, the shear strength depends on the stress concentration effects associated with the presence of fibres and voids, and on the strength of the interfacial bond. At low v_f the stress concentration factor is relatively insensitive to v_f but it rises rapidly when v_f is greater than 0.60. For composites with brittle matrices, the stress concentration effect will lead to values of $\hat{\tau}_{12}$ lower than the shear strength of the matrix alone. The effect is more pronounced when fibre bunching occurs. The strength of a composite with more flexible matrices is approximately the same as for the pure matrix because any stress concentrations are relaxed by local deformation processes. The same arguments apply to the effect of voids and weakly bonded interfaces. Because of the complexities, no serious attempts have been made to model this mode of failure.

8.4 THERMAL AND MOISTURE EFFECTS

As already mentioned, carbon and aramid fibres have a very small, or even slightly negative, coefficient of longitudinal thermal expansion (α_f). One consequence of this is that residual stresses are set up in a unidirectional composite as it cools from the curing temperature. In the transverse direction these stresses can be a significant fraction of the failure stress of the matrix. Hence, any calculation which attempts to predict failure should include these thermal effects.

A second consequence of the fibre and matrix having different expansion coefficients is that the composite has different coefficients in the longitudinal (α_1) and transverse (α_2) directions. Based on the assumption that $v_f \simeq v_m$ the following simple expressions may be derived:

$$\alpha_1 = \frac{1}{E_{11}}(\alpha_f E_f v_f + \alpha_m E_m v_m), \tag{8.20}$$

$$\alpha_2 = (1+v_f)\alpha_f v_f + (1+v_m)\alpha_m v_m - \alpha_1 v_{12}. \tag{8.21}$$

The anisotropy of thermal expansion $(\alpha_1 \neq \alpha_2)$ will cause residual thermal stresses in laminates, in addition to those in each ply mentioned above.

Resin matrices will absorb moisture and therefore swell during normal operating conditions. Some fibres, such as Kevlar, also absorb water. The water up-take in a resin matrix is usually by a process of Fickian diffusion and the swelling can be characterized by an absorption coefficient (β), which is directly equivalent to the thermal expansion coefficient.

For a homogeneous material it can be shown that β (defined as the strain per unit change of moisture content) is given by

$$\beta = \frac{1}{3}\frac{\rho}{\rho_w}, \tag{8.22}$$

where ρ is the density of the material and ρ_w is the density of water.

For a unidirectional PMC, reinforced with carbon or glass fibres, the swelling coefficient in the fibre direction (β_1) may be taken as zero. The transverse coefficient (β_2) depends on the expansion of the matrix, β_m (defined as in equation 8.22), and may be taken as

$$\beta_2 = \frac{\rho_c}{\rho_m}(1+v_m)\beta_m, \tag{8.23}$$

where the suffices m and c refer to matrix and composite repectively.

8.5 SUMMARY

We have seen that micromechanics models are successful in predicting the longitudinal modulus in tension and compression, and reasonably successful in predicting the corresponding strengths. Whilst models exist for predicting transverse and shear stiffness, they do not compare well with experimental data. There are no satisfactory methods for predicting transverse and shear strengths.

The deficiencies in micromechanics models means that we cannot at the moment predict laminate performance from only a knowldge of fibre and matrix properties. However, we can successfully use macromechanics to predict laminate behaviour provided we use, as 'input', values of stiffness and strength obtained from experiments on representative unidirectional samples.

FURTHER READING

General

Hull, D. (1981) *An Introduction to Composite Materials*, Cambridge University Press.

PROBLEMS

8.1   A unidirectional glass fibre-reinforced plastic is to be modified by replacing some of the glass fibres by carbon fibres, the total fibre volume fraction remaining unchanged at 0.50. Calculate the fraction of carbon fibres needed to double the original longitudinal modulus.
Assume the following properties: glass fibres, $E = 70$ GPa; carbon fibres, $E = 300$ GPa; matrix, $E = 5$ GPa.

8.2   Calculate the shear modulus (G_{12}) of a unidirectional composite containing 65% by volume of fibres. Assume both fibre and matrix to be isotropic, with Poisson's ratio of 0.3. Take $E_f = 250$ GPa and $E_m = 5$ GPa.

Self-assessment questions

Indicate whether statements 1 to 4 are true or false.

1.   The longitudinal modulus of a unidirectional composite is determined solely by the fibre modulus.

(A) true
(B) false

2. The fibre modulus has a stronger influence on longitudinal modulus than it does on transverse modulus.

(A) true
(B) false

3. The shear modulus of a unidirectional composite (G_{12}) is strongly dependent on the shear modulus of the matrix.

(A) true
(B) false

4. Brittle fibres have high values of the Weibull modulus, m.

(A) true
(B) false

For each of the statements of questions 5 and 6, one or more of the completions given are correct. Mark the correct completions.

5. The longitudinal modulus of a unidirectional composite

(A) is larger than the fibre modulus,
(B) is directly proportional to fibre volume fraction,
(C) is calculated assuming the fibres to have identical strength,
(D) is calculated assuming the failure strain of the fibres to be greater than that of the matrix,
(E) is calculated assuming that stress is proportional to strain.

6. Transverse properties are more difficult to predict than longitudinal properties because

(A) fibres are not uniformly packed,
(B) they are dependent on matrix properties,
(C) there is a lack of data on appropriate fibre properties,
(D) the strain is constant,
(E) the coefficient of expansion is different for fibre and matrix.

Each of the sentences in questions 7 to 11 consists of an assertion followed by a reason. Answer:

(A) if both assertion and reason are true statements and the reason is a correct explanation of the assertion,
(B) if both assertion and reason are true statements but the reason is not a true explanation of the assertion,
(C) if the assertion if true but the reason is a false statement,
(D) if the assertion is false but the reason is a true statement,
(E) if both the assertion and reason are false statements.

7.   When calculating the longitudinal modulus of unidirectional FRP the matrix can be neglected *because* the fibre modulus is much greater than the matrix modulus.

8.   Having the fibre modulus will halve the composite's transverse modulus *because* transverse modulus is proportional to fibre modulus.

9.   The major Poisson's ratio of a unidirectional composite is given by a Rule of Mixtures expression *because* transverse strains are neglected.

10.   For practical FRPs the composite's longitudinal tensile stress is directly proportional to fibre stress *because* the fibre modulus is much greater than the matrix modulus.

11.   The longitudinal and transverse expansion coefficients of CFRP are the same *because* the fibre expansion coefficient is larger than that of the matrix.

ANSWERS

Problems

8.1   0.163.
8.2   5.32 GPa.

Self-assessment

1. B;   2. A;   3. A;   4. B;   5. A,E;   6. A,C;   7. A;   8. E;   9. C;
10. A;   11. E.

Strength of unidirectional composites and laminates

9

9.1 INTRODUCTION

In Chapter 7 we established the stress–strain relationships for an individual lamina, or ply, and for a laminate. We can use the constitutive equations to calculate the stresses in each ply when we know the values of the loads acting on the laminate. By comparing these stresses with a corresponding limiting value we can decide whether, or not, the laminate will fail when subjected to the service loads.

There are several ways to define failure. The obvious one is when we have complete separation, or fracture; clearly, then, the component can no longer support the loads acting on it. However, a more general definition would be 'when the component can no longer fulfil the function for which it was designed'.

Such a definition includes total fracture but could also include excessive deflection as seen when a laminate buckles (basically a *stiffness* rather than a *strength* limit), or even just matrix cracking. The latter could constitute failure for a container because any contents would be able to leak through the matrix cracks in the container's walls.

As for isotropic materials (see Chapter 6) a failure criterion can be used to predict failure. A large number of such criteria exist, no one criterion being universally satisfactory. We shall start by considering a single ply before moving on to discuss laminates.

As before we take the lamina to be a regular array of parallel continuous fibres, perfectly bonded to the matrix. We saw in the last chapter that there are five basic modes of failure of such a ply: longitudinal tensile or compressive, transverse tensile or compressive, or shear. Each of these modes would involve detailed failure mechanisms associated with fibre, matrix or interface failure. Some typical strengths of PMCs are shown in Table 9.1 (see also Chapter 5).

We can regard the strengths in the principal material axes (parallel and transverse to the fibres) as the fundamental parameters defining failure. When a ply is loaded at an angle to the fibres, as it is when it is part of a multidirectional laminate, we have to determine the stresses in the principal directions and compare them with the fundamental values.

Table 9.1 Typical strength of unidirectional PMC laminates ($v_f = 0.5$) (values in MPa)

Material	Longitudinal tension	Longitudinal compression	Transverse tension	Transverse compression	Shear
Glass–polyester	650–950	600–900	20–25	90–120	45–60
Carbon–epoxy	850–1500	700–1200	35–40	130–190	60–75
Kevlar–epoxy	1100–1250	240–290	20–30	110–140	40–60

9.2 STRENGTH OF A LAMINA

Strength can be determind by the application of failure criteria, which are usually grouped into three classes: *limit criteria*, the simplest; *interactive criteria* which attempt to allow for the interaction of multiaxial stresses; and *hybrid criteria* which combine selected aspects of limit and interactive methods. In this text we shall only discuss criteria that fit into the first two classes.

9.2.1 Limit criteria

(a) *Maximum stress criterion*

The maximum stress criterion consist of five sub-criteria, or limits, one corresponding to the strength in each of the five fundamental failure modes. If any one of these limits is exceeded, by the corresponding stress expressed in the principal material axes, the material is deemed to have failed.

In mathematical terms we say that failure has occurred if

$$\sigma_1 \geqslant \hat{\sigma}_{1T} \quad \text{or} \quad \sigma_1 \leqslant \hat{\sigma}_{1C} \quad \text{or} \quad \sigma_2 \geqslant \hat{\sigma}_{2T} \quad \text{or} \quad \sigma_2 \leqslant \hat{\sigma}_{2C} \quad \text{or} \quad \tau_{12} \geqslant \hat{\tau}_{12}. \quad (9.1)$$

(recalling that a compressive stress is taken as negative so, for example, failure would occur if $\sigma_2 = -200$ MPa and $\hat{\sigma}_{2T} = -150$ MPa).

(b) *Maximum strain criterion*

The maximum strain criterion merely substitutes strain for stress in the five sub-criteria. We now say that failure has occurred if

$$\varepsilon_1 \geqslant \hat{\varepsilon}_{1T} \quad \text{or} \quad \varepsilon_1 \leqslant \hat{\varepsilon}_{1C} \quad \text{or} \quad \varepsilon_2 \geqslant \hat{\varepsilon}_{2T} \quad \text{or} \quad \varepsilon_2 \leqslant \hat{\varepsilon}_{2C} \quad \text{or} \quad \gamma_{12} \geqslant \hat{\gamma}_{12}. \tag{9.2}$$

As when calculating stiffness, it is important that we can deal with the situation in which the fibres are not aligned with the applied stresses. We illustrate this by considering the simple case of a single stress σ_x, inclined at an angle θ to the fibres (Figure 9.1). We now use equations 7.6 to obtain the stresses in the principal material directions. Putting $\sigma_y = \tau_{xy} = 0$ in those equations we obtain

$$\sigma_1 = \sigma_x \cos^2 \theta, \quad \sigma_2 = \sigma_x \sin^2 \theta, \quad \tau_{12} = -\sigma_x \sin \theta \cos \theta. \tag{9.3}$$

We then apply equations 9.1 to determine whether failure has occurred. We are seeking the value of σ_x to cause failure and we see from equation 9.3 that there are three possible results, i.e.

$$\sigma_x = \hat{\sigma}_{1T}/\cos^2 \theta - \text{fibre failure,}$$

or

$$\sigma_x = \hat{\sigma}_{2T}/\sin^2 \theta - \text{transverse failure,}$$

or

$$\sigma_x = -\hat{\tau}_{12}/\sin \theta \cos \theta - \text{shear failure;}$$

clearly the smallest value of σ_x will be the failure stress.

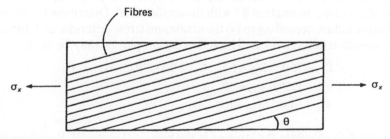

Figure 9.1 Uniaxial stress, σ_x, inclined at an angle θ to the fibres of a unidirectional composite.

The effect on the value of σ_x at failure as θ is varied as illustrated in Figure 9.2. We see that each mode of failure is represented by a separate curve. Fibre failure is most likely when θ is small, transverse (either matrix or interface) failure when θ approaches $90°$, and shear failure at intermediate angles.

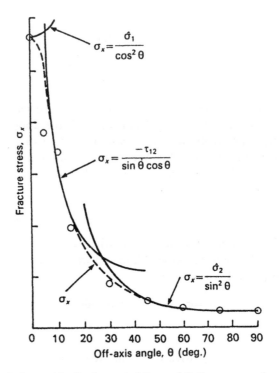

Figure 9.2 Variation with θ of σ_x at failure: full lines – maximum stress limit criterion; dotted line – Tsai–Hill interactive criterion.

Example 9.1

A tensile specimen of a unidirectional composite is prepared such that the fibres make an angle of $5°$ with the applied load. Determine the stress to cause failure according to (a) the maximum stress criterion, and (b) the maximum strain criterion. The following properties may be used:

$$E_{11}=76.0, \quad E_{22}=5.5, \quad G_{12}=2.35 \text{ GPa}; \quad \nu_{12}=0.33;$$

$$\hat{\sigma}_{1T}=1250, \quad \hat{\sigma}_{2T}=30, \quad \hat{\tau}_{12}=50,$$

$$\hat{\sigma}_{1C}=1000, \quad \hat{\sigma}_{2C}=100 \quad \text{MPa}.$$

(a) We first calculate the stresses in the principal material directions as described above, i.e.

$$\sigma_1=\sigma_x\cos^2\theta, \quad \sigma_2=\sigma_x\sin^2\theta, \quad \tau_{12}=-\sigma_x\sin\theta\cos\theta$$

where, here, $\theta=5°$.

Our conditions for failure become (there are no compressive stresses in this instance)

$$\sigma_1=\sigma_x\cos^2\theta\geqslant\hat{\sigma}_{1T}, \quad \text{or} \quad \sigma_2=\sigma_x\sin^2\theta\geqslant\hat{\sigma}_{2T},$$

or

$$\tau_{12} = -\sigma_x \sin\theta \cos\theta \geqslant \hat{\tau}_{12}.$$

Taking the equality we get three values of σ_x, the smallest of which is the desired result.

Substituting for θ and the given strengths gives

$$\sigma_x = 1250/\cos^2 5 \qquad = 1260 \text{ MPa},$$

or

$$\sigma_x = 30/\sin^2 5 \qquad = 3949 \text{ MPa},$$

or

$$\sigma_x = 50/\sin 5 \cos 5 = 575 \text{ MPa}.$$

Hence we have a failure stress of 575 MPa, the failure being in a shear mode.

(b) We can determine the direct strains in the principal directions using equation 7.1, noting that they depend on both stresses, i.e.

$$\begin{bmatrix} \varepsilon_1 \\ \varepsilon_2 \end{bmatrix} = \begin{bmatrix} \dfrac{1}{E_{11}} & -\dfrac{v_{21}}{E_{22}} \\ -\dfrac{v_{12}}{E_{11}} & \dfrac{1}{E_{22}} \end{bmatrix} \begin{bmatrix} \sigma_1 \\ \sigma_2 \end{bmatrix}$$

or, using the results from part (a)

$$\varepsilon_1 = \frac{\sigma_1}{E_{11}} - \frac{v_{21}}{E_{22}}\sigma_2 = \left(\frac{\cos^2\theta}{E_{11}} - \frac{v_{21}\sin^2\theta}{E_{22}} \right)\sigma_x$$

and

$$\varepsilon_2 = \frac{-v_{12}\sigma_1}{E_{11}} + \frac{\sigma_2}{E_{22}} = \left(\frac{-v_{12}\cos^2\theta}{E_{11}} + \frac{\sin^2\theta}{E_{22}} \right)\sigma_x.$$

The shear strain is obtained quite simply as

$$\gamma_{12} = \frac{\tau_{12}}{G_{12}} = \frac{-\sigma_x \sin\theta \cos\theta}{G_{12}}.$$

We now need to determine the failure strains from the data given above, assuming the material to have a linear stress–strain relationship, i.e.

$$\hat{\varepsilon}_{1T} = 1250/76 \times 10^3 = 0.0164,$$
$$\hat{\varepsilon}_{2T} = 30/5.5 \times 10^3 = 0.0055,$$
$$\hat{\gamma}_{12} = 50/2.35 \times 10^3 = 0.0218,$$
$$\hat{\varepsilon}_{1C} = -1000/76 \times 10^3 = -0.0132,$$
$$\hat{\varepsilon}_{2C} = -100/5.5 \times 10^3 = -0.0182.$$

The conditions for failure will again give three values for σ_x. Clearly in this problem the values for the shear mode of failure will be the same for both criteria. Also ε_2 will be compressive so we should use $\hat{\varepsilon}_{2C}(-0.0182)$ as our limiting strain.

Corresponding to ε_1 we get

$$\sigma_x = \hat{\varepsilon}_{1T} \bigg/ \left(\frac{\cos^2 \theta}{E_{11}} - \frac{v_{21} \sin^2 \theta}{E_{22}} \right) = 1259 \text{ MPa},$$

and corresponding to ε_2 we get

$$\sigma_x = \hat{\varepsilon}_{2C} \bigg/ \left(\frac{-v_{12} \cos^2 \theta}{E_{11}} + \frac{\sin^2 \theta}{E_{2C}} \right) = 6210 \text{ MPa}.$$

So, finally, the failure stress is seen to be 575 MPa, as given by the maximum stress criterion.

Although they are simple to use, limit criteria do not agree well with experimental data unless the fibre angle is close to 0° or 90°. This is because at intermediate angles there will be a stress field in which both σ_1 and σ_2 can be significant. These stresses will interact and affect the failure load, a situation which is not represented in criteria where a mode of failure is assumed not to be influenced by the presence of other stresses.

Also, the maximum stress and maximum strain criteria will give different predictions when the stress–strain relation is nonlinear. This will certainly be the case for shear deformations and hence the assumption of linearity made in the last example is seen to be invalid. In such cases the maximum strain criterion generally gives better agreement with experiment than does the maximum stress criterion.

9.2.2 Interactive criteria

Interactive criteria, as the name suggests, are formulated in such a way that they take account of stress interactions. The objective of this approach is to allow for the fact that failure loads when a multi-axial stress state exists in the material may well differ from those when only a uniaxial stress is acting.

There are many such criteria, of varying complexity, their success in predicting failure often being confined to one fibre/resin combination subjected to a well defined set of stresses (e.g., a tube under internal pressure). The *Tsai–Hill* criterion which has proven to be successful in a wide variety of circumstances is the only method that will be discussed here.

The Tsai–Hill criterion was developed from Hill's anisotropic failure criterion which, in turn, can be traced back to the von Mises yield criterion

(Chapter 6). In its most general form the Tsai–Hill criterion defines failure as

$$\left(\frac{\sigma_1}{\hat{\sigma}_1}\right)^2 - \frac{\sigma_1\sigma_2}{\hat{\sigma}_1^2} + \left(\frac{\sigma_2}{\hat{\sigma}_2}\right)^2 + \left(\frac{\tau_{12}}{\hat{\tau}_{12}}\right)^2 \geqslant 1. \tag{9.4}$$

Because it is usually small compared with the others, the second term $(\sigma_1\sigma_2/\hat{\sigma}_1^2)$ is often neglected. The modified form of the criterion is then

$$\left(\frac{\sigma_1}{\hat{\sigma}_1}\right)^2 + \left(\frac{\sigma_2}{\hat{\sigma}_2}\right)^2 + \left(\frac{\tau_{12}}{\hat{\tau}_{12}}\right)^2 \geqslant 1. \tag{9.4a}$$

The values of strength used in equation 9.4 or 9.4a are chosen to correspond to the nature of σ_1 and σ_2. So if σ_1 is tensile $\hat{\sigma}_{1T}$ is used, if σ_2 is compressive $\hat{\sigma}_{2C}$ would be used, and so on.

It should be noted that only one criterion has to be satisfied, as opposed to the five sub-criteria of the limit methods. Thus, only one value is obtained for the failure stress. Another point to bear in mind is that the *mode of failure* is not indicated by the method, unlike with the limit criteria. This latter issue has an influence on how we predict the failure of laminates, as we shall see later. For a unidirectional composite subjected to uniaxial stress parallel to a principal direction the Tsai–Hill and maximum stress criteria will give the same failure stress.

As with the limit criteria we are interested in the case of off-axis loading, i.e. stress and fibres not aligned. To illustrate this we again take the simple case of a single stress σ_x acting at θ to the fibres. Substituting the stresses of equation 9.3 in equation 9.4 gives, at failure,

$$\left(\frac{\sigma_x \cos^2\theta}{\hat{\sigma}_1}\right)^2 - \frac{\sigma_x^2 \cos^2\theta \sin^2\theta}{\hat{\sigma}_1^2} + \left(\frac{\sigma_x \sin^2\theta}{\hat{\sigma}_2}\right)^2 + \left(\frac{\sigma_x \sin\theta \cos\theta}{\hat{\tau}_{12}}\right)^2 = 1. \tag{9.5}$$

The way in which σ_x, obtained from the latter equation, varies with θ is shown by the dotted curve in Figure 9.2. Note that there is a continuous variation, rather than three separate curves as obtained with the limit criterion.

Example 9.2
Repeat Example 9.1 using the Tsai–Hill criterion.

We use equation 9.5 in conjunction with the appropriate strengths, i.e.

$$\left(\frac{\sigma_x \cos^2\theta}{1250}\right)^2 - \frac{\sigma_x^2 \cos^2\theta \sin^2\theta}{1250^2} + \left(\frac{\sigma_x \sin^2\theta}{30}\right)^2 + \left(\frac{\sigma_x \sin\theta \cos\theta}{50}\right)^2 = 1.$$

Substituting for $\theta = 5°$ we find at failure

$$\sigma_x = 520\,\text{MPa}.$$

Notice that this does not agree with the prediction of the maximum stress criterion, due to the allowance made here for the interactive effect of the stresses. Examination of the relative magnitudes of the terms in the equation shows that the last is dominant. This indicates that failure would probably be in shear (as predicted by the maximum stress criterion).

Example 9.3

Determine whether the rotated lamina of Example 7.5 will fail when subjected to the loading shown, according to (a) the maximum strain criterion, (b) the Tsai–Hill criterion. The uniaxial strengths of the composite may be taken as

$$\hat{\sigma}_{1T}=2100, \quad \hat{\sigma}_{1C}=1800, \quad \hat{\sigma}_{2T}=35, \quad \hat{\sigma}_{2C}=210, \quad \hat{\tau}_{12}=90 \text{ MPa}.$$

From that earlier example the stresses and strains in the principal material directions were found to be

$$\sigma_1=-3.25, \quad \sigma_2=1.75, \quad \tau_{12}=-5.75 \text{ MPa},$$

and

$$\varepsilon_1=-27.36\times10^{-6}, \quad \varepsilon_2=202.38\times10^{-6}, \quad \gamma_{12}=-809.86\times10^{-6}.$$

(a) We first need to find the failure strains corresponding to the given uniaxial stresses. We need

$$\hat{\varepsilon}_{1C}=-1800/138\times10^3=0.0130=-13000\times10^{-6},$$
$$\hat{\varepsilon}_{2T}=35/8.96\times10^3=0.0039=3900\times10^{-6},$$
$$\hat{\gamma}_{12}=90/7.1\times10^3=0.0127=12700\times10^{-6}.$$

(assuming linear elasticity and the moduli used in Example 7.5). Taking the ratios of limiting (failure) strain to calculated strain we get

$$\hat{\varepsilon}_{1C}/\varepsilon_1=1300/27.36 \quad =475.15,$$
$$\hat{\varepsilon}_{2T}/\varepsilon_2=3900/202.38 \quad = 19.27,$$
$$\hat{\gamma}_{12}/\gamma_{12}=12700/809.86= 15.68.$$

So, we see that failure has not occurred, the strain closest to failure being the shear strain.

(b) We substitute the calculated stresses and the appropriate strengths into the Tsai–Hill criterion (equation 9.4):

$$\left(\frac{\sigma_1}{\hat{\sigma}_1}\right)^2-\frac{\sigma_1\sigma_2}{\hat{\sigma}_1^2}+\left(\frac{\sigma_2}{\hat{\sigma}_2}\right)^2+\left(\frac{\tau_{12}}{\hat{\tau}_{12}}\right)^2\geq1,$$

that is

$$\left(\frac{-3.25}{-1800}\right)^2-\frac{(-3.25\times1.75)}{(-1800)^2}+\left(\frac{1.75}{35}\right)^2+\left(\frac{-5.75}{90}\right)^2\geq1 \text{ for failure}.$$

When evaluated, the left-hand side of the last expression is found to be 0.00658. Clearly failure has not occurred.

We have shown that application of a failure criterion will tell us whether, or not, failure will occur for a given set of stresses. It is clearly also of interest to be able to predict the magnitude of the stress, or stresses, that will cause failure. We can achieve this by simple factoring. If we are dealing with a limit criterion we merely examine the smallest ratio of limit-to-calculated strains and scale accordingly.

Thus, for example, in Example 9.3(a) we had a ratio of 15.68 (from the shear strains). In other words we need a strain 15.68 times the calculated value to cause failure.

Using an interactive criterion is slightly more complicated. Suppose we scale our set of stresses σ_1, σ_2 and τ_{12}, (and the associated Cartesian stresses σ_x, σ_y and τ_{xy}) by the factor R. Then equation 9.4 becomes

$$\left(\frac{R\sigma_1}{\hat{\sigma}_1}\right)^2 - \frac{R\sigma_1 R\sigma_2}{\hat{\sigma}_1^2} + \left(\frac{R\sigma_2}{\hat{\sigma}_2}\right)^2 + \left(\frac{R\tau_{12}}{\hat{\tau}_{12}}\right)^2 = 1 \quad \text{at failure,}$$

that is

$$\left(\frac{\sigma_1}{\hat{\sigma}_1}\right)^2 - \frac{\sigma_1\sigma_2}{\hat{\sigma}_1^2} + \left(\frac{\sigma_2}{\hat{\sigma}_2}\right)^2 + \left(\frac{\tau_{12}}{\hat{\tau}_{12}}\right)^2 = \frac{1}{R^2}. \tag{9.6}$$

R is clearly a measure of the available strength in the system. It is often referred to as the *Reserve Factor*.

Let us again use Example 9.3 as an illustration. In part (b) of that example we evaluated the left-hand side of equation 9.6 as 0.00658, i.e.

$$\frac{1}{R^2} = 0.00658,$$

or

$$R = 12.33,$$

i.e. we can increase all the stresses by this factor before failure occurs.

9.3 STRENGTH OF A LAMINATE

9.3.1 Initial failure

Suppose we take a cross-ply laminate (0/90° lay-up) and apply an increasing load in the direction of the 0° fibres. At a relatively low load cracks will be seen in the matrix parallel to the fibres in the 90° plies. These cracks will increase in density until a saturation state is reached. At this point the 90° plies contribute virtually no stiffness to the laminate in the 0° direction, a fact that is shown by the change in slope of the load–extension curve for such a laminate (Figure 9.3). The commencement of transverse ply cracking is known as *initial failure* or *first ply failure*.

Similar behaviour would be seen in an angle-ply laminate ($\pm\theta$ lay-up), with initial failure indicated by cracks parallel to the fibres. These cracks

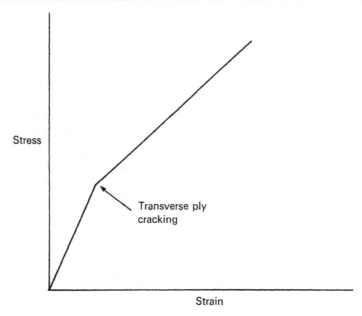

Figure 9.3 Change in slope of stress–strain curve for a cross-ply laminate at onset of transverse (90°) ply cracking.

would be caused by interlaminar shear at low values of θ and by the transverse tension at high values of θ.

It is possible to predict initial failure of laminate by combining *Classical Laminate Theory* with a failure criterion. Clearly, the choice of criterion is crucial and, as already stated, there are many available, each one often being relevant only to a very specific situation (loading and geometry).

We start with the plate constitutive equations (equation 7.20), i.e.

$$\begin{bmatrix} N \\ M \end{bmatrix} = \begin{bmatrix} A & B \\ B & D \end{bmatrix} \begin{bmatrix} \varepsilon^0 \\ k \end{bmatrix}. \tag{9.7}$$

Solution of equations 9.7 will give the plate mid-plane strains (ε^0) and plate curvatures (k) for a known set of forces (N) and moments (M).

$$\begin{bmatrix} \varepsilon^0 \\ k \end{bmatrix} = \begin{bmatrix} A & B \\ B & D \end{bmatrix}^{-1} = \begin{bmatrix} A^1 & B^1 \\ B^1 & D^1 \end{bmatrix} \begin{bmatrix} N \\ M \end{bmatrix}, \tag{9.8}$$

where

$$A^1 = A^* + B^*[D^*]^{-1}[B^*]^t,$$

$$B^1 = B^* - [D^*]^{-1},$$

$$D^1 = [D^*]^{-1},$$

$$A^* = A^{-1},$$

$$B^* = A^{-1}B,$$

$$D^* = D - BA^{-1}B.$$

The plate strains are expressed in the plate x–y axes (see Figure 7.7). Thus the strains in each ply can be found from the transformations of equation 7.7, i.e.

$$\bar{\varepsilon}_{12} = \mathbf{T}\bar{\varepsilon}_{xy}. \tag{9.9}$$

Finally the ply stresses are obtained from the stiffness matrix equation 7.5, i.e.

$$\sigma_{12} = \mathbf{Q}\varepsilon_{12} \tag{9.10}$$

By applying, on a ply-by-ply basis, a selected failure criterion the occurrence of failure can be determined.

Example 9.4
The unidirectional CFRP material considered in Example 9.3 (and Example 7.5) is used to form a 6-ply (0.75 mm thick) laminate with a $(\pm 30/0^\circ)_s$ lay-up. Determine the uniaxial in-plane stress to cause first ply failure according to the maximum stress criterion.

Using the elastic properties given in Example 7.5 ($E_{11} = 138$ GPa, $E_{22} = 8.9$ GPa, $G_{12} = 7.1$ GPa, $v_{12} = 0.30$) we can assemble the **A** matrix (**B** = 0 because the stacking sequence is symmetric and **D** is not needed because the loading is in-plane). Then, applying the constitutive equations (9.7) we can calculate the membrane strains and, finally, using the transformations (equation 9.9) and stress–strain relationships (equation 9.10) we get the ply stresses in the principal directions.

Suppose, here, we apply a load $N_x = 10^5$ N/m (parallel to the 0° fibres). The resulting strains are $\varepsilon_x^0 = 0.1561\%$, $\varepsilon_y^0 = -0.1615\%$, $\gamma_{xy}^0 = 0$. The ply stresses are illustrated below (values in MPa).

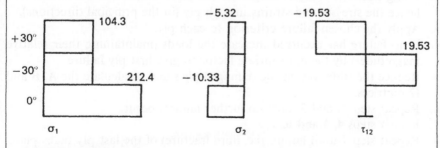

We now compare the above stresses with the appropriate strengths from Example 9.3, i.e.

$$\hat{\sigma}_{1T} = 2100, \quad \hat{\sigma}_{2C} = 210, \quad \hat{\tau}_{12} = 90 \text{ MPa}.$$

The corresponding minimum reserve factors are:

longitudinal tension (0° ply), $R = 2100/212.4 = 9.89$;

transverse compression (0° ply), $R = 210/10.33 = 20.33$;

shear ($\pm 30°$ plies), $R = 90/19.53 = 4.61$.

We see that if we increase the applied load by 4.61 (to 4.61×10^5 N/m) the first ply failure will be by shear in the $\pm 30°$ plies.

9.3.2 Final failure and strength

The final, or ultimate, failure load of an angle ply laminate is often coincident with, or only slightly higher than, the load to cause initial failure. This is not necessarily the case for other lay-ups and final failure can be at a considerably higher load than that to cause first ply failure.

It is clear that once a ply has sustained failure its stiffness in certain directions will have been reduced. However, unless the damaged ply has completely delaminated from the rest of the laminate it will still contribute to the overall stiffness of the plate. The magnitude of this contribution depends on the amount of damage, the fibre/matrix combination and the nature of the loading on the ply.

In general an iterative method is adopted, successively applying the approach described in section 9.3.1 until final failure has occurred. As a starting point the relative value of the forces and moments (N_x, N_y, N_{xy}, M_x, M_y, M_{xy}) acting on the laminate will be known from, say, a structural analysis of the component being designed. Alternatively a parametric study of a range of lay-ups could be carried out for an arbitrary set of load values. The steps taken are as follows.

1. Apply, in the previously-determined ratio, a small value of the loads to the laminate under consideration.
2. Using laminate analysis determine the plate strains and curvatures and hence the stresses and strains in each ply (in the principal directions).
3. Apply the chosen failure criterion to each ply.
4. If no failure has occurred increase the loads (maintaining their relative magnitudes) by the appropriate factor to give first ply failure.
5. Reduce the stiffnesses in the damaged plies and recalculate the A, B and D matrices.
6. Repeat steps 2 and 3 until no further failures occur.
7. Repeat steps 4, 5 and 6.
8. Repeat step 7 until failure (i.e. fibre fracture) of the last ply takes place. This defines the ultimate strength of the laminate.

It is when applying failure criteria in the above procedure that the choice between limit and interactive methods becomes important. As we have seen, the former tells us the *mode* of failure. Thus, when we come to step 5, we know which stiffnesses to reduce if we use either the maximum stress or maximum strain criterion. If, on the other hand, we were to use the Tsai–Hill criterion we would have to infer the mode of failure from the relative magnitudes of the terms in equation 9.4.

Even when we know the failure mode it is still necessary to select the reduction factor for the stiffness terms; at the moment there is no universally accepted approach. Suppose, for example, that step 4 of our procedure indicates transverse tensile failure in a particular ply. One approach would be to put the corresponding stiffness, E_{22}, together with v_{21}, to zero. An alternative would be to reduce the original values by, say, 50% or 90%; some researchers have found the former appropriate to CFRP and the latter to GFRP. The simplest, and least realistic, approach would be to completely disregard the damaged layer. In this case, when recalculating the **A**, **B** and **D** matrices, the laminate would appear to have empty space in place of these layers.

9.4 ADDITIONAL FACTORS

9.4.1 Hygrothermal effects

It was noted in an earlier chapter that some high performance fibres, notably carbon and Kevlar, have small, or even negative, coefficients of thermal expansion. As a consequence, in a unidirectional ply, when cooled to room temperature after curing, the fibres will be in compression and the matrix in tension (parallel to the fibres).

It is easy to see that a complex state of stress will exist in a laminate at room temperature because, in addition to the effect described above, the relatively large transverse contraction of a free ply will be constrained by other plies having differently oriented fibres. These so-called residual thermal stresses can be a significant fraction of the failure stress of the matrix. It is, therefore, vital that such issues are included in any strength prediction procedure.

Similar effects are seen due to the absorption of moisture by a polymer matrix. The resultant swelling is akin to thermal expansion and can be treated in the same way by modification of the laminated plate constitutive equations. Such hygrothermal topics will not be addressed here.

9.4.2 Edges

Classical laminate theory (CLT) applies only to plates that are infinitely long and wide. In other words it ignores edges. In many real situations laminates will have edges, for example a plate containing a hole or a plate of finite width.

At such edges the assumptions of CLT break down and the in-plane stresses $(\sigma_x, \sigma_y, \sigma_{xy})$ alone are found not to satisfy local equilibrium on the stress-free boundaries. The through-thickness direct (σ_z) and shear (τ_{xz}, τ_{yz}) stresses (Figure 9.4) can be calculated from advanced applied mechanics theories or from finite element analysis. It is found that within one plate thickness from the edge these stresses can be sufficiently high to exceed the (low) through-thickness strengths. Both their magnitude and sense (tension or compression) are determined by the laminate's stacking sequence. Hence it is possible to minimize their effect and reduce the chance of failure, by interlaminar shear or tension, initiating at an edge.

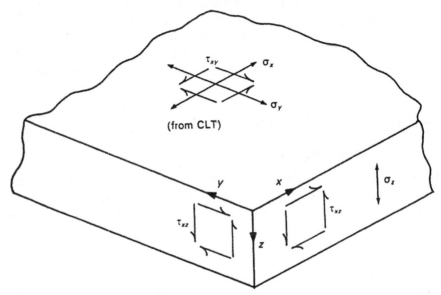

Figure 9.4 Stresses at free edge of a laminate: σ_z, τ_{xz}, τ_{yz} (not predicted by CLT), in the thickness (z) direction.

A simple illustration of this situation is given by a cross-ply laminate when loaded in tension parallel to the 0° plies. With a $(0/90°)_s$ stacking sequence the through-thickness direct stress on the edges parallel to the load is tensile, whilst for a $(90/0°)_s$ stacking sequence the stress is compressive. Clearly the former stacking sequence is to be preferred.

9.5 SUMMARY

In this chapter we have considered ways of calculating the strength of a lamina or a laminate when subjected to a set of loads (forces and moments) that cause in-plane stresses. The general approach is to obtain from CLT the stresses, or strains, in the principal directions of each ply and to compare these with some limiting values. The comparison is made via a failure criterion.

A large number of failure criteria have been developed, although many of them apply only in restricted circumstances (a given fibre/matrix combination and/or set of loads). Here we looked at the two limit criteria, maximum stress and maximum strain, and one interactive criterion, the Tsai–Hill method.

Initial, or first-ply, failure of a laminate can readily be predicted using a criterion, but the way of predicting final failure is still open to debate. Finally, a full analysis should include hygrothermal effects and, where appropriate, the additional stresses that arise at edges.

FURTHER READING

General

Agarwal, B. D. and Broutman, L. J. (1990) *Analysis and Performance of Fiber Composites*, 2nd edn., Wiley.

Rowlands, R. E. (1985) Strength (failure) theories and their experimental correlation, in *Handbook of Composites*, Vol. 3, (eds G. C. Sih and A. M. Skudra), North Holland, Amsterdam, pp. 71–125.

PROBLEMS

9.1 A unidirectional lamina, with the properties given below, is subjected to a shear stress as illustrated in the figure. Determine the magnitude of this stress to cause failure according to the Tsai–Hill criterion when

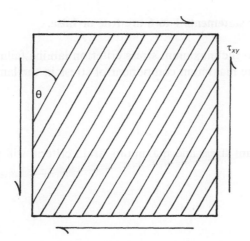

$\theta = 10°$, $30°$ and $60°$. $E_{11} = 130.0$, $E_{22} = 10.0$, $G_{12} = 5.0$ GPa; $v_{12} = 0.3$; $\hat{\sigma}_{1T} = 1500$, $\hat{\sigma}_{2T} = 50$, $\hat{\sigma}_{1C} = 1300$, $\hat{\sigma}_{2C} = 120$, $\hat{\tau}_{12} = 80$ MPa.

9.2   A 4-ply laminate with $(\pm 45°)_s$ lay-up is subjected to the set of loads shown in the figure. Each ply is 0.125 mm thick with the elastic properties used in Example 7.2 and the strengths of Problem 9.1. Determine the reserve factor appropriate to the given loads according to the maximum stress criterion.

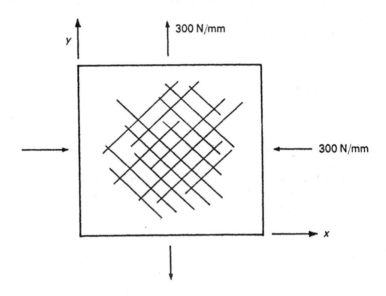

SELF-ASSESSMENT QUESTIONS

Indicate whether statements 1 to 6 are true or false.

1.   According to the maximum stress criterion lamina failure is determined by the absolute maximum component of stress in the lamina.

   (A) true
   (B) false

2.   The maximum stress and maximum strain criteria will predict the same failure loads.

   (A) true
   (B) false

3. A hybrid stress criterion is used for composites containing more than one fibre type (i.e. a hybrid composite).

    (A) true
    (B) false

4. An interactive stress criterion cannot directly predict the mode of failure.

    (A) true
    (B) false

5. An interactive criterion will always predict failure stresses different to those predicted by the maximum stress criterion.

    (A) true
    (B) false

6. When predicting the final failure of a laminate it is necessary to know the failure mode of individual plies.

    (A) true
    (B) false

For each of the statements of questions 7 and 8, one or more of the completions given are correct. Mark the correct completions.

7. Maximum strain criterion

    (A) can be obtained from the maximum stress criterion by dividing each term by an appropriate stiffness,
    (B) comprises five sub-criteria,
    (C) cannot predict failure stresses,
    (D) will give the same prediction as the maximum stress criterion for a lamina in a uniaxial stress state when the stress is parallel to the fibres,
    (E) gives a better prediction than the maximum stress criterion when the stress–strain relation shows significant nonlinearity.

8. Tsai–Hill criterion

    (A) is only applicable if the direct stresses are tensile,
    (B) cannot predict the mode of failure,
    (C) gives better prediction than does a limit criterion for a unidirectional lamina when the fibres are not aligned with the applied stress,
    (D) cannot be used to predict the final failure of a laminate,
    (E) cannot be used to obtain a reserve factor.

Each of the sentences in questions 9 to 15 consists of an assertion followed by a reason. Answer:

(A) if both assertion and reason are true statements and the reason is a correct explanation of the assertion,

(B) if both assertion and reason are true statements but the reason is not a true explanation of the assertion,

(C) if the assertion is true but the reason is a false statement,

(D) if the assertion is false but the reason is a true statement,

(E) if both the assertion and reason are false statements.

9. A lamina is deemed to have failed when the fibres fracture *because* the fibres carry the highest stresses.

10. When predicting the failure of an off-axis lamina it is necessary to calculate the stresses in the principal directions *because* these stresses are always greater than the applied stresses.

11. The maximum stress criterion will always predict failure in tension *because* the longitudinal tensile strength of a unidirectional ply is greater than the corresponding compressive strength.

12. The Tsai–Hill criterion gives a more accurate prediction for off-axis loading *because* it does not predict the mode of failure.

13. Initial failure of a cross-ply laminate can only be predicted by the Tsai–Hill criterion *because* it corresponds to transverse ply cracking.

14. Prediction of laminate failure requires an iterative approach *because* ply stiffnesses are modified as failures occur.

15. Classical Laminate Theory cannot predict failure of finite width laminates *because* it ignores the existence of through-thickness stresses.

ANSWERS

Problems

9.1   82.7, 104.4, 104.4 MPa.
9.2   0.133.

Self-assessment

1. B;   2. B;   3. B;   4. A;   5. B;   6. A;   7. B,D,E;   8. B,C;   9. B;   10. C;
11. D;   12. B;   13. D;   14. A;   15. A.

Short fibre composites | 10

10.1 INTRODUCTION

In previous chapters we have discussed the mechanical properties of continuous fibre composites. As the reader will see, for various reasons, many composites are not reinforced by continuous fibres but by short fibres. The properties of *short fibre composites* are very different from those of their continuous fibre counterparts and in this chapter a number of models for the mechanical performance of short fibre composites will be presented. Most examples of the application of these models, and of the effects of processing, will be polymer matrix systems as there is such a wealth of information available on these compared with metal and ceramic matrix composites.

10.2 REASONS FOR USING SHORT FIBRE COMPOSITES

The reader will recall from Chapter 2 on reinforcements that, as a general rule, continuous fibres are more expensive than other forms of reinforcement. Furthermore the manufacturing processes for continuous fibre reinforced composites tend to be slow and inflexible. Let us look at these two points in more detail for PMCs.

Hand lay-up, filament winding, autoclave and vacuum bag processing techniques, generally associated with continuous fibre reinforced thermoset composites, are suitable for short runs or one-off requirements of high-performance high-priced products. Such techniques are however restrictive when very large numbers of articles are required. For many applications the development of peak strength or stiffness properties are not the main requirement. Where this is the case, there is frequently a desire to manufacture the product in numbers which make the processing techniques employed for continuous fibres prohibitively lengthy, and expensive, in terms of machine time and man-hours. The use of pultrusion to produce continuous fibre high performance articles is a fairly rapid process, but can only be applied where a constant profile is required. For large numbers of articles with complex shapes, injection, compression and transfer mould-ing are favoured.

The price to be paid for the use of such mass production techniques is a shortening of fibre length. This reduction in fibre length is partly due to the requirements of the processing technique, but some processes which involve mechanical shearing and mixing actions also promote considerable fibre breakage. Fibre damage is particularly noticeable for injection moulding, extrusion and mixing of polyester moulding compounds.

10.3 FIBRE LENGTH

At this stage it should be clearly stated that there is not a fibre length which is constant for all materials and below which the fibres may be termed 'short'. Instead the fibre length should be considered relative to a parameter known as the *critical fibre length*. In section 10.6 it will be shown that the critical fibre length is a function of the matrix and the reinforcement and as such varies considerably from composite to composite. It is therefore possible for fibres of, say, 5 mm length to be classified as short in one system and not in another. However there is always a clear distinction between continuous fibre composites and any other type. The behaviour of very short fibres is dominated by end effects and they do not therefore act as good reinforcing agents.

Discontinuous fibres are normally supplied by manufacturers in standard lengths for different processing routes. Typical length parameters are in-dicated for a number of well-known polymeric materials and processes in Table 10.1. Note that the processed fibre length may be significantly less than the preprocessed fibre length.

Table 10.1 Typical length parameters for discontinuous fibres

Material		*Preprocessed fibre length* (mm)	*Processed fibre length* (mm)
Polyester–Epoxy–CSM		50	50
Polyester	SMC	25	25
moulding compounds	BMC	12	<4
Injection moulding thermoplastics		3	<3

Given that a typical fibre may have a diameter of approximately 10 μm, it is clear that high levels of length degradation are required to reduce them to 'particles'. This is just as well since, as shown by Table 10.1, processing techniques such as injection moulding have a devastating effect on fibre length. For example, the micrograph in Figure 10.1(a) shows fibres separated from a polymer matrix by burning. The composite had been injection moulded, and marked differences in fibre length are apparent. The initial length of the fibres prior to moulding was 6 mm. The histogram in Figure 10.1(b) shows the distribution of fibre length for a sample taken from an injection moulded thermoset. Initial fibre length was again 6 mm, and the histogram shows that in this sample none of the fibres remained unscathed during the processing operation. The degree of fibre length degradation depends on several process parameters such as screw design, shear rates, melt viscosity and fibre volume fraction. Other processes such as extrusion can be just as damaging to fibres. For example, extruding glass fibres of initial length 6 mm and of weight fraction 28% in a polypropylene matrix through a circular die of diameter 3 mm results in a mean fibre length of only 0.5 mm, although some fibres as long as 2 mm may be found.

It is clear from the above discussion that dependent upon the type of material used, and the method chosen to process it to its final shape, a wide variety of fibre lengths will be present. Whilst fibres even down to 50 µm in

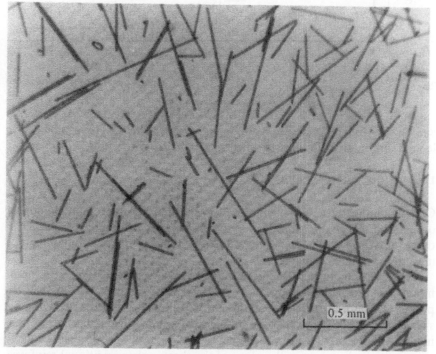

Figure 10.1 (a) Light micrograph showing the wide variation in fibre length after injection moulding. The fibres have been separated from the thermoplastic matrix; (b) histogram of the lengths of fibres extracted from a thermoset injection moulding: initial fibre length 6 mm. (Source: Pennington, 1979.)

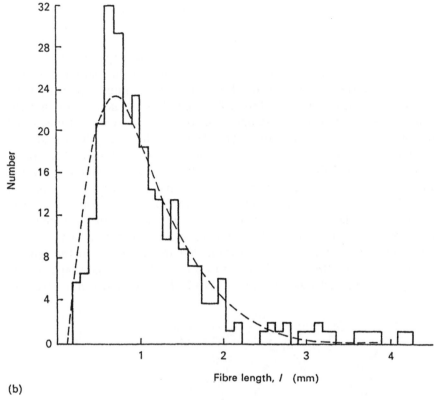

(b)

Figure 10.1 *Contd.*

length may retain some ability to reinforce, it is the fact that actual fibre length and its distribution are uncertain that can cause design problems.

10.4 FIBRE ORIENTATION

Of equal importance to the length of fibres is their orientation. Although the orientation effects for short fibre systems are not in general as marked as those described in Chapter 8 for off-axis loading of unidirectional composites, they are not negligible.

The fibre orientation depends on the processing route. When continuous fibres are used the lay-up can be controlled to give predictable end properties for the composite in terms of stiffness and strength. Chopped strand mat (CSM) and sheet moulding compounds (SMC), because of their nature and the methods of processing into panel form, give properties which are essentially isotropic in the plane of the sheet, unless additional reinforcement is added. The properties however are very different normal to the plane of the sheet and composites with this type of anisotropy are sometimes referred to as 2−D or *in-plane randomly orientated* composites.

On the other hand, where fibres are shorter, and processing methods involve flow of material in a mould, changes in fibre orientation throughout a moulding are inevitable. Such is the case for bulk moulding compounds (BMC) and for the wide range of reinforced thermoplastics currently available. Where thick or variable sections and several injection points are involved, the orientation of the fibres may be impossible to predict. In any case the properties of the material could differ markedly from area to area within the same moulding.

Changes in fibre orientation are related to a number of factors, such as the geometrical properties of the fibres, viscoelastic behaviour of the fibre-filled matrix, mould design and the change in shape of the material produced by the processing operation. In many processing operations the polymer melt, or charge, undergoes both elongational (or extensional) flow and shear flow. The effect of these flow processes on fibre orientation is shown in Figure 10.2 for simple two-dimensional deformation. During extensional flow the fibres rotate towards the direction of extension. With large extensions a high degree of alignment may be produced. In shear flow

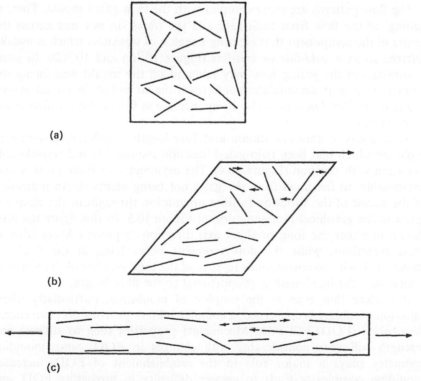

(a)

(b)

(c)

Figure 10.2 Schematic diagram of the changes in fibre orientation during flow: (a) initial random distribution; (b) rotation during shear flow; (c) alignment during elongational flow. (Source: Hull, 1981.)

there is a similar tendency for fibres to rotate towards the direction of shear. The viscosity of the matrix affects fibre orientation mainly through its influence on mould filling, which determines the distribution of elongational and shear flow fields.

Typical flow patterns obtained when injection moulding single-gated and double-gated moulds are presented in Figure 10.3 for short glass fibre-reinforced polypropylene. First let us consider the single-gated mould. As the material passes through the gate it experiences large elongational and compressive fields. Solidification first takes place at the mould surfaces, forming a skin. The mould is then filled by material which flows through the core region to the advancing flow front (Figure 10.3(a)). A velocity profile is established within the core, and the deformtion field in the region of the solidifying skin involves significant elongational flow. Solidification of the core occurs when the mould is full, under completely different conditions to the skin, to give the flow pattern shown in Figure 10.3(b). It is usually found that there is a pronounced preferred orientation of the fibres parallel to the flow direction in the outer layers, and a more random distribution in the core.

The flow patterns are more complex with the twin-gated mould. There is 'jetting' of the flow from each gate and the fibres do not mix across the centre of the component thus creating a region of weakness which is usually referred to as a *weld-line* or *knit-line* (Figure 10.3(c) and 10.3(d)). In some circumstances the jetting flow may rebound off the mould wall facing the gates and cause an accumulation of fibres at the far end of the mould as seen in Figure 10.3(c). There may also be an increase in fibre concentration at the gate mouths.

An analysis of fibre orientation and fibre length at different positions in a plaque which had been twin-gated injection moulded found considerable variation with position (Figure 10.4). The asymmetry of these plots is due, presumably, to the flows from the gates not being identical. An indication of the extent of the variation in fibre orientation throughout the plaque is given in the graphical representation of Figure 10.5. In this figure the lines drawn in either the long or short axis directions represent fibres lying in these directions, while the dots represent fibres lying in the thickness direction. Each line corresponds to 10% of the total number of fibres in that region, and the line length is proportional to the fibre length.

It is clear that even in the simplest of mouldings, particularly where injection moulding is concerned, a complex inhomogeneous *fibre orientation distribution* (FOD) will exist. Mechanical properties such as stiffness and strength will vary considerably with changes in FOD. Since moulding geometry plays a major role in the establishment of FOD, increased moulding complexity leads to greater difficulty in predicting FOD, and consequent uncertainty as to the mechanical properties in any section of the component. The prediction of FOD precisely, in anything but the simplest

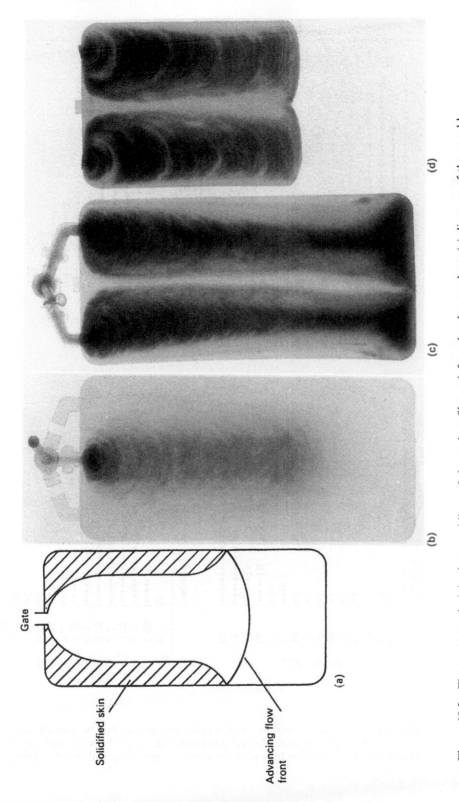

Figure 10.3 Flow patterns in injection mouldings of short glass fibre-reinforced polypropylene: (a) diagram of the mould filling process; (b) single-gated mould; (c and d) twin-gated moulds. (Source: Akay and Barkley, 1985.)

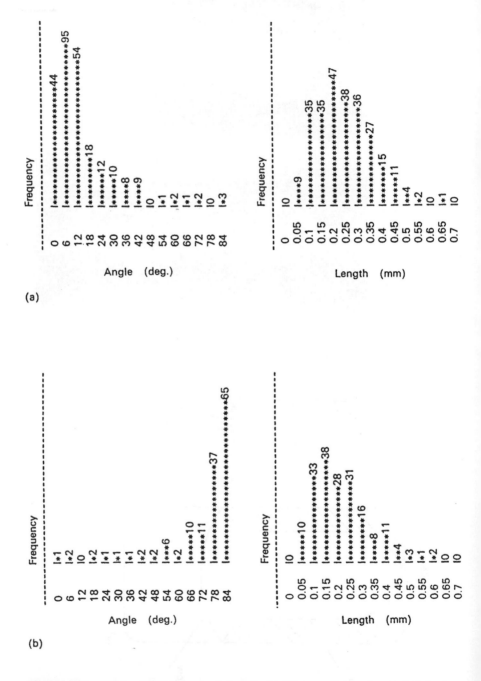

Figure 10.4 Fibre orientation and length histograms for twin-gated injection moulded short glass fibre-reinforced polypropylene: (a) at the weld, mid-way along plaque; (b) at the weld on far-wall from gate. (Source: Akay and Barkley, 1985.)

Gate   Gate

Figure 10.5 A graphical representation of the fibre orientation and length with position in a twin-gated injection moulded plaque. Dots represent fibres lying in thickness direction. Length of line is proportional to fibre length. (Source: Akay and Barkley, 1985.)

geometries, is unlikely to be a possibility in the foreseeable future. Even if it were possible to establish the point-to-point variation in the plane and through the thickness of a moulding, the complexity of the stress analysis required to cope with such anisotropy and inhomogeneity, whilst not impossible, would be extremely daunting and probably not justified for the type of component under discussion. Satisfactory prediction of mechanical performance can usually be achieved with simpler methods employing averaged properties.

10.5 STRESS AND STRAIN DISTRIBUTION AT FIBRES

In the analysis of continuous fibre composites in Chapter 7 any effects associated with fibre ends were neglected. As the *aspect ratio* (l/D where l and D are fibre length and diameter respectively) of the fibres decreases the end effects become progressively more significant and the efficiency of the fibres in stiffening and reinforcing the matrix decreases. As well as affecting stiffness and strength, fibre ends play an important role in the fracture of short fibre composites, and may also contribute to fracture processes in continuous fibre composites, since the long fibres may break down into discrete lengths.

Let us consider a single fibre of length l embedded in a matrix of lower modulus, and aligned with the loading direction. If we assume that the fibre is well bonded to the matrix, the stress applied to the matrix will be transferred to the fibre across the interface. The matrix and the fibre will experience different tensile strains because of their different moduli; in the region of the fibre ends the strain in the fibre will be less than that in the matrix, as indicated in Figure 10.6(b). As a result of this strain difference, shear stresses are induced around the fibres in the direction of the fibre axis, and the fibre is stressed in tension. The shear strength of the fibre–matrix interface is relatively low, typically of the order of 20 MPa, although it can exceed 50 MPa, for a PMC. However the surface area of the fibre is large, so that, given sufficient length, the fibre can carry a significant load, even up to the fibre fracture load.

It has been shown analytically that the stress distribution along a fibre aligned parallel to the loading direction of the matrix may be represented as in Figure 10.7. The main assumptions made in this analysis are that the fibre and matrix only deform elastically and that the interface is thin and gives good bonding between the fibre and the matrix. According to the analysis the tensile stress is zero at the fibre ends, and a maximum at the centre of the fibre. Conversely the shear stress around the fibre is a maximum at fibre ends, and for a sufficiently long fibre falls to almost zero in the centre. It is this variation of shear stress ('shear effect') that causes the build-up of tensile stress in the fibre. The analysis has been supported by results from photoelasticity and laser Raman spectroscopy experiments, e.g., Figure 10.8 shows *in situ* axial strain measurements obtained from a polydiacetylene fibre at different levels of matrix strain; the strain distribution in the fibre is of the same form as the tensile stress distribution given in Figure 10.7.

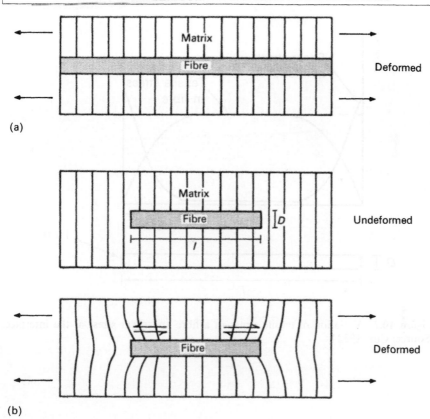

Figure 10.6 Effect of deformation on the strain around a fibre in a low modulus matrix: (a) continuous fibre; (b) short fibre.

The theoretical analysis and the experimental data demonstrate that regions towards the ends of fibres do not carry the full load. It follows that the average stress in a short fibre is less than that in a continuous fibre subjected to the same external loading and hence continuous fibre reinforcement is more efficient. The reinforcing efficiency of fibres also depends on interface strength since load transfer requires a strong interfacial bond. The large shear stresses at fibre ends can produce undesirable effects such as

(a) interfacial shear debonding,
(b) cohesive failure of matrix or fibre, and
(c) matrix yielding.

10.6 CRITICAL FIBRE LENGTH AND AVERAGE FIBRE STRESS

The maximum strain that can be achieved in a fibre is that applied to the matrix, ε_m, and in that situation the tensile stress in the fibre, σ_{Tf}, is

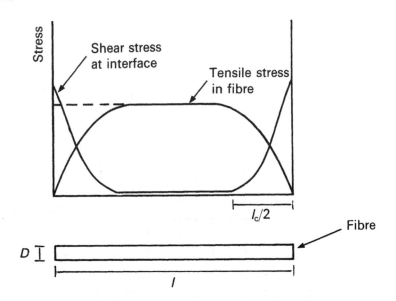

Figure 10.7 Variation of tensile stress in a fibre and shear stress at the interface. (Source: Cox, 1952.)

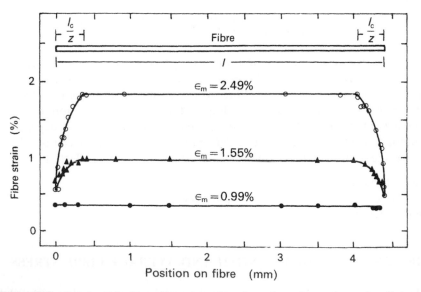

Figure 10.8 Axial strain in a fibre as a function of position along the fibre at different loadings. (Source: Young *et al.*, 1985.)

given by

$$\sigma_{Tf} = \varepsilon_m E_f, \tag{10.1}$$

where E_f is the Young's modulus of the fibre and it is assumed that the fibre has deformed elastically.

As indicated in Figures 10.7 and 10.8 for a given loading on a composite the maximum strain and stress occur away from the fibre ends. As the 'shear effect' builds up from the fibre ends, the stress carried increases until it reaches a maximum, for the particular applied load, of σ_{Tf} ($=\varepsilon_m E_f$ from equation 10.1) at $l_c/2$.

Should the applied load be increased then ε_m will be larger resulting in a higher stress in the fibre; the maximum stress which can be obtained in the fibre is of course the fracture stress $\hat{\sigma}_{Tf}$ (Figure 10.9(a)). In order to achieve this level of stress in the fibre the fibre length must be at least equal to a critical value l_c known as the *critical fibre length*. l_c may be defined as the minimum fibre length for a given diameter which will allow tensile failure of the fibre rather than shear failure of the interface, i.e., the minimum length of fibre required for the stress to reach the fracture stress of the fibre (Figure 10.9(b)). The critical fibre length may be determined by considering a force balance in the fibre when the fibre stress is $\hat{\sigma}_{Tf}$:

$$\text{tensile force in fibre} = \hat{\sigma}_{Tf}\, \pi D^2/4, \tag{10.2}$$

$$\text{shear force at interface} = \tau \pi D l_c/2, \tag{10.3}$$

by equating these two quantities:

$$l_c = \hat{\sigma}_{Tf} D/2\tau. \tag{10.4}$$

For calculation purposes, values for the interfacial shear strength or the matrix shear strength are used for τ in equation 10.4. Typical values for l_c and l_c/D, which is called the *critical aspect ratio*, are given for a range of composites in Table 10.2.

Owing to the less effective end portions, the average stress in a discontinuous fibre will be lower than the maximum stress that could be achieved with a continuous fibre. The value of the average stress depends on the stress distribution in the ends of the fibres and upon the fibre length. Figure 10.10 indicates the forces acting on half a fibre and, assuming a linear stress variation in the fibre, we have

$$F_1 = \sigma_{Tf}\pi D^2/4, \tag{10.5}$$

$$F_2 = [\sigma_{Tf} + d\sigma_{Tf}]\pi D^2/4, \tag{10.6}$$

$$F_3 = \tau \pi D\, dx. \tag{10.7}$$

For equilibrium

$$F_1 + F_3 = F_2, \tag{10.8}$$

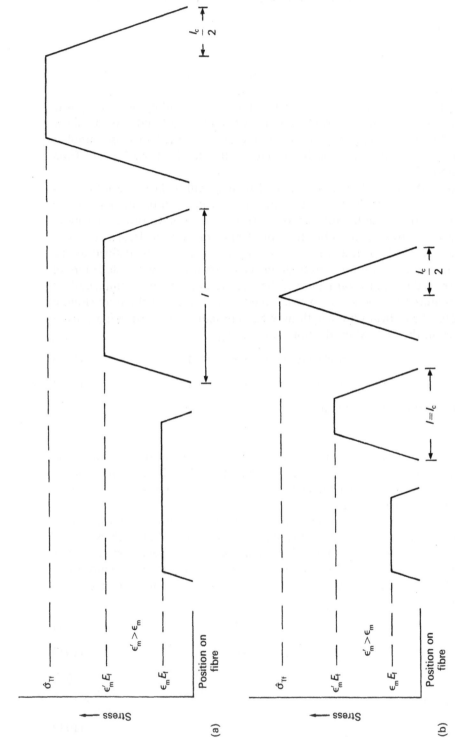

Figure 10.9 Effect of load and fibre length on the tensile (axial) stress distribution in a fibre: (a) $l > l_c$; (b) $l = l_c$. ε_m and ε'_m are two different matrix strains with $\varepsilon'_m > \varepsilon_m$ and $\hat{\sigma}_{Tf}$ is the fibre tensile strength.

Table 10.2 Typical values for the critical length (l_c) and the critical aspect ratio (l_c/D)

Matrix	Fibre	$l_c(mm)$	l_c/D
Ag	Alumina	0.4	190
Cu	Tungsten	38	20
Al	Boron	1.8	20
Epoxy	Boron	3.5	35
Epoxy	Carbon	0.2	35
Polycarbonate	Carbon	0.7	105
Polyester	Glass	0.5	40
Polypropylene	Glass	1.8	140
Alumina	SiC	0.005	10

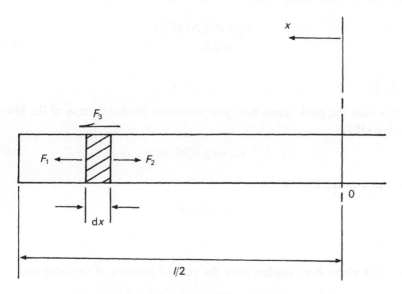

Figure 10.10 The force balance on half a fibre.

from which, by integration and simplification

$$\sigma_{Tf} = 4\tau(l/2 - x)/D. \tag{10.9}$$

Note that x is measured from the mid-length of the fibre and that τ has been assumed to be constant.

There are three possibilities for fibre length:

(a) $l < l_c$,
(b) $l = l_c$,
(c) $l > l_c$.

The stress–fibre length diagrams corresponding to these three possibilities are shown in Figure 10.11.

(a) $l < l_c$

The stress never reaches that sufficient to break the fibre and other mechanisms such as matrix failure and fibre pull-out will occur. The peak stress occurs at the centre of the fibre $(x = 0)$, and substituting for $x = 0$ in equation 10.9 gives

$$\sigma_{Tf} = 2\tau l/D. \tag{10.10}$$

The average fibre stress, $\bar{\sigma}_{Tf}$ is determined by taking the area under the stress–fibre length graph and dividing by the fibre length, so that

$$\bar{\sigma}_{Tf} = (l/2[2\tau l/Dl])/l$$
$$= \tau l/D. \tag{10.11}$$

(b) $l = l_c$

In this case the peak stress may just reach the fracture stress of the fibre at $x = 0$, so that

$$\sigma_{Tf} = \hat{\sigma}_{Tf} = 2\tau l_c/D, \tag{10.12}$$

and the average fibre stress

$$\bar{\sigma}_{Tf} = \tau l_c/D. \tag{10.13}$$

(c) $l > l_c$

The peak stress here applies over the central portion of the fibre and

$$\hat{\sigma}_{Tf} = 2\tau l_c/D. \tag{10.14}$$

The average fibre stress is given by

$$\bar{\sigma}_{Tf} = [(l - l_c) + (l_c/2)]\hat{\sigma}_{Tf}/l,$$

$$\bar{\sigma}_{Tf} = [1 - (l_c/2l)]\hat{\sigma}_{Tf}. \tag{10.15}$$

It can be seen from the last equation, and from Figure 10.12, that in order to obtain an average fibre stress close to the maximum fibre stress applicable to a continuous fibre, which is $\hat{\sigma}_{Tf}$, the fibres must be considerably longer than the critical length, e.g., according to equation 10.15 when $l = 5l_c$ the average fibre stress $\bar{\sigma}_{Tf} = 0.9\ \hat{\sigma}_{Tf}$.

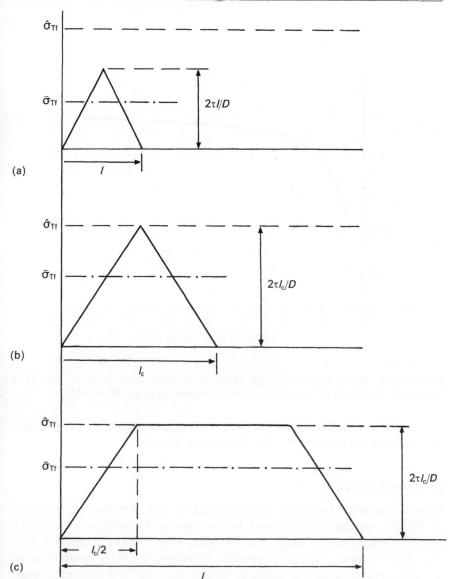

Figure 10.11 Tensile stress distribution in short fibres: (a) $l<l_c$; (b) $l=l_c$; (c) $l>l_c$. See equations 10.10 to 10.15 for the maximum stress ($\hat{\sigma}_{Tf}$) and the average stress ($\bar{\sigma}_{Tf}$) given in this figure.

10.7 STIFFNESS AND STRENGTH

Having obtained an understanding of the concept of critical fibre length and how it determines the average stress in a fibre, we are in a position to study the mechanical properties of short fibre reinforced composites. For the

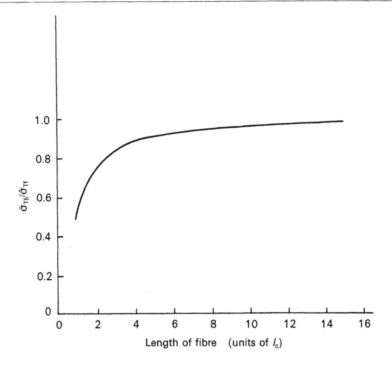

Figure 10.12 Ratio of the average stress in a short fibre ($\bar{\sigma}_{Tf}$) to that in a continuous fibre ($\hat{\sigma}_{Tf}$) as a function of length according to equation 10.15.

consideration of stiffness and strength of these composites, it is convenient to sub-divide such materials into three main classes.

(a) *Aligned systems*

These composites have the fibres aligned in one direction. In the case of PMCs this may be achieved by wet lay-up techniques with special methods employed to align the chopped fibres. Good alignment may also be obtained with compression, injection or extrusion/pultrusion methods, provided care is taken to ensure that the flow geometry aligns the fibres.

(b) *2-D composites*

These are materials with in-plane randomly oriented fibres, as in chopped strand mats and sheet moulding compounds. For CSM the reinforcing unit is the fibre bundle. This is also true to a lesser extent with SMC where the fibre distribution and bundle diameter may be influenced by the 'size' applied to the fibres.

(c) *Variable fibre orientation*

In practice it is extremely difficult to produce a composite with a completely random distribution of fibres. For example, although the fibres may be random in injection moulded thermoplastics and in bulk moulding compounds usually there is some preferred orientation of the fibres owing to variations in flow fields. Indeed, large differences in orientation may occur from one part of the moulding to another which can have a significant effect on performance.

10.7.1 Stiffness

The effective modulus of short fibre reinforced composites is bound to be lower than that for long fibre materials since the reinforcing efficiency is lower. The general case has a three-dimensional distribution of fibre orientations and a distribution of fibre lengths and, as yet, there is no satisfactory description of elastic properties in terms of these parameters. There are however reasonable models available for aligned and some non-aligned systems which we will now discuss.

(a) *Aligned systems*

For unidirectionally aligned materials with fibres of length l, a modified rule of mixtures incorporating a length efficiency parameter, η_L, is often used:

$$E_{11} = \eta_L E_f v_f + E_m (1 - v_f),   \tag{10.16}$$

where

$$\eta_L = 1 - [\tanh((1/2)\beta l)/((1/2)\beta l)],   \tag{10.17}$$

and

$$\beta = [8G_m / E_f D^2 \log_e(2R/D)]^{1/2}.   \tag{10.18}$$

E_{11} is the Young's modulus parallel to the fibres, i.e., in the longitudinal direction. $2R$ is the interfibre spacing and G_m is the shear modulus of the matrix.

When $E_f \gg E_m$ the effective modulus of a short fibre material is given by

$$E_{(short)} = \eta_L E_{(continuous)}$$

Theoretical values of η_L for several fibre lengths, calculated from equation 10.17 and assuming a strong interface, are given in Table 10.3 for two typical systems. For long fibres η_L approaches unity and hence equation 10.16 approximates to the simple law of mixtures (equation 8.5). The efficiency appears to be good provided $l > 1.0$ mm, which by comparison with the l_c value of 0.2 mm for carbon–epoxy (Table 10.2) suggests that l must be greater than about $5l_c$ for short fibres to be effective.

SHORT FIBRE COMPOSITES

Table 10.3 Values of Length Efficiency Parameter η_L. (Source: Hull, 1981)

Material	Fibre length, l(mm)	Fibre diameter, $D(\mu m)$	Volume fraction, v_f	η_L
Carbon–epoxy	0.1	8	0.3	0.20
	1.0	8	0.3	0.89
	10.0	8	0.3	0.99
Glass–nylon	0.1	11	0.3	0.21
	1.0	11	0.3	0.89
	10.0	11	0.3	0.99

Experimental results for aligned discontinuous carbon fibre/epoxy composites are in good general agreement with the theoretical predictions in that the experimental data of Table 10.4 indicate that no significant increase in efficiency is obtained by increasing fibre length from 1.0 to 6.0 mm. However the experimental efficiencies presented in Table 10.4, which were determined from comparison of the theoretical modulus for continuous fibres with the experimental modulus for the short fibre composite, are slightly less than the theoretical efficiencies given in Table 10.3 for similar length fibres. This demonstrate the importance of the orientation of the fibres as it is thought that the slight reduction in efficiency is due to imperfect fibre alignment.

Table 10.4 Experimental efficiency values for aligned short fibre carbon–epoxy. Theoretical E_{11} for continuous fibre composite calculated assuming $E_f = 390$GPa. (Source: Dingle, 1974)

Fibre length, l (mm)	Volume fraction, v_f	Theoretical E_{11} for continuous fibres (GPa)	Experimental E_{11} for discontinuous fibres (GPa)	η_L
1	0.49	194	155	0.80
4	0.32	128	112	0.87
6	0.42	167	141	0.84

An alternative expression for the longitudinal modulus has been developed as an extension to the Halpin–Tsai equations. The Halpin–Tsai equations enable the moduli (E_{11}, E_{22}, G_{12}, etc.) of aligned continuous fibre composites to be estimated and the modified expression for the Young's modulus parallel to short aligned fibres is

$$E_{11} = E_m(1 + \Psi\eta v_f)/(1 - \eta v_f), \tag{10.19}$$

where

$$\eta = [(E_f/E_m) - 1]/[(E_f/E_m) + \Psi], \tag{10.20}$$

and Ψ is a shape factor for the fibres, which in this case is equivalent to $2l/D$.

Previously it was suggested that the efficiency of short fibres is reduced by misalignment of the fibres. The magnitude of this effect can be estimated by measuring the modulus of a well aligned short fibre composite as the direction of stressing is changed. Typical results are illustrated in Figure 10.13 where the modulus of glass-reinforced epoxy composites are plotted against the angle between the stress axis and the direction of alignment of the fibres. The fall in the modulus as misorientation angle increases is significant for the short fibre composites but not as great as the theoretical prediction for a continuous fibre composite.

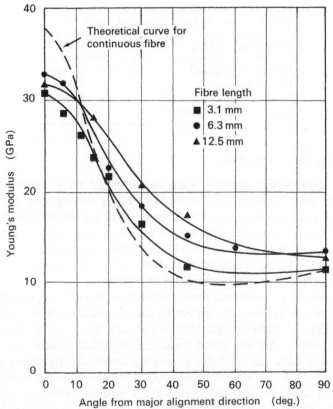

Figure 10.13 Young's modulus as a function of the angle between the stress axis and the direction of alignment of fibres in short glass-fibre reinforced epoxies. (Source: Kaier *et al.*, 1975, 1977, 1978.)

(b) *Non-aligned systems*

For non-aligned systems there is a distribution of fibre orientation and the reader will not be surprised to learn that the reinforcing efficiency of the fibres is reduced further. An additional orientation efficiency factor, η_o, is introduced into equation 10.16 to account for the further reduction in efficiency:

$$E_c = \eta_o \eta_L E_f v_f + E_m(1 - v_f). \tag{10.21}$$

Values of η_o have been calculated for simple fibre orientation distributions assuming elastic deformation of the matrix and fibres, and equality of strains. These results are given in Table 10.5 and show, for example, that the contribution from the fibres is reduced by almost a half when the orientation is changed from random-in-plane to three-dimensional random.

Table 10.5 Orientation efficiency factor η_o for several systems. (Source: Krenchel, 1964)

Orientation of fibres	η_o
Aligned-longitudinal	1
Aligned-transverse	0
Random in-plane (2–D)	0.375
Three-dimensional random	0.2

(c) *Variable fibre orientation*

For less than perfect systems, such as produced by injection moulding, where the fibre orientation distribution varies throughout the product, greater care is required in predicting and measuring properties and there is still no accepted practice. The measurement of mechanical properties on standard tensile bars alone is clearly not representative since in any other component the fibres' alignment may differ or the stresses may be applied at a different angle from the direction of alignment.

In an attempt to overcome these problems a 'bounds' approach has been suggested which involves the measurement of typical extremes of behaviour for the different fibre orientations which are likely to be found in the moulded article. An end-gated injection moulded tensile specimen can provide such data; the modulus measured longitudinally giving the upper bound and the lower bound being indicated by transverse measurements. Upper and lower bound 100-second *creep modulus* data are presented in Figure 10.14 (100-second creep modulus is the stress to produce a strain of 0.10% in 100 s). It is clear from Figure 10.14 that the use of such data in standard isotropic stress analyses can lead to significant over or under design, since the bounds are widely separated.

It has been suggested that in fairly complex three-dimensional components, high degrees of fibre alignment in one direction throughout a moulding are unlikely. The more likely situation is that layers of alignment in one direction are balanced by changes in preferred orientation, either through the thickness or from place to place in the moulding. It follows that

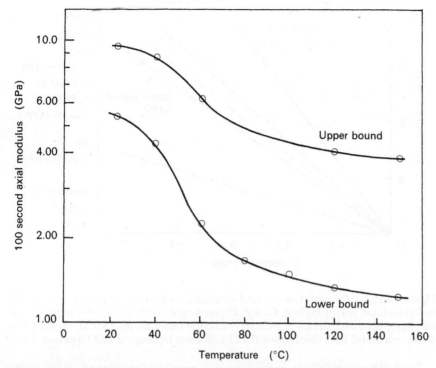

Figure 10.14 100 s isochronous creep modulus at 0.1% strain against temperature for a glass filled polyamide (nylon) showing upper and lower bounds. (Source: Darlington and Smith, 1978.)

the general deformation behaviour is some average of the response of the regions of differing fibre orientation over the area of interest. The use of creep modulus data which correspond to a fibre orientation which is randomly oriented in the plane of the moulding (RITP) has been proposed for design purposes. Such an approach appears to give quite acceptable agreement between experimental data and finite element predictions for a fairly complex moulding. Figure 10.15 shows some typical data, including upper and lower bound results which show significant divergence from the experimental data. Note the excellent agreement between experimental and predicted data for the unreinforced polypropylene.

10.7.2 Strength

(a) *Aligned systems*

The concept of average stress in a fibre was discussed earlier in section 10.6. Here, this concept is developed to embrace aligned fibre composites making use of the rule of mixtures.

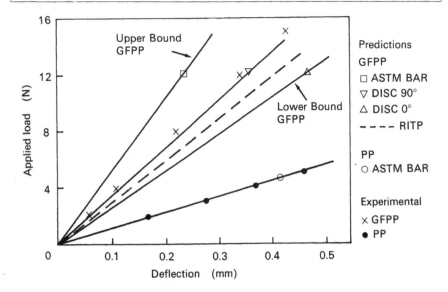

Figure 10.15 100 s isochronous load–deflection behaviour of polypropylene, PP, and glass filled polypropylene, GFPP. Experimental results for GFPP show reasonable agreement with the finite element prediction based on data for an in-plane randomly orientated fibre distribution (RITP). (Source: Darlington and Upperton, 1985.)

Essentially two failure conditions are possible for aligned short brittle fibre composites, dependent upon l, τ, $\hat{\sigma}_{Tf}$ and $\hat{\sigma}_{Tm}$. These are either fibre pull-out or fibre fracture. Fibre pull-out will occur if $l < l_c$, the stress build up in the fibres being insufficient to cause fibre fracture (Figure 10.11(a)). The average fibre stress is given in equation 10.11, and the longitudinal strength of the composite $\hat{\sigma}_{1T}$, i.e., parallel to the fibres, may be written from the rule of mixtures, such that

$$\hat{\sigma}_{1T} = (\tau l / D)v_f + \hat{\sigma}_{Tm}(1 - v_f), \tag{10.22}$$

where $\hat{\sigma}_{Tm}$ is the strength of the matrix.

On the other hand when $l > l_c$ the condition for fracture of the fibres is exceeded, and the stress pattern in the fibres is that shown in Figure 10.11(c) with the average fibre stress given by equation 10.15. Substitution of the average fibre stress into the rule of mixtures equation gives an expression for the strength of the composite:

$$\hat{\sigma}_{1T} = \hat{\sigma}_{Tf}[1 - (l_c/2l)]v_f + \sigma'_m(1 - v_f), \text{ or}$$

$$\hat{\sigma}_{1T} = 2\tau l_c / D[1 - l_c/2l]v_f + \sigma'_m(1 - v_f), \tag{10.23}$$

where σ'_m is the stress on the matrix at the failure strain of the fibres. Using this expression we can see how the ratio of the strength of an aligned short fibre composite to the strength of an aligned continuous fibre composite varies as a function of fibre length. Figure 10.16 shows that for short fibres, say $l < 5l_c$, the strength of the short fibre composite is significantly less than that of a continuous fibre composite with the same volume fraction of fibres. However, the strength of the short fibre composite increases with length of fibres for a given volume fraction, and for lengths above about $10l_c$ the difference between the strengths of the two composites becomes small.

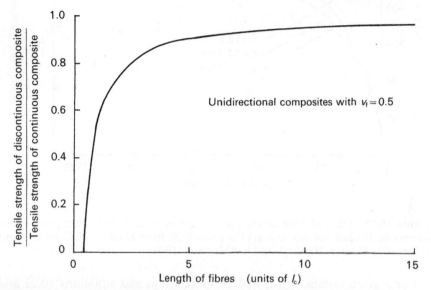

Figure 10.16 Ratio of the tensile strengths of composites containing discontinuous and continuous fibres as a function of fibre length.

Experimental data are available which confirm the deductions drawn from equation 10.23. For example Figure 10.17 presents some results from a glass fibre–epoxy composite, where $\hat{\sigma}_{Tf} = 1.8\,\text{GPa}$, $\hat{\sigma}_{Tm} = 91.5\,\text{MPa}$, $D = 12.7\,\mu\text{m}$ and $v_f = 0.26$. The experimental points are fitted by the theoretical line at $l_c = 12.7\,\text{mm}$. It is interesting to note that the strength of this discontinuous fibre material is equivalent to a continuous fibre composite at $l = 10l_c$. Also at a fibre length of $l = l_c$, the strength of the discontinuous fibre material is approximately 50% of the value achieved by the continuous fibre material, which is in accordance with equation 10.23, provided the contribution to strength from the matrix is negligible.

Distributions of fibre length may be tackled by adding the contributions from fibres in different fibre length ranges, for example above and below l_c. The simplest approach is to use the summation of the first terms in equations 10.22 and 10.23 for the contributions from the fibres of length below and above l_c respectively; this gives

$$\hat{\sigma}_{1T} = \sum_{l_i < l_c} (\tau l/D)v_i + \sum_{l_j < l_c} (2\tau l_c/D)(1 - l_c/2l_j)v_j + \sigma'_m(1 - v_f) \qquad (10.24)$$

where $\Sigma_i v_i + \Sigma_j v_j = v_f$.

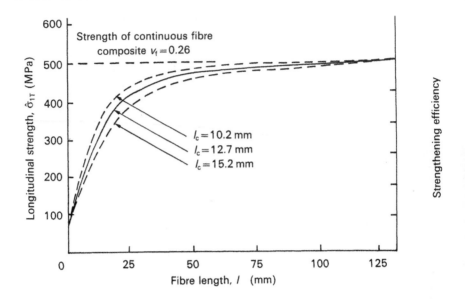

Figure 10.17 Effect of fibre length on strength of aligned short fibre glass–epoxy composite. Dashed and full lines are the predictions from equation 10.23 for different values of l_c. (Source: Hancock and Cuthbertson, 1970.)

For a given system, τ, l, l_c and D are constant and equations 10.22 and 10.23 predict that the longitudinal tensile strength of a composite is directly proportional to fibre volume fraction. Equation 10.24 indicates that this is also the case when there is a variation in fibre length provided the length distribution remains unchanged with v_f. A linear increase in $\hat{\sigma}_{1T}$ with increasing v_f has been reported for PMCs and MMCs and some typical results are shown in Figure 10.18. The intercept on the stress axis depends on the values for σ'_m or $\hat{\sigma}_{Tm}$ and therefore changes with test temperature.

Nonlinear stress–strain curves may be the consequence of a number of phenomena including creep of a resin matrix, plastic deformation of a metal matrix and distribution of fibre length. We are concerned with the last of these and a distribution in fibre length yields a stress–strain curve the slope of which decreases with increasing strain. In other words, the efficiency of the fibres decreases as the strain on the composite increases, as shown in Figure 10.9. At low strains all fibres contribute to load carrying but the contribution from the shorter fibres decreases with a given strain increment and as the strain increases only the longer fibres continue to contribute fully. This phenomenon is illustrated in Figure 10.19 for a nylon 6.6–glass fibre material.

Variations in tensile strength with angle between the fibre and loading axes for aligned discontinuous fibre composites are similar to those for continuous fibre materials (Figure 10.20). The models proposed for the

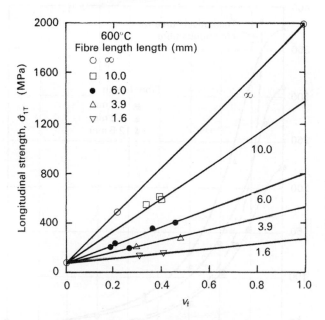

Figure 10.18 Graph showing a linear relationship between tensile strength and volume fraction of 0.2 mm diameter tungsten wire in copper tested at 600 °C. Numbers are the fibre lengths in mm. (Source: Kelly and Tyson, 1965.)

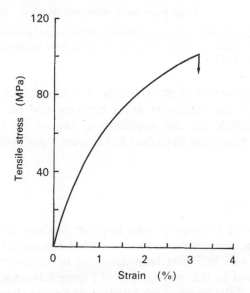

Figure 10.19 Typical stress–strain curve for aligned short glass fibre-reinforced nylon showing breakdown in linear relationship. (Source: Bowyer and Bader, 1972.)

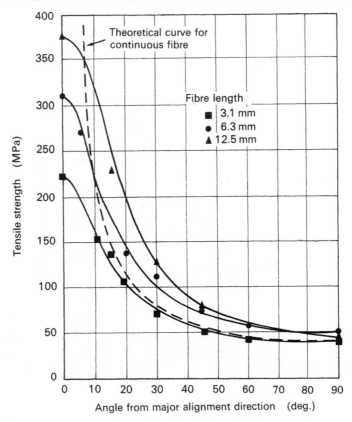

Figure 10.20 Tensile strength as a function of the angle between the stress axis and the direction of alignment of fibres in short glass fibre-reinforced epoxies. (Source: Kaier *et al*; 1975, 1977, 1978.)

orientation dependence of the strength of unidirectional continuous composites, such as the maximum stress criterion and the stress interaction (Tsai–Hill) criterion, are also applicable to aligned short fibre composites. These models have been described in Chapter 8 and will not be discussed further.

(b) *2-D composites*

An example of a composite with in-plane random fibres is the porous carbon–carbon composites, known as *carbon bonded carbon fibre* (CBCF). The structure of CBCF has been described in an earlier chapter and the reader is referred to the micrographs of Figure 4.24 which show a random arrangements of the fibres in the xy plane and some alignment of the fibres in the planes normal to that plane, i.e., zx and zy planes. The orientation of the fibres is quantified in the orientation histograms of Figure 10.21; note

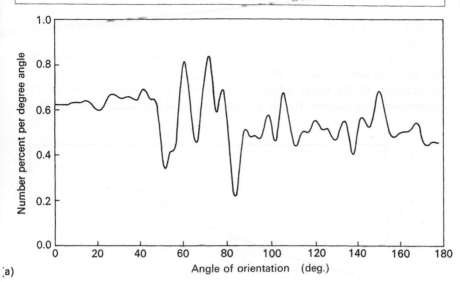

(a)

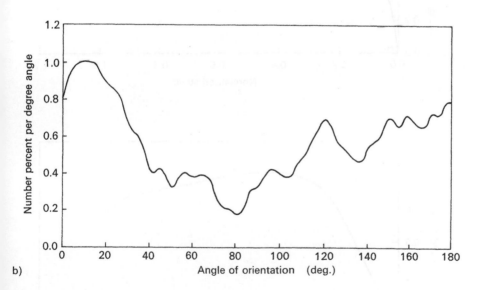

b)

Figure 10.21 Orientation histograms for carbon bonded carbon fibre material: (a) random arrangement in xy plane; (b) some preferred orientation in the zx or zy plane. (Source: Davies, 1992.)

the even distribution of orientations in the xy plane and the high frequency of the 0–30° and 150–180° orientations in the zx or zy planes.

We have already learned (section 4.4.3) that the strength is lower if a crack is propagating normal to the z-direction, in other words propagating between the randomly orientated planes. Let us now look at the stress–strain curves obtained in compression and correlate these with the microstructure. When the compressive axis is parallel to the z-direction there is a plateau in the stress–strain curve at intermediate stresses (Figure 10.22(a)). In region AB of the curve the deformation is reversible and is mainly associated with elastic deformation of the fibres which interlink the xy planes of random fibre orientation. The plateau, BC, is a consequence of permanent damage to the

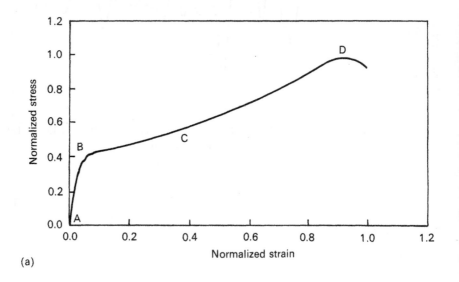

(a)

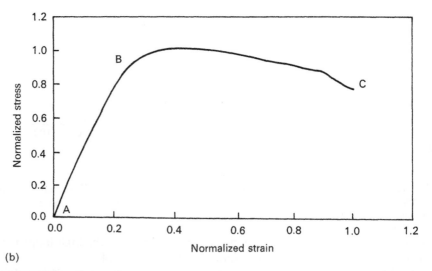

(b)

Figure 10.22 Stress–strain curves obtained from compression tests on carbon bonded carbon fibre material: (a) compressive axis parallel to z-direction; (b) compressive axis parallel to x- or y-direction. (Source: Davies, 1992.)

interlinking fibres whereas region CD is dominated by elastic deformation of the fibres and matrix in the xy planes. As can be seen from Figure 10.22(b) the stress–strain curve of a sample compressed along the x- or y-direction has fewer distinct regions of deformation: region AB is due to elastic deformation of the xy planes and BC is associated with failure within these planes. These results demonstrate the difference in mechanical properties within and perpendicular to the xy planes of random fibres.

We now turn our attention to some PMC examples of 2-D composites. In CSM the reinforcing unit becomes the fibre bundle and, to a lesser extent, this is also true of SMC. Each bundle in CSM contains typically 400 fibres; the bundles are distributed fairly randomly in the plane of sheet material giving a composite with near isotropy in the plane of the sheet. The effective aspect ratio in CSM is about $l/20D$, where l is the length of the bundle and D is the fibre diameter, compared with l/D for single fibres of the same length. It follows that pull-out of the bundles occurs at much lower stresses than for single fibres, given the same interface properties for fibres and bundles.

A typical stress–strain curve for CSM is shown in Figure 10.23, the volume fraction of fibres being 0.17 and matrix failure strain about 2%. Three stages of progressive failure can be identified. Linear behaviour below

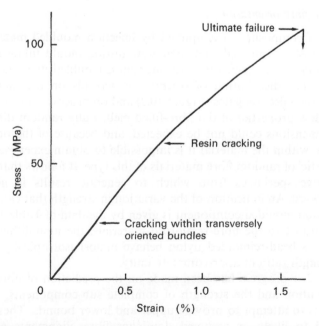

Figure 10.23 Stress–strain curve from a tensile test on CSM polyester laminate. (Source: Johnson, 1979.)

0.3% strain is followed by a slight deviation due to cracking within transversely oriented bundles. At higher strains cracking occurs in bundles oriented at other angles to the applied stress. The more rapid change of slope above 0.7% strain is due to cracking between fibre bundles. Further application of strain results in debonding, shear cracking and resin fracture until complete failure at 1.5–2.0% strain.

There is no satisfactory theory for the strength prediction of in-plane random fibre composites, although empirical approaches have been attempted by adapting the rule of mixtures. Using this approach the tensile strength of a composite, $\hat{\sigma}_{Tc}$, is given by

$$\hat{\sigma}_{Tc} = K_1 \eta_o \hat{\sigma}_{Tf} v_f, \qquad (10.25)$$

where η_o is the orientation efficiency referred to previously (see equation 10.21) and K_1 is the strengthening efficiency of short fibre bundles and is defined as

$$K_1 = \frac{[\hat{\sigma}_{1T(short)} - \sigma'_m(1 - v_f)]}{[\hat{\sigma}_{1T(continuous)} - \sigma'_m(1 - v_f)]} \qquad (10.26)$$

There are few values for K_1 and η_o available, and these are not established with any degree of certainty, and hence equation 10.26 is not widely used for the prediction of strength in undocumented 2-D composite systems.

(c) Variable fibre orientation

This class of composite is exemplified by injection moulded materials. We have previously mentioned that melt flow during mould filling results in orientation of fibres which varies throughout a moulded item and leads to anisotropy. The distribution of orientation depends on the number and position of the injection gates (Figure 10.3) and on process conditions which affect the flow properties of the fibre-filled melt. Fully random distribution in three dimensions could not be expected, and because of the orientation distribution within the material it is impossible to obtain experimental data 'characteristic' of random fibre materials of this type. It follows that selecting representative specimens from which to generate results is not at all straightforward. An indication of the variation in strength that can be found in an injection moulded component is given by the data of Table 10.6. Note that, in contrast to the fibre-reinforced materials, the monolithic matrices and the glass bead-reinforced nylon behave almost isotropically, i.e., they have a strength ratio of approximately unity.

Until further progress has been made on the prediction of fibre orientation distribution and the strength of complete sub-components, the safest procedure is to attempt to provide upper and lower bounds. These bounds correspond to likely or assumed dominant fibre alignments along and transverse to applied stress fields. Design is then based on the lower bound

Table 10.6 Flexural strength of specimens cut with their lengths parallel and perpendicular to the direction of flow in an injection moulded component (P. J. Cloud and F. C. McDowell, 1979)

Matrix	Reinforcement	Flexural strength (MPa)		Strength ratio parallel to perpendicular
		Parallel	Perpendicular	
Nylon		110	115	0.96
Nylon 6.6	15% glass fibre	180	150	1.20
Nylon 6.6	30% glass fibre	240	175	1.37
Nylon 6.6	40% glass beads	130	140	0.93
Nylon 6.6	40% carbon fibre	300	190	1.58
Polycarbonate		100	105	0.95
Polycarbonate	30% glass fibre	190	145	1.31

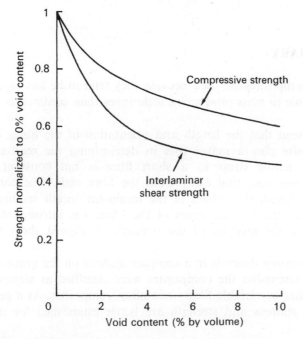

Figure 10.24 Graph illustrating the marked effect of void content on the strength of GRP. (Source: Potter, 1989.)

unless experience suggests this to be unduly pessimistic. The results of this type of approach to design are likely to be somewhat conservative, and may in some circumstances lead to substantial over-design, partly from the choice of the lower bound for design, but also stemming from the likely choice of failure criterion used. In some circumstances, such as failure due to bending, an elastic–plastic failure analysis appears to be more appropriate than the, perhaps more normal, maximum stress or maximum strain criteria.

(d) *Defects*

So far the only structural features which we have considered are the volume fraction, length and the orientation of the fibres. It is worth recalling that composites, in common with other materials, usually contain defects which may markedly affect mechanical performance. For example in fibre-reinforced polymers a relatively low concentration of small voids can significantly degrade strength as illustrated by typical results for GRP in Figure 10.24. These results extend up to 10% void content but normal good practice should keep the void content to less than 5% and more sophisticated techniques, which are used for the production of critical components, can reduce the content to about 0.5%. The detection of defects in composites by non-destructive testing will be the subject of the final chapter of this book.

10.8 SUMMARY

Short fibre composites are used because they tend to be less expensive and more amenable to mass-production techniques than continuous fibre composites.

We have seen that the length and orientation of the fibres in a short fibre composite play a major role in determining the mechanical properties. The tensile stress in a short fibre is not constant with the maximum stress occurring away from the fibre ends. The concept of a critical fibre length (l_c), which is the minimum length required for the stress to reach the fracture stress of the fibre, was introduced and used particularly in the analysis of the strength of aligned short fibre composites.

Fibre orientation depends in a complex manner on the processing route. For ease of discussion the composites were classified as aligned systems, 2-D composites or variable fibre orientation composites. As a general rule, analyses for stiffness and strength are better established for the aligned systems.

FURTHER READING

General

Chawla, K. K. (1987) *Composite Materials – Science and Engineering*, Springer-Verlag, New York.
ASM (1987) *Engineered Materials Handbook, Vol. 2, Composites*. ASM International.

Specific

Akay, M. and Barkley, D. (1985) *Composite Structures*, Vol. 3, p. 269.

Bowyer, W. H. and Bader, M. (1972) *J. Mat. Sci.*, **7**, 1315.

Cloud, P. J. and McDowell, F. C. (1979) *The Effects of Fibre Orientation on Physical Properties*, Plastic Des. Forum.

Cox, H. L. (1952) *Brit. J. Appl. Phys.*, **3**, 72.

Darlington, M. W. and Upperton, P. H. (1985) *Plastics and Rubber Int.*, **10**, 35.

Darlington, M. W. and Smith, G. R. (1978) *Plastics and Rubber: Materials and Applications*, Vol. 3, p. 97.

Davies, I. J. (1992) *Ph.D. Thesis*, Imperial College of Science, Technology and Medicine, London University.

Dingle, L. E. (1974) Aligned discontinuous carbon fibre composites. *Proc. 4th Int. Conf. on Carbon Fibres, their Composites and Applications*, Plastics Institute, London.

Hancock, P. and Cuthbertson, R. C. (1970) *J. Mat. Sci.*, **5**, 762.

Hull, D., (1981) *An Introduction to Composite Materials*, Cambridge University Press.

Johnson, A. F. (1979) *Engineering Design Properties of GRP*, British Plastics Federation, London.

Kaier, L., Narkis, M. and Ishai, O. (1975) *Polym. Eng. Sci.*, **15**, 525, 532; (1977) *Polym. Eng. Sci.*, **17**, 234; (1978) *Polym. Eng. Sci.*, **18**, 45.

Kelly, A. and Tyson, W. R. (1965) Fibre-strengthened materials, in *High Strength Materials*, (ed. V. F. Zackay), Wiley and Sons. p. 578.

Pennington, D. (1979) *Ph.D. Thesis*, University of Liverpool.

Krenchel, H. (1964) *Fibre Reinforcement*, Akademisk Forlag, Copenhagen.

Potter, R. T. (1989) *Concise Encyclopedia of Composite Materials*, (ed. A. Kelly), Pergamon Press, p. 248.

Young, R. J., Galiotis, C. and Batchelder, D. N. (1985) Measurement of fibre strain in all-polymer composites, *6th Int. Conf. on Def., Yield and Fract.*, Cambridge, UK.

PROBLEMS

10.1. Define the critical fibre length.

An aligned short fibre composite consists of 40 vol. % carbon fibres of length 2 mm and diameter 7 µm in a polycarbonate matrix. The tensile strength of the fibres and the shear strength of the fibre–matrix interface are 2.5 GPa and 12.5 MPa respectively. Calculate the critical fibre length, l_c, and then estimate the longitudinal tensile strength of the composite given that the stress on the matrix at the failure strain of the fibres is 30 MPa.

10.2 Explain why the stiffness of a discontinuous fibre composite is less than that of a continuous fibre composite with the same constituents and same proportion of reinforcement.

Calculate the longitudinal Young's modulus, E_{11}, for an aligned short fibre carbon–epoxy composite with 42 vol. % reinforcement given that Young's modulus of the fibre and matrix are 390 GPa and 5.5 GPa respectively and that the length efficiency parameter, η_L, is 0.84.

SELF-ASSESSMENT QUESTIONS

Indicate whether statements 1 to 7 are true or false.

1. A short fibre composite is one with a fibre length less than 10 mm.

    (A) true
    (B) false

2. In the context of short fibre composites, the aspect ratio is defined as l^2/D and thus has units of length. (l and D are fibre length and diameter respectively.)

    (A) true
    (B) false

3. If the fibre length is less than l_c, where l_c is the critical fibre length, the tensile stress in the fibre never reaches the fibre fracture stress.

    (A) true
    (B) false

4. 2-D composites have in-plane randomly orientated fibres.

    (A) true
    (B) false

5. If the Young's modulus of the fibres is much greater than that of the matrix the effective modulus of an aligned short fibre composite is given by
$E_{(\text{short})} = \eta_L E_{(\text{continuous})}$

    (A) true
    (B) false

6. In contrast to the behaviour of aligned continuous fibre composites, the tensile strength of an aligned discontinuous (short) fibre composite is independent of the angle between the fibre and the loading axis.

    (A) true
    (B) false

7. In variable fibre orientation composites, such as injection moulded materials, the safest procedure is to obtain upper and lower bounds for strength and then base design on the lower bound unless experience suggests this is unduly pessimistic.

    (A) true
    (B) false

For each of the statements of questions 8 to 11, one or more of the completions given are correct. Mark the correct completions.

8. The main reasons for the widespread use of short fibre composites are

(A) it is easier to align short fibres than continuous fibres,
(B) short fibres are generally less expensive than continuous fibres,
(C) short fibre composites are always isotropic,
(D) manufacturing processes for continuous fibre composites tend to be slow and inflexible,
(E) the longitudinal strength of an aligned short fibre composite is greater than that of an aligned continuous fibre composite, with the same volume fraction of fibres.

9. If the fibre length $l > l_c$, where l_c is the critical length, the average fibre stress is

(A) less than the fibre fracture stress,
(B) more than the fibre fracture stress,
(C) proportional to the fibre fracture stress,
(D) independent of l,
(E) increases with increasing l.

10. Young's modulus E_{11} parallel to the fibres in an aligned short fibre composite is given by

(A) $E_{11} = \eta_L E_f v_f - E_m(1 - v_f)$ where v is volume fraction, subscripts f and m refer to fibre and matrix respectively and η_L is the length efficiency parameter,
(B) $E_{11} = \eta_L E_f v_f + E_m(1 - v_f)$,
(C) $E_{11} = \eta_L E_f v_f + E_m(v_m)$,
(D) a modified law of mixtures incorporating a length efficiency parameter,
(E) a modified maximum stress criterion incorporating a length efficiency parameter.

11. The longitudinal strength $\hat{\sigma}_{1T}$ of an aligned short brittle fibre composite

(A) is greater than that of an aligned continuous fibre composite,
(B) is less than that of an aligned continuous fibre composite,
(C) is approximately half that of an aligned continuous fibre composite for $l = l_c$ where l and l_c are fibre length and critical fibre length respectively,
(D) is approximately twice that of an aligned continuous fibre composite for $l = l_c$ where l and l_c are fibre length and critical fibre length respectively,
(E) approaches that of an aligned continuous fibre composite when $l > 10l_c$,
(F) approaches that of a random short fibre composite when $l > 10l_c$.

Each of the sentences in questions 12 to 15 consists of an assertion followed by a reason. Answer:

(A) if both assertion and reason are true statements and the reason is a correct explanation of the assertion,

(B) if both assertion and reason are true statements but the reason is not a true explanation of the assertion,

(C) if the assertion is true but the reason is a false statement,

(D) if the assertion is false but the reason is a true statement,

(E) if both the assertion and reason are false statements.

12. A weld-line may be found in a component which has been injection moulded using a twin-gated mould *because* jetting of the flow from each gate may occur thus restricting mixing of the fibres across the centre of the component.

13. The critical length l_c may readily be calculated from micrographs *because* l_c is simply the average length of the fibres aligned with the stress axis.

14. The average stress in a short fibre is less than the maximum stress that can be achieved in a continuous fibre *because* the stiffness of short fibres is usually a factor of four less than that of a continuous fibre.

15. Porous carbon–carbon composite, known as carbon bonded carbon fibre (CBCF), is a good example of an aligned short fibre composite *because* there is a random arrangement of the fibres in the xy plane and some alignment of the fibres in planes zx and xy normal to that plane.

16. Select the correct word from each of the groups given in italics in the following passage.

The shear stress around a short fibre is (*zero/maximum/minimum*) at the fibre ends and (*almost zero/maximum/minimum*) in the centre. The tensile stress in a fibre is (*zero/maximum/minimum*) at the fibre ends, and (*zero/maximum/minimum*) at the centre of the fibre. This means the reinforcing efficiency of short fibres (*remains constant/increases/decreases*) with decreasing fibre length. It is often useful to consider fibre length in units of the critical fibre length which is the (*minimum/maximum*) fibre length required for the stress to reach the fracture stress of the fibre.

ANSWERS

Problems

10.1 $l_c = 0.7$ mm, therefore $l > l_c$. Longitudinal strength of composite is 843 MPa.

10.2 141 GPA.

Self-assessment

1. B; 2. B; 3. A; 4. A; 5. A; 6. B; 7. A; 8. B,D; 9. A,C,E; 10. B,C,D; 11. B,C,E; 12. A; 13. E; 14. C; 15. D; 16. maximum, almost zero, zero, maximum, decreases, minimum.

11 Fracture mechanics and toughening mechanisms

11.1 INTRODUCTION

This chapter first gives a brief introduction to fracture mechanics and then discusses the various toughening mechanisms that may be operative in composite materials.

Concerning the fracture mechanics, we shall assume that the materials behave in a linear elastic fashion so that linear elastic fracture mechanics (LEFM) may be employed. The whole subject is based on the concept that there are always cracks present and that we seek to describe the conditions under which they will propagate. We shall do this in terms of energy per unit area, G_C, to create the new crack surfaces and to produce any associated deformation and the closely related parameter K_C, the critical stress intensity factor. The easiest approach to the subject, and the one that we shall follow, is to first consider the elastic energy in the system.

11.2 ENERGY ANALYSIS

Elastic strain energy is stored in a body when it is deformed elastically. This energy is released as the crack grows through the body and is used for the creation of new surfaces associated with the crack.

Consider an elastic body having a thickness B and containing a crack of length a. If we apply a load P and the body deforms elastically, a linear load–deflection curve, as shown in Figure 11.1., is obtained. From this curve we may determine the *compliance C* which is defined as

$$C = u/P \tag{11.1}$$

where u is the deflection.

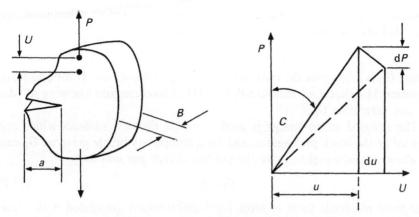

Figure 11.1 General loading on a body of thickness B and crack length a resulting in a linear load–deflection curve (full line). The dotted load–deflection curve is for the same body with a crack of length $a+\mathrm{d}a$.

If the crack grows an amount $\mathrm{d}a$ then the crack area increases by $B\,\mathrm{d}a$. Let us also suppose the load changes by $\mathrm{d}P$ and the displacement by $\mathrm{d}u$. We may now write expressions for the various energies as follows.

Initial energy stored, $U1=(1/2)Pu$, i.e., the area under the initial (full line in Figure 11.1) load–deflection curve.

Final energy stored, $U2=(1/2)(P+\mathrm{d}P)(u+\mathrm{d}u)$, i.e., the area under the dotted load–deflection curve of Figure 11.1.

External work done, $U3=(P+\mathrm{d}P/2)\,\mathrm{d}u$.

Energy released, $\mathrm{d}U = U1 + U3 - U2$, which on substitution for $U1$, $U2$ and $U3$ gives

$$\mathrm{d}U =(1/2)Pu+(P+\mathrm{d}P/2)\,\mathrm{d}u-(1/2)(P+\mathrm{d}P)(u+\mathrm{d}u)$$
$$=(1/2)(P\,\mathrm{d}u-u\,\mathrm{d}P),$$

if we neglect the product of small quantities.

We now define the *strain energy release rate G* as

$$G=\frac{1}{B}\frac{\mathrm{d}U}{\mathrm{d}a}=\frac{1}{2B}\left(P\frac{\mathrm{d}u}{\mathrm{d}a}-u\frac{\mathrm{d}P}{\mathrm{d}a}\right) \qquad (11.2)$$

From equation 11.1 we have, after differentiating with respect of a,

$$P^2\frac{\mathrm{d}C}{\mathrm{d}a}=P\frac{\mathrm{d}u}{\mathrm{d}a}-u\frac{\mathrm{d}P}{\mathrm{d}a} \qquad (11.3)$$

and on substituting into equation 11.2,

$$G=\frac{P^2}{2B}\frac{\mathrm{d}C}{\mathrm{d}a}=\frac{u^2}{2BC^2}\frac{\mathrm{d}C}{\mathrm{d}a}=\frac{U1}{BC}\frac{\mathrm{d}C}{\mathrm{d}a}. \qquad (11.4)$$

Thus if we know C as a function of a, from experiment or calculation, we may find dC/da. Furthermore, at fracture we say that

$$G = G_C,$$

where G_C is known as the *critical strain energy release rate*. It follows from equation 11.4 that if we measure P, u or $U1$ at fracture, and knowing dC/da, we can determine G_C.

The released strain energy is used to create the new surfaces which are formed as the crack propagates, and for a completely brittle material G_C can be shown to be equal to twice the surface energy per unit area, γ:

$$G_C = 2\gamma \tag{11.5}$$

For most materials there is some local deformation associated with crack propagation and this also requires energy, hence G_C is greater than 2γ.

Numerous forms of equation 11.4 are available for different specimen geometries where C and dC/da are written in terms of dimensions of the specimen and Young's modulus E. For example consider the *double cantilever beam*, DCB, specimen shown in Figure 11.2. The two arms may be treated, to a first approximation, as two slender cantilevers which may be analysed using conventional beam theory to give

$$u = \frac{8a^3P}{EBD^3}$$

hence

$$C = \frac{8a^3}{EBD^3} \text{ and } \frac{dC}{da} = \frac{24a^2}{EBD^3},$$

and therefore, from equation 11.4

$$G = \frac{12P^2a^2}{EB^2D^3} = \frac{3ED^3u^2}{16a^4} = \frac{U1}{B}\left(\frac{3}{a}\right). \tag{11.6}$$

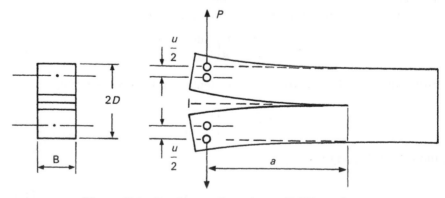

Figure 11.2 Double cantilever beam (DCB) specimen.

Note that G increases with crack length, a, for a fixed P and decreases for a fixed u. It is possible to design specimens to give G independent of a so that G, for example, is constant for a fixed load. One such geometry is the *double torsion* specimen for which,

$$G = \frac{3(1+v)l^2 P^2}{2EB^3 D B_{\mathrm{C}}},$$

where P, E and v are load, Young's modulus and Poisson's ratio respectively, l and B are specimen dimensions as defined in Figure 11.3, and B_{C} is the crack thickness when there are side grooves to guide the crack.

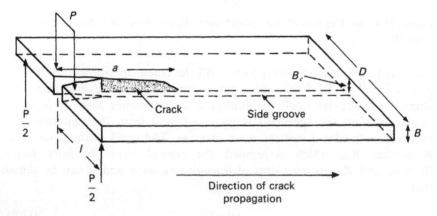

Figure 11.3 Double torsion, DT, specimen (with side grooves to guide crack) for which G is independent of crack length a.

11.3 LOCAL STRESSES

In some configurations it is not possible to compute C and we must thus look at local stresses around the crack tip. The crack acts as a stress concentrator giving high stresses at the crack tip as shown in Figure 11.4. It is possible to use elasticity theory to calculate the stresses at any point (r, θ) in the vicinity of the crack tip, and the results of such an analysis are

$$\sigma_{yy} = \frac{K}{(2\pi r)^{1/2}} \cos \tfrac{1}{2}\theta (1 + \sin \tfrac{1}{2}\theta \sin \tfrac{3}{2}\theta),$$

$$\sigma_{xx} = \frac{K}{(2\pi r)^{1/2}} \cos \tfrac{1}{2}\theta (1 - \sin \tfrac{1}{2}\theta \sin \tfrac{3}{2}\theta), \tag{11.7}$$

$$\tau_{xy} = \frac{K}{(2\pi r)^{1/2}} (\cos \tfrac{1}{2}\theta \sin \tfrac{1}{2}\theta \cos \tfrac{3}{2}\theta),$$

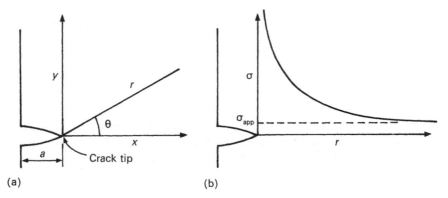

Figure 11.4   (a) The coordinate system used; (b) the stress as a function of r at a crack tip.

where K is the *stress intensity factor*. All the crack tip stresses are directly proportional to K. Note also that, according to these equations, the stresses are singular (tend to infinity) as r approaches zero. K increases as the load on the specimen is increased, and from the equations this means that the local stresses must increase. The crack propagates when K reaches K_C, which is termed the *critical stress intensity factor*. Both G_C and K_C are measures of *fracture toughness* and it can be shown that

$$K_C^2 = EG_C \tag{11.8a}$$

for isotropic materials and plane stress conditions. The corresponding equation for plane strain is

$$K_C^2 = \frac{EG_C}{(1 - v^2)}. \tag{11.8b}$$

In fact there is very little difference between equations 11.8a and 11.8b as Poisson's ratio v is less than unity, typically around 0.25, therefore v^2 is only about 0.06.

For an infinite plate containing a crack of length $2a$ as shown in Figure 11.5 we have that

$$K^2 = \pi \sigma^2 a \tag{11.9}$$

and it follows that

$$G = \frac{\pi \sigma^2 a}{E} \tag{11.10}$$

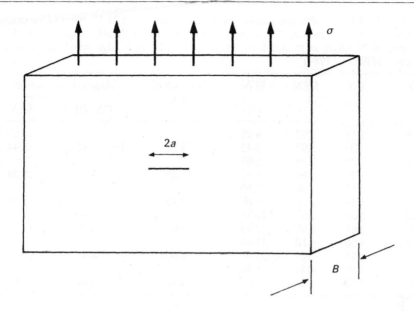

Figure 11.5 Infinite plate containing a crack of length $2a$.

for this case. Note that at fracture, that is when the crack propagates, σ is the fracture stress σ_F and $G = G_C$. If we now substitute for G_C from equation 11.5 we obtain

$$\sigma_F = \left(\frac{2\gamma E}{\pi a}\right)^{1/2},$$ (11.11)

which is the well known *Griffith* equation.

π is the calibration factor for the infinite plate but often the crack is not small in comparison with the specimen and then the more general form of equation 11.9 has to be used:

$$K^2 = Y^2 \sigma^2 a,$$ (11.12)

where Y is a calibration factor that depends on the crack length and specimen dimension D. Values for Y are available for most geometries and Table 11.1 gives some examples.

For anisotropic materials equation 11.8a is replaced by

$$G_C = K_C^2 \left(\frac{S_{11}S_{22}}{2}\right)^{1/2} \left[\left(\frac{S_{22}}{S_{11}}\right)^{1/2} \left(\frac{2S_{12} + S_{33}}{2S_{11}}\right)\right]^{1/2},$$ (11.13)

where S_{11}, etc. are the orthotropic elastic compliances (Chapter 7). The Y^2 functions are approximately the same as for the isotropic case.

Table 11.1 Finite width correction factors: Y^2 in equation 11.12, where CN–centre notched plate in tension, CT–compact tension, DEN–double edge notched in tension, SEN–single edge notch in tension, SENB–single edge notch in bending. (Source: Williams, 1984)

a/D	CN	DEN	SEN	Tan formula CN	Approximate method CN, DEN	Approximate method SEN
0.05	3.17	3.97	4.05			
0.1	3.20	3.99	4.42	3.18	3.32	3.94
0.15	3.25	3.99	5.05			
0.2	3.30	3.98	5.92	3.25	3.55	5.28
0.25	3.37	3.96	7.10			
0.3	3.48	3.96	8.69	3.40	3.85	7.71
0.35	3.62	3.97	10.86			
0.4	3.81	4.01	13.94	3.63	4.28	12.56
0.45	4.06	4.10	18.44			
0.5	4.38	4.23	25.16	4.00	4.91	23.76
0.55	4.80	4.43	35.36			
0.6	5.32	4.71	51.03	4.59	5.94	55.09

a/D	CT $H/D=0.6$	Pure bending	SENB $S/D=4$	SENB $S/D=8$
0.05	–	3.60	3.28	3.44
0.1	–	3.43	3.05	3.24
0.15	–	3.41	2.97	3.19
0.2	–	3.49	3.02	3.26
0.25	–	3.68	3.16	3.42
0.3	112.4	3.97	3.41	3.69
0.35	121.0	4.38	3.79	4.08
0.4	132.3	4.97	4.33	4.65
0.45	153.8	5.81	5.11	5.46
0.5	182.3	7.03	6.26	6.65
0.55	246.5	8.82	7.95	8.40
0.6	302.8	11.47	10.49	11.00

11.4 OTHER PARAMETERS

11.4.1 Damage zone

A volume of damage occurs at the crack tip due to the high stresses (Figure 11.6). If the stress to give damage is σ_D (for debonding, etc.) then from equation 11.7 we have the size of the *damage zone*, r_D, as

$$r_C = \frac{K_C^2}{2\pi\sigma_D^2}. \tag{11.14}$$

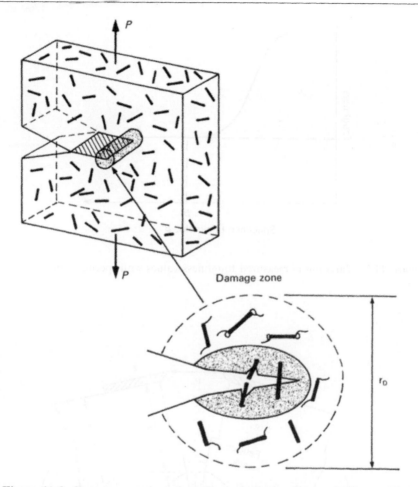

Figure 11.6 Damage zone in a short-fibre composite. (Source: Williams, 1990.)

For valid K_C and G_C values (i.e., for LEFM to apply), r_D must be much smaller than the dimensions of the specimen, B and D. Anomalously high values for toughness are obtained for small specimens as shown in Figure 11.7. The general criterion adopted for a fracture toughness to be considered to be valid is that

$$B, D \text{ and } a > 2.5(K_C/\sigma_D)^2. \tag{11.15}$$

Clearly the tougher the material for a given strength the larger the specimen required for a valid determination of K_C or G_C.

11.4.2 Crack opening displacement (COD)

The distance apart of the faces of a crack, which is termed the *crack opening displacement* (COD), is another useful parameter. The crack opening

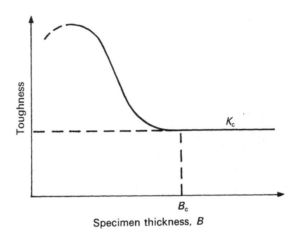

Figure 11.7   Variation of measured toughness values with specimen dimensions.

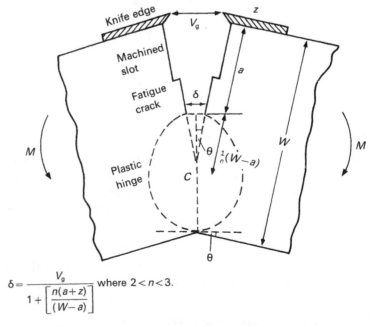

$$\delta = \frac{V_g}{1 + \left[\dfrac{n(a+z)}{(W-a)}\right]} \quad \text{where } 2 < n < 3.$$

Figure 11.8   Relationship between displacement at the crack tip, δ, and the displacement at the crack mouth, v_g, as measured using knife-edges on the surface of a bend specimen. (Source: Knott, 1973.)

displacement at the crack tip is usually given the symbol δ. On loading a body with a crack, the crack will open prior to crack extension until a critical value for δ is reached, at which point the crack will propagate. The critical displacement at the crack tip, δ_C, is a measure of toughness which is often used for materials such as metals, that yield prior to fracture. The relationship between δ_C and the previously discussed fracture toughness parameters G_C and K_C is

$$\delta_C = G_C/\sigma_D = K_C^2/(E\sigma_D). \tag{11.16}$$

It is not normally feasible to measure δ at the crack tip directly. The usual procedure is to measure the displacement, V_g, of the crack faces at the mouth of the crack and to calculate δ at the crack tip from V_g as illustrated in Figure 11.8.

11.5 FRACTURE INITIATION

In this section we will show how the concepts introduced previously may be used to determine the fracture toughness at crack initiation. We may determine K_C by taking a series of specimens with various crack lengths, a, and then testing them so that the stress at fracture σ_F can be found. Using equation 11.12 we can plot $\sigma_F^2 Y^2$ versus a^{-1} (or $\sigma_F Y$ versus $a^{-1/2}$) to obtain a straight line from which we can calculate K_C, as the gradient is K_C^2. For DCB specimens we calibrate directly to obtain dC/da. An example is given in Figure 11.9 for a random glass fibre filled polyester resin for which K_C is $5.5\ \mathrm{MPa\,m}^{1/2}$ and G_C approximately $3000\ \mathrm{J/m}^2$. Note that for polyesters without fibres the toughness is much lower, $K_C = 0.6\ \mathrm{MPa\,m}^{1/2}$, giving $G_C = 120\ \mathrm{J/m}^2$.

11.6 IMPACT

Many mechanical properties are strain rate sensitive and, as a general rule, materials become more brittle the faster the application of load. The toughness at the fast strain rate corresponding to impact may be assessed using standard impact tests, such as the *Izod-type* test, on specimens of fixed size and geometry. Such tests yield only semi-quantitative data in that the values obtained are not true material parameters but only apply for the specimen size and geometry, and testing conditions employed.

A better approach is based on the energy analysis described in section 11.2 and involves instrumented impact testing. We return to equation 11.4 and use

$$U1 = GBD\phi, \tag{11.17}$$

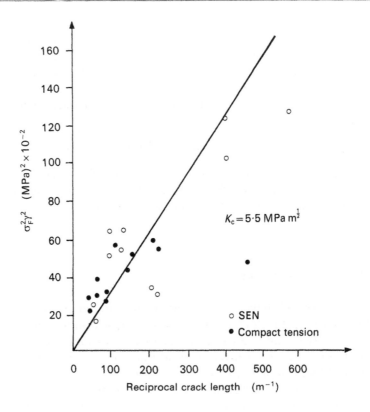

Figure 11.9 Determination of K_c for a random glass fibre filled polyester resin from the slope of a plot of $\sigma_F^2 Y^2$ versus a^{-1}. (Courtesy J. G. Williams.)

where $\phi = C/(\mathrm{d}C/\mathrm{d}(a/D))$ and is the energy calibration factor for impact (it is the impact equivalent to Y which was discussed earlier). Table 11.2 gives values for ϕ for three different impact specimen geometries that are shown, together with some other geometries, in Figure 11.10. We then plot $U1$, the impact fracture energy, against $BD\phi$ and calculate G_C from the slope as shown for polypropylene reinforced with 20% glass fibres in Figure 11.11. Note that there is a positive intercept on the energy axis from kinetic energy.

11.7 SLOW CRACK GROWTH AND CRACK OPENING MODES

Slow crack growth can occur at K values less than K_C. For all materials there is a unique K versus crack velocity (v) relationship for a given test environment. This relationship is often of the form

$$v = AK^n,$$ (11.18)

Table 11.2 Energy calibration factor ϕ (see equation 11.17) for three impact specimen geometries, where a = crack length, L = specimen length, D = specimen width. (Source: Williams, 1984)

	Charpy energy calibration factor, ϕ				
a/D	$L/D=4$	$L/D=6$[a]	$L/D=8$	$L/D=10$[a]	$L/D=12$[a]
0.05	1.383	1.996	2.612	3.231	3.846
0.1	0.782	1.101	1.424	1.748	2.070
0.15	0.576	0.789	1.005	1.224	1.438
0.2	0.469	0.624	0.781	0.942	1.100
0.25	0.402	0.520	0.639	0.761	0.880
0.3	0.354	0.446	0.538	0.633	0.724
0.35	0.318	0.389	0.461	0.535	0.605
0.4	0.287	0.343	0.398	0.455	0.510
0.45	0.260	0.302	0.345	0.388	0.430
0.5	0.234	0.266	0.298	0.331	0.362
0.55	0.210	0.234	0.257	0.280	0.304
0.6	0.187	0.205	0.222	0.239	0.255

[a]from interpolated Y values

	Izod energy calibration factor, ϕ				
a/D	$L/D=4$	$L/D=6$	$L/D=7$	$L/D=9$	$L/D=11$
0.05	1.738	1.990	2.125	2.500	–
0.1	1.060	1.165	1.230	1.360	1.570
0.15	0.748	0.834	0.874	0.953	1.050
0.2	0.600	0.642	0.670	0.730	0.810
0.25	0.511	0.540	0.555	0.600	0.679
0.3	0.452	0.480	0.489	0.519	0.587
0.35	0.410	0.438	0.447	0.473	0.525
0.4	0.387	0.410	0.420	0.441	0.478
0.45	0.370	0.391	0.398	0.419	0.440
0.5	0.360	0.379	0.385	0.399	0.411

	SEN tension energy calibration factor, ϕ			
a/D	$L/D=4$	$L/D=8$	$L/D=16$	$L/D=24$
0.04	12.50	24.99	49.96	74.93
0.08	5.93	11.83	23.62	35.41
0.1	4.57	9.09	18.14	27.18
0.2	1.77	3.46	6.83	10.21
0.3	0.87	1.64	3.18	4.71
0.4	0.48	0.84	1.55	2.27
0.5	0.28	0.44	0.76	1.08
0.6	0.18	0.25	0.38	0.51
0.8	0.12	0.13	0.15	0.17

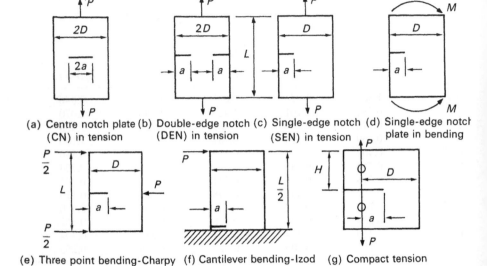

(a) Centre notch plate (CN) in tension   (b) Double-edge notch (DEN) in tension   (c) Single-edge notch (SEN) in tension   (d) Single-edge notch plate in bending

(e) Three point bending-Charpy   (f) Cantilever bending-Izod   (g) Compact tension

Figure 11.10   Some common test geometries.

where n is a constant. The relationship between v and K may be determined by a number of techniques but the most convenient use constant G or K specimens such as employed in the double torsion, DT, test (Figure 11.3). As slow crack growth tends to be sensitive to the environment, v versus K plots are often used for environmental testing. Figure 11.12 shows data from DT tests on zirconia-toughened alumina tested in different liquids. It can be seen that at a given K the crack velocity can vary by as much as two orders of magnitude depending on the liquid environment.

It is possible to predict the lifetime of a component subjected to a service stress of σ from equation 11.18. Lifetime t_f can be considered to consist of the time to initiate a crack, t_i, plus the time to grow that crack to the critical size at which unstable crack growth occurs, t_p, that is

$$t_f = t_i + t_p.$$

In most components there are inherent flaws and so the time to initiate a crack tends to zero and may be neglected. We therefore have

$$t_f = t_p.$$

The time to grow an inherent flaw of size a_0 to the critical flaw size, a_c, may be obtained by integration of equation 11.18:

$$t_f = t_p = \int_{a_0}^{a_c} \frac{1}{A K_C^n} \, da.$$

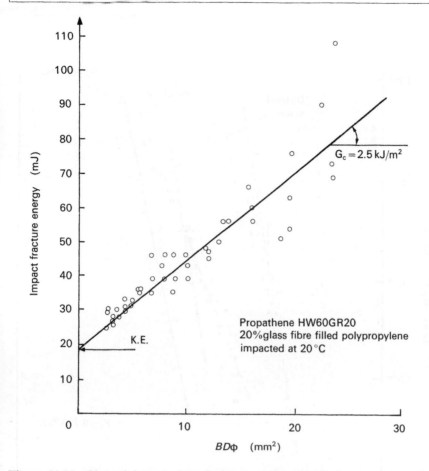

Figure 11.11 Plot of impact data (using equation 11.17) from polypropylene reinforced with 20% glass fibres to obtain G_c. (Courtesy J. G. Williams.)

We can substitute for K from equation 11.12 to give

$$t_f = t_p = \int_{a_0}^{a_C} \frac{1}{A Y^n \sigma^n a^{n/2}} \, da,$$

which yields on integration:

$$t_f = \frac{1}{(A Y^n \sigma^n)} \frac{2}{(n-2)} (a_0^{((2-n)/2)} - a_C^{((2-n)/2)}).$$

However usually $u_C \gg a_0$ and $n \geqslant 10$, therefore the a_C term can be ignored and the lifetime is given by

$$t_f = \frac{1}{(A Y^n \sigma^n)} \frac{2}{(n-2)} a_0^{((2-n)/2)}$$

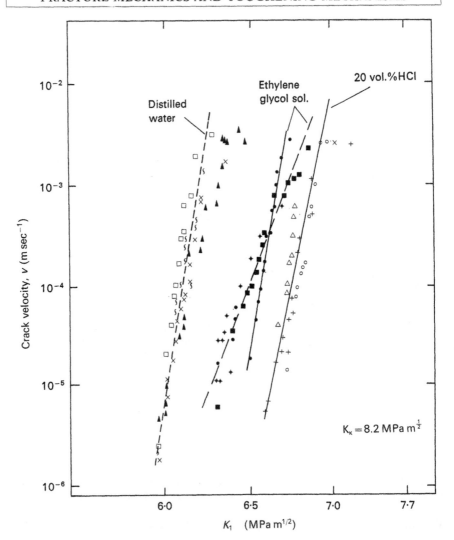

Figure 11.12 Plot of crack velocity versus K obtained from double torsion tests on zirconia-toughened alumina in different liquid environments. (Source: Thompson and Rawlings, 1992.)

Similar tests can be performed in fatigue loading to obtain the crack growth rate per cycle, da/dN, as a function of the stress intensity factor range ΔK, i.e. $da/dN - \Delta K$ curves as exemplified by the results for polymer matrix composite presented in an earlier chapter (Figure 5.23). It is also possible to integrate the lines to predict lifetimes in the manner just described for static loading.

The reader may have noticed that so far in this chapter the critical stress intensity factor and the critical strain energy release rate have been specified by K_C and G_C respectively, whereas in some previous chapters the designations K_{IC} and G_{IC} have appeared. The subscript I simply specifies a specific *crack mode*. In mode I (Figure 11.13(a)), which is termed the *opening mode*, the crack faces are displaced normal to the plane of the crack. There are also modes II and III so we can have K_{IIC} and K_{IIIC}. Mode II is known as the *shearing* or *edge-sliding* mode (Figure 11.13(b)) and mode III as the *antiplane strain* or *tearing* mode (Figure 11.13(c)).

The fracture of a composite material can be complex even when tested using relatively simple geometries and stressing as found in the tensile test. As shown in Figure 11.14, all three crack modes may be operative during a fracture test on a composite material.

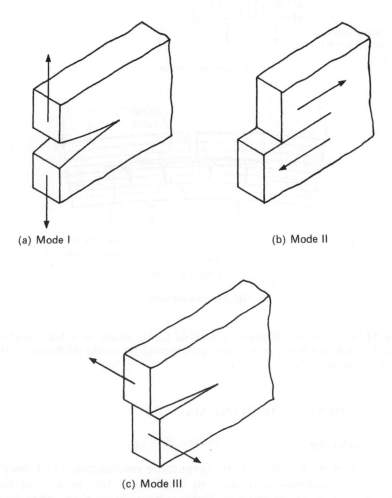

(a) Mode I   (b) Mode II

(c) Mode III

Figure 11.13 Crack modes: (a) I, opening mode; (b) II, shearing or edge-sliding mode; (c) III, antiplane strain or tearing mode.

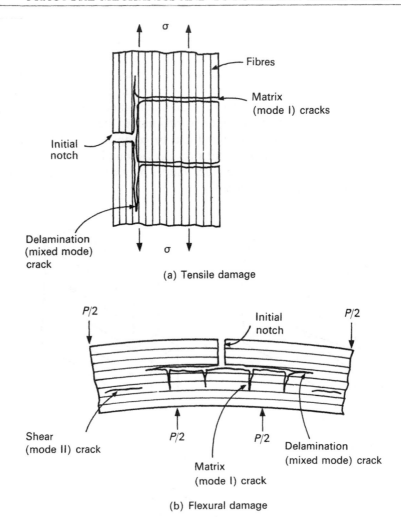

Figure 11.14 Schematic diagram of possible failure modes in a high toughness uniaxially reinforced brittle matrix composite during: (a) tensile; (b) flexural testing. (Source: Evans and Marshall, 1990.)

11.8 TOUGHENING MECHANISMS

11.8.1 Introduction

In this section we will discuss the toughening mechanisms which may be operative in a composite. It must be emphasized that in a given composite more than one mechanism is likely to be operating although one mechanism may dominate. As will be seen in the following sections the effectiveness of the toughening mechanisms depends to varying degrees on:

(a) size, morphology and volume fraction of the reinforcement;
(b) interfacial bond;
(c) properties (e.g., mechanical, thermal expansion) of the matrix and the reinforcement; and
(d) phase transformations.

In view of the sensitivity of the toughening mechanisms to these details of the composite, it is not surprising that the dominant mechanism is not the same for all composites. In some composites the situation is so complex that it is difficult to determine which is the dominant mechanism.

11.8.2 Crack bowing

Any reinforcement in the stress field at the tip of a propagating crack perturbs the crack front causing a reduction in the stress intensity in the matrix. This of course hinders the propagation of the crack. We can identify two dominant perturbations, namely crack bowing and crack deflection.

Crack bowing originates from the resistance of the reinforcement to fracture and results in a *nonlinear* crack front (Figure 11.15). The crack bows between the reinforcements, shown as spherical particles in Figure 11.15, leading to a decrease in the stress intensity K along the bowed section in the matrix, whilst producing an increase in K at the reinforcement. As the crack front becomes more bowed so K at the reinforcement increases until it reaches the fracture toughness of the reinforcement. The crack then breaks through the reinforcement and advances through the composite.

The crack bowing toughening increases with the volume fraction of reinforcement and depends on the magnitude of the crack–reinforcement interaction, which in turn is determined by the separation and toughness of the reinforcement (Figure 11.16). We can see from Figure 11.16 that the morphology of the reinforcement is also important – the extent of the bowing and hence the enhancement of toughness is greater the higher the aspect ratio.

11.8.3 Crack deflection

The interaction between the reinforcement and the crack front can cause the crack to be deflected and to become *non-planar*. The deflection may be a *tilting* or *twisting* motion around the reinforcement (Figures 11.17(a) and 11.17(b)). We learnt earlier that there are three modes of crack opening (Figure 11.13) and deflection alters the mode. Usually the crack initially propagates in mode I, the opening mode, but this is not so once deflection has occurred; tilting produces a mixture of modes I and II (shearing) and

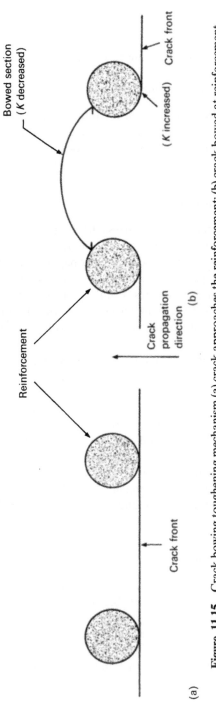

Figure 11.15 Crack bowing toughening mechanism (a) crack approaches the reinforcement; (b) crack bowed at reinforcement.

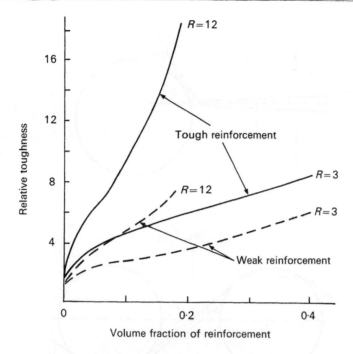

Figure 11.16 Theoretical plots demonstrating the effects of volume fraction, toughness and aspect ratio (R) of a disc-shaped reinforcement on crack bowing toughening. (Source: Faber and Evans, 1983.)

twisting produces a mixture of modes I and III (tearing). In either case crack propagation is hindered, although it is thought that crack deflection toughening arises mainly from twisting rather than tilting of the crack.

A number of crack–reinforcement interactions, such as the interaction of the crack with the residual stress fields due to differences in the thermal expansion coefficients or elastic moduli between the matrix and the reinforcement, can cause deflection. As for the crack bowing mechanism, high aspect ratio reinforcements are the most effective (Figures 11.17(c) and 11.18) and the toughening increases with increasing reinforcement content, although for crack deflection toughening there appears to be little extra benefit above about 0.2 volume fraction (Figure 11.18).

11.8.4 Debonding

The process of debonding, shown in Figure 11.19, creates new surfaces in the composite and therefore requires energy. Although the surface energy per unit area is small, the total area of new surface can be large hence the total energy associated with new surfaces is significant. Nevertheless other energy-requiring processes must also be occurring as experimental debonding

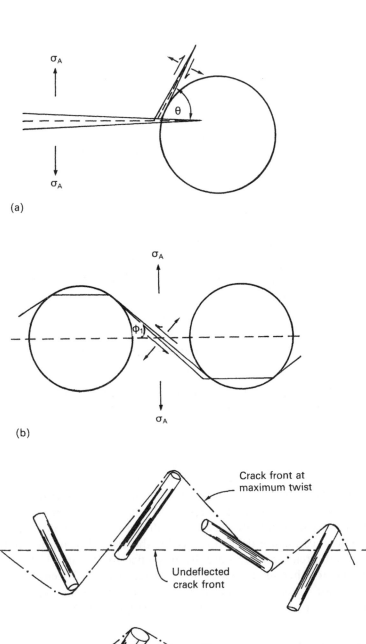

Figure 11.17 Crack deflection toughening: (a) tilt of crack front; (b) twist of crack front; (c) effect of aspect ratio on twist. (Source: Faber and Evans, 1983.)

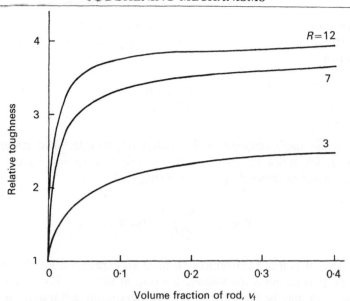

Figure 11.18 Theoretical plots demonstrating the effects of volume fraction and aspect ratio of a rod-shaped reinforcement on crack deflection toughening. (Source: Faber and Evans, 1983.)

energies are found to be greater than the total energy needed for the new surfaces. We can calculate the energy of debonding per fibre W_D by equating it to the strain energy released in the fibre as the stress relaxes. Consider that debonding takes place over a length l of a fibre of diameter D. Then

$$W_D = W \times (\text{volume of fibre debonded}),$$

where W is the strain energy in the fibre per unit volume and may be expressed by an equation of the form $\sigma^2/(2E_f)$. Hence on substituting for the volume of the debonded section of the fibre we obtain for W_D:

$$W_D = \frac{\pi D^2}{8E_f} \int_0^l \left(\sigma - \frac{4\tau l^2}{D} \right) dl.$$

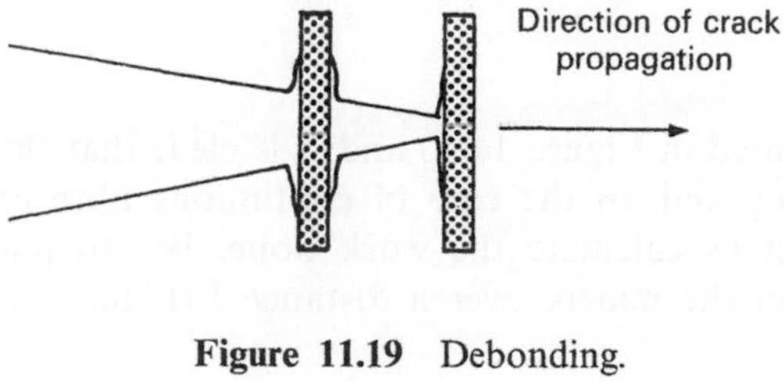

Direction of crack propagation

Figure 11.19 Debonding.

The maximum work occurs if the tensile stress in the fibre reaches the fracture stress $\hat{\sigma}_{Tf}$, which is the situation when l is half the critical length, l_c, that is when

$$l = \frac{l_c}{2} = \frac{\hat{\sigma}_{Tf} D}{4\tau},$$

where τ is the shear strength of the matrix or the interfacial strength (see Chapter 10 for more information on l_c). Substituting for l gives for the maximum work of debonding per fibre ($W_{D(max)}$):

$$W_{D(max)} = \frac{\pi D^2 \hat{\sigma}_{Tf}^2 l_c}{48 E_f} \quad \text{per fibre.} \tag{11.19}$$

Sometimes it is more convenient to have an expression for the energy of debonding that includes the volume fraction of fibres v_f. For aligned fibres equation 11.19 may be modified to give the maximum debonding energy per unit cross-sectional area of composite as

$$W_{D(max)} = \frac{\hat{\sigma}_{Tf}^2 l_c v_f}{12 E_f} \quad \text{per unit area.} \tag{11.20}$$

From this analysis we can see that to maximize the toughening due to debonding we need a large volume fraction of strong fibres with a weak fibre–matrix interface, as a low τ gives a large l_c. An analysis of debonding in terms of critical strain energy release rates leads to a similar conclusion as it is found that debonding rather than fibre failure only occurs if

$$\frac{(G_C)_i}{(G_C)_f} \leq \frac{1}{4},$$

where the subscripts i and f refer to fibre–matrix interface and fibre respectively.

11.8.5 Pull-out

Pull-out is illustrated in Figure 11.20 and it is clear that this cannot occur without debonding and, in the case of continuous fibre composites, also fibre fracture. Let us calculate the work done, W_p, in pulling a fibre of diameter D out of the matrix over a distance l (Figure 11.21(a)). A force

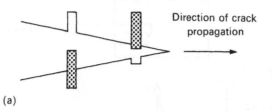

Direction of crack
propagation

(a)

(b)

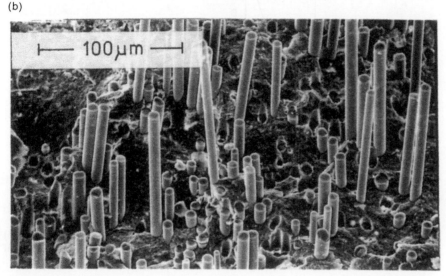

├── 100 µm ──┤

Figure 11.20 Pull-out: (a) schematic diagram; (b) fracture surface of 'Silceram' glass-ceramic reinforced with SiC fibres. (Courtesy H. S. Kim, P. S. Rogers and R. D. Rawlings.)

is required to pull out the fibre as normal frictional forces have to be overcome and also, as the stress on the fibre has relaxed due to debonding, Poisson's expansion of the fibre takes place and 'jams' the fibre in the matrix.

$$W_P = (\text{average force}) \times (\text{distance})$$

$$= \left(\frac{\pi D l \tau}{2}\right) \times (l) = \frac{\pi D \tau l^2}{2}.$$

This equation assumes that shear stress is constant during pull-out. Now if the embedded length to be pulled out is greater than the critical length, the tensile stress in the fibre will reach the fracture stress and the fibre will break. Therefore the maximum length of fibre that can be pulled out is $l_c/2$.

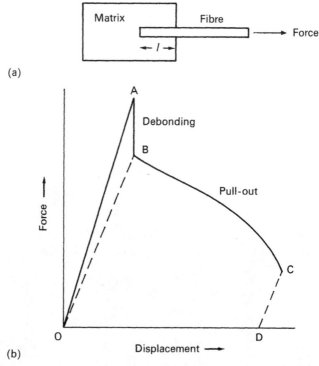

Figure 11.21 Pulling a fibre out of the matrix: (a) experimental arrangement; (b) force–displacement curve (energy of debonding is given by area OAB and that of pull-out by area OBCD).

Substituting $l = l_c/2$ gives us for the maximum work to pull out a fibre, $W_{P(max)}$ as

$$W_{P(max)} = \frac{\pi D \tau l_c^2}{8} = \frac{\pi D^2 \hat{\sigma}_{Tf} l_c}{16} \text{ per fibre.} \tag{11.21}$$

We now have the information to be able to compare the energies required for debonding and pull-out. The ratio of equations 11.19 and 11.21 gives

$$\frac{W_{P(max)}}{W_{D(max)}} = \frac{3E_f}{\hat{\sigma}_{Tf}}$$

and reference to the mechanical property data for fibres in Chapter 2 demonstrates that this ratio is always greater than unity, in other words, the energy of pull-out is greater than the energy of debonding. The difference in this energy can be considerable, for example substituting values for Young's modulus and strength from Tables 2.7 and 2.8 for carbon (pitch P-75S) and SiC Nicalon fibres yields $W_{P(max)}/W_{D(max)}$ ratios of 743 and 341 respectively. Experiments on pulling single fibres from a brittle matrix produce force–displacement curves of the form illustrated in Figure 11.21(b). These confirm that pull-out is a more significant toughening mechanism as the energy of debonding is given by the area OAB whereas the energy of pull-out is given by the much larger area OBCD.

So far we have discussed pull-out in general terms and have not considered any differences that might exist between continuous and discontinuous fibre composites. In the latter, when the conditions are favourable for pull-out, all the fibres with ends within a distance $l_c/2$ of the cross-section at which failure occurs will pull out. If the length of the fibres, l, is greater than l_c then only a fraction of the fibres, given by l_c/l, will pull out. Under these circumstaces the energy of pull-out is

$$W_P = \left(\frac{l_c}{l}\right)\frac{(\pi D\tau l_c^2)}{12} \quad \text{per fibre.}$$

On the other hand, when l is less or equal to l_c, the lengths of fibres pulled out will range from 0 to $l_c/2$ and the average work per fibre is

$$W_P = \frac{(\pi D\tau l^2)}{24} \quad \text{per fibre.}$$

It can be seen that for both cases the energy of pull-out is less than the maximum energy of pull-out given by equation 11.21.

Pull-out also occurs in continuous fibre composites as variations in strength along the length of fibres allows fracture to occur. The fibres can then be considered to be of a certain length and equations similar to those for discontinuous composites have been proposed.

If fibre fracture precedes pull-out there will be an additional contribution to toughness as a loaded debonded fibre dissipates energy when it fails. The energy dissipated may be estimated from the stresses in the fibre and matrix immediately before and after fracture. The contribution to toughness is directly proportional to v_f/E_f and a complex function of the stresses in the fibre and the debond length.

11.8.6 Wake toughening

When this form of toughening is associated with fibre reinforcement it is also known as *fibre bridging*. Let us consider the situation illustrated in Figure 11.22(a) where debonding has taken place but not all of the fibres in the crack wake have fractured and some bridge the faces of the crack. As the crack opens under the action of the applied stress, some of the stress will be transferred to the fibres which will deform elastically. The stresses in the bridging fibres are viewed as crack closure tractions which reduce the stresses at the crack tip. There is a corresponding reduction in the stress intensity factor at the crack tip and hence crack propagation is hindered.

The interesting feature of this toughening mechanism is that it is occurring in the crack wake and not at the damage, or process, zone at the crack

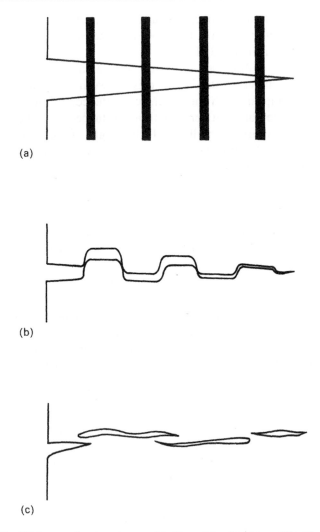

Figure 11.22 Wake toughening due to: (a) fibres (also known as fibre bridging); (b) interlocking of grains of ceramic; (b) bridging of intact material across the crack faces.

tip. The damage zone remains approximately constant in size as the crack grows therefore the contribution to the toughness from toughening mechanisms which take place in this zone is also constant. In contrast as a crack extends, its wake region increases in size and hence the contribution from wake toughening mechanisms increases until a steady state toughness value is reached. This increasing resistance to crack growth as the crack extends is shown for fibre bridging in the plot of toughness against crack extension, which is termed a *crack resistance curve* (R-curve), of Figure 11.23. Note that

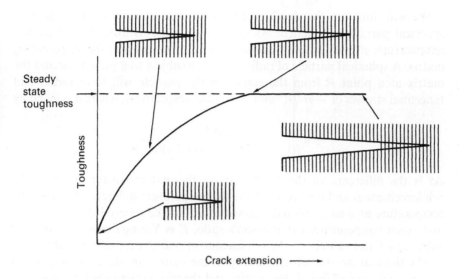

Figure 11.23   Crack resistance curve (R-curve) due to fibre bridging.

the steady state toughness corresponds to a constant wake structure of a certain number of bridging fibres in a zone called the *bridging zone*. It is possible in small specimens, or in components of composites with large bridging zones, for failure to occur before the steady state situation is reached.

We have concentrated on fibre bridging but any microstructural feature that restricts the relative movement of the crack faces in the crack wake can lead to wake toughening. Two such mechanisms, which could be operative either in a monolithic material or in the matrix of a composite, are given in Figures 11.22(b) and 11.22(c).

11.8.7 Microcrack toughening

The toughness of a material can be enhanced by the presence of micro-cracks. The strain energy of the primary crack is lowered since the microcracks enter the stress field at its crack tip and interactions cause crack blunting, branching and deflection.

In brittle non-cubic crystal matrices microcracking can occur because of the differences in the coefficients of thermal expansion between neighbour-ing grains; this source of microcracking is not specific to composite materials. The main cause of microcracking in composites is the thermal stresses which arise because of the differences in the coefficients of thermal expansion between the matrix and the reinforcement. We have already touched upon these thermal stresses in previous chapters (see for example sections 3.4.2 and 4.1) but it is appropriate at this point to study them in more depth.

We will illustrate the development of thermal stresses by looking at a spherical particle in a matrix. As the composite cools from the fabrication temperature, stresses develop within the reinforcement and the surrounding matrix. A spherical particle of radius r will be subject to a pressure σ and the matrix at a point R from the centre of the particle will have radial and tangential stresses of $-\sigma r^3/R^3$ and $+\sigma r^3/2R^3$ respectively, where σ is given by

$$\sigma = \frac{\Delta\alpha\Delta T}{[(1+v_m)/(2E_m)] + [(1-2v_f)/E_f]}.$$

$\Delta\alpha$ is the difference in the coefficients of thermal expansion between the reinforcement α_f and matrix α_m, ΔT is the temperature difference between the temperature at which residual stresses could not be relieved during cooling and room temperature, v is Poisson's ratio, E is Young's modulus and the subscripts f and m refer to the reinforcement and matrix respectively.

The thermal stresses which develop in the matrix are the most important as far as microcracking of the matrix, and therefore microcrack toughening, is concerned. For composites with $\alpha_f > \alpha_m$, σ is negative and hence there is a tangential compressive stress and a radial tensile stress in the matrix. Crack propagation is favoured when the crack faces are opened by a tensile stress, consequently circumferential microcracks are formed if the stress is sufficient (Figure 11.24(a)). Conversely, for $\alpha_f < \alpha_m$ we find the tensile stress in the matrix is tangential thus causing radial microcracking provided the stress is high enough (Figure 11.24(b)). Radial cracking can be especially damaging if uncontrolled.

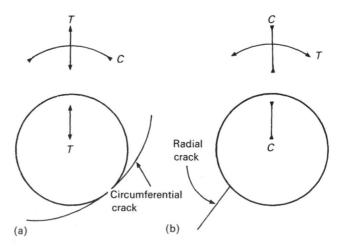

Figure 11.24 Stress distribution and microcrack formation around spherical particles when: (a) $\alpha_f > \alpha_m$; (b) $\alpha_f < \alpha_m$. C and T refer to compressive and tensile stresses respectively.

The volume change associated with phase transformations can also be a source of microcracks as described in Chapter 4 for zirconia-toughened alumina, ZTA (section 4.4.1). No matter what the mechanism of microcrack formation the density and size of the microcracks are critical in determining the toughness. As shown in Figure 4.17 for ZTA, if the microcracking is too extensive the toughness falls. Figure 4.17 also demonstrates that although the presence of microcracks might result in an increase in toughness this is obtained at the expense of strength as the microcracks lead to larger flaws, i.e., if we apply equation 11.12 at fracture we find that the increase in flaw size a outweighs the increase in K_C so σ_F falls.

11.8.8 Transformation toughening

A *metastable* reinforcement may transform to a lower energy state when it becomes influenced by the stress field at the tip of the primary crack (Figure 11.25). The toughness is increased as energy is absorbed ahead of the primary crack owing to the transformation. Also the transformation induces

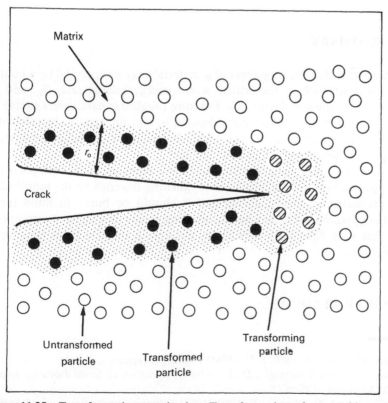

Figure 11.25 Transformation toughening. Transformation of metastable particles at the crack tip gives a zone, of width r_o, of transformed particles in the crack wake.

compressive stresses on the crack faces behind the crack tip. The classic example of transformation toughening in a composite is ZTA when the zirconia particles are retained in the metastable tetragonal state after fabrication. As previously discussed in section 4.4.1 the particles transform to the monoclinic structure and give an increase in toughness ΔK_{TT} of

$$\Delta K_{TT} = 0.3 v_{zirc} \Delta \varepsilon E_m r_0^{1/2},$$

where v_{zirc} is the volume fraction of the metastable zirconia particles, $\Delta \varepsilon$ is the unconstrained strain accompanying the transformation, E_m is Young's modulus of the alumina matrix and r_0 the width of the zone in the crack wake containing the transformed particles (Figure 11.25). As some of the toughening is occurring behind the tip of the crack we might expect that transformation toughening produces similar R-curve behaviour to wake toughening. Also, unlike microcrack toughening, an improvement in strength may be associated with transformation toughening (Figure 4.18).

11.9 SUMMARY

We have seen that the toughness of a material may be specified by a number of parameters such as the critical strain energy release rate, G_C, and the critical stress intensity factor, K_C. Fracture toughness parameters enable the critical flaw size in a component subjected to a given stress to be calculated. Slow crack growth under static, or fatigue, loading may also be analysed to obtain an estimate for component lifetime.

The various toughening mechanisms have been discussed with particular emphasis on how the magnitude of toughening depends on the morphology and properties of the reinforcement. It should be borne in mind that a number of mechanisms may be contributing to the toughness of a composite and that it is often very difficult to determine the dominant toughening mechanism.

FURTHER READING

General

Ewalds, H. L. and Wanhill, J. H. (1984) *Fracture Mechanics*, Arnold.

Rooke, D. P. and Cartwright, D. J. (1976) *Compendium of Stress Intensity Factors*, HM Stationery Office.

Sih, G. C. and de Oliveira, F. (eds) (1984) Fracture mechanics methodology, in *Engineering Applications of Fracture Mechanics*, Vol. 1, Nijhoff.

Specific

Evans, A. G. and Marshall, D. B. (1990) Mechanical behaviour of ceramic matrix
   composites, in *Fibre Reinforced Ceramic Composites*, (K. S. Mazdiyasni), Noyes
   Publication, p. 1.
Faber, K. T. and Evans, A. G. (1983) *Acta Metall.*, **31**, 565.
Knott, J. F. (1973) *Fundamentals of Fracture Mechanics*, Butterworths.
Thompson, I. and Rawlings R. D. (1992) *J. Mat. Sci.*, **27**, 2823, 2831.
Williams, J. G. (1984) *Fracture Mechanics of Polymers*, Ellis Harwood/Wiley.
Williams, J. G. (1990) *Proc. Inst. Mech. Eng.*, **204**, 209.

PROBLEMS

11.1.   Define the strain energy release rate, G.
For the standard double cantilever beam specimen shown in Figure 11.2, G
was shown to be

$$G = \frac{12P^2a^2}{EB^2D^3}$$

Determine the corresponding expression for G for the 'constant G' tapered
DCB illustrated below for which the taper half angle $\alpha = 11°$. (*Hint*: for G
to be independent of crack length a^2/D^3 must be equal to a constant. Let
the constant be k and calculate D_0k for any value of a/D_0 in the range
5 to 20.)

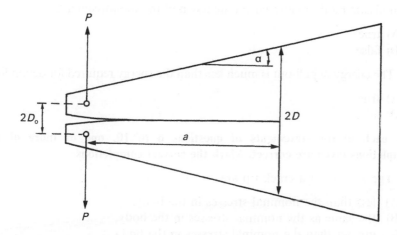

11.2.   Discuss the effect of specimen size on the applicability of linear elastic
fracture mechanics.

The critical stress intensity factor determined for a brittle matrix composite using a double cantilever beam specimen (Figure 11.2) with $D = 30$ mm, $a = 5$ mm and $B = 6$ mm was 8 MPa m$^{1/2}$. If the stress to give damage, σ_D, in the composite was 400 MPa, is the K_C determination valid?

SELF-ASSESSMENT QUESTIONS

Indicate whether statements 1 to 5 are true or false.

1. It is impossible to devise a fracture toughness specimen that has strain energy release rate independent of crack length.

   (A) true
   (B) false

2. The crack opening displacement (COD) at the crack tip is usually given the symbol δ.

   (A) true
   (B) false

3. The fracture of a composite material is complex and all three crack opening modes (modes I, II and III) may be operative.

   (A) true
   (B) false

4. Crack deflection can be caused by the interaction of the crack with the residual stress fields due to differences in the thermal expansion coefficients or in elastic moduli between the matrix and the reinforcement.

   (A) true
   (B) false

5. The energy of pull-out is much less than the energy required for debonding.

   (A) true
   (B) false

For each of the statements of questions 6 to 10, one or more of the completions given are correct. Mark the correct completions.

6. The stresses at a crack tip are

   (A) less than the nominal stresses in the body,
   (B) the same as the nominal stresses in the body,
   (C) greater than the nominal stresses in the body,
   (D) proportional to $K^{1/2}$ where K is the stress intensity factor,
   (E) proportional to K where K is the stress intensity factor,

(F) proportional to $r^{-1/2}$ where r is the distance from the tip,

(G) independent of r.

7. In the figure below σ_F is the fracture stress and Y is a calibration factor. The slope of the plot

(A) is Young's modulus, E,

(B) is the compliance, C,

(C) is C^2,

(D) enables the fracture toughness to be calculated,

(E) is K_C,

(F) is K_C^2,

(G) enables the crack velocity to be calculated.

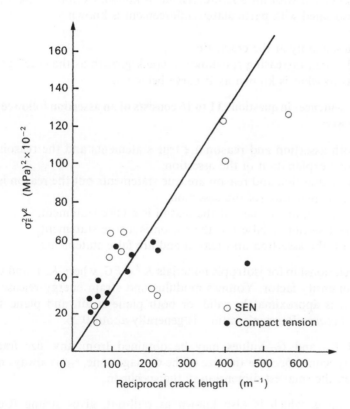

8. Slow crack growth

(A) occurs at stress intensity values K of less than K_C,

(B) occurs at stress intensity values greater than K_C,

(C) is sensitive to the environment,

(D) proceeds at a velocity v which is independent of K,

(E) proceeds at a velocity which is proportional to K^n where n is a constant,

(F) proceeds at a velocity which is proportional to $\exp(-K)$.

9. Toughening by crack bowing

(A) results in a non-planar crack,
(B) results in a nonlinear crack front,
(C) is greater the smaller the aspect ratio of the reinforcement,
(D) is greater the larger the volume fraction of reinforcement,
(E) is independent of the mechanical properties of the reinforcement.

10. Wake toughening

(A) if associated with fibre reinforcement is known as fibre bridging,
(B) if associated with particulate reinforcement is known as crack deflection,
(C) occurs mainly at the crack tip,
(D) leads to an increasing resistance to crack growth as the crack grows,
(E) leads to what is known as R-curve behaviour.

Each of the sentences in questions 11 to 16 consists of an assertion followed by a reason. Answer:

(A) if both assertion and reason are true statements and the reason is a correct explanation of the assertion,
(B) if both assertion and reason are true statements but the reason is not a true explanation of the assertion,
(C) if the assertion is true but the reason is a false statement,
(D) if the assertion is false but the reason is a true statement,
(E) if both the assertion and reason and are false statements.

11. The relationship for isotropic materials $K^2 = EG$, where K, E and G are the stress intensity factor, Young's modulus and strain energy release rate respectively, is approximately valid for both plane strain and plane stress conditions *because* Poisson's ratio v is generally about 0.25.

12. Valid K_C and G_C values may be obtained from any size fracture toughness specimen *because* the size of the damage zone, r_D, is always much smaller then the smallest dimension of the specimen.

13. Debonding, which is also known as pull-out, gives strong R-curve behaviour *because* strong fibres with a weak fibre–matrix interface are needed to maximize the toughening due to debonding.

14. Microcrack toughening, which is the breaking up of the primary crack into a large number of microcracks, is a most desirable toughening mechanism *because* it also produces a significant increase in strength.

15.   When the relative values for the coefficient of thermal expansion of the matrix and a spherical reinforcement is $\alpha_f > \alpha_m$ there is the possibility of radial microcracking *because* the thermal tensile stresses in the matrix are radial.

16.   ZTA, which is alumina reinforced with zirconia particles of the appropriate size and composition, undergoes transformation toughening *because* the alumina matrix transforms from the equilibrium condition to a metastable state under the action of the high stresses at a crack tip.

Fill in the missing words in the sentences. Each dash represents a letter.

17.   When a body is deformed elastically stress divided by strain gives the _ _ _ _ _ _ _   _ _   _ _ _ _ _ _ _ _ _ _ and displacement divided by load the _ _ _ _ _ _ _ _ _ _ _ .

18.   The subscript I in K_{IC} specifies a specific crack mode. In mode I, which is termed the _ _ _ _ _ _ _ mode, the crack faces are displaced _ _ _ _ _ _ to the plane of the crack.

19.   The _ _ _ _ _ _ torsion specimen geometry is designed to give G independent of crack length so that G is _ _ _ _ _ _ _ _ for a fixed load.

20.   The tougher the material for a given strength the _ _ _ _ _ _ the specimen required for a valid determination of K_C or G_C, that is for _ _ _ _ _   _ _ _ _ _ _ fracture mechanics to be valid.

21.   A force is required to pull out a fibre as normal _ _ _ _ _ _ _ _ _ _ forces have to be overcome and also, as the stress on the fibre has relaxed due to _ _ _ _ _ _ _ _ _, Poisson's expansion of the fibre takes place and 'jams' the fibre in the matrix.

22.   A _ _ _ _ _ _ _ _ _ _ reinforcement may transform to a _ _ _ _ _ energy state when it enters the stress field at the tip of the primary crack.

ANSWERS

Problems

11.1. $D_0 k \approx 3.6$; $G \approx \dfrac{43P^2}{EB^2 D_0}$.

11.2.   valid as $2.5(K_C/\sigma_D)^2 = 1 \times 10^{-3}$ m.

Self-assessment

1. B; 2. A; 3. A; 4. A; 5. B; 6. C,E,F; 7. D,F; 8. A,C,E; 9. B,D; 10. A,D,E; 11. A; 12. E; 13. D; 14. E; 15. D; 16. C; 17. modulus of elasticity, compliance; 18. opening, normal; 19. double, constant; 20. larger, linear elastic; 21. frictional, debonding; 22. metastable, lower.

Impact resistance 12

12.1 INTRODUCTION

All structures and components are likely to sustain impacts during service. It is vital that the reduction in performance caused by a 'normal' impact is not so large as to make the item unsafe. Fibre-reinforced polymer composites, especially CFRP, are very susceptible to accidental impact damage and to reductions in compression strength under hot/wet conditions. As a consequence design limits are set at about 0.4% strain, less than one third of the fibres' capability. In this chapter we shall look at some of the factors affecting impact performance of PMCs and, particularly, of CFRP.

Traditionally, techniques such as Izod and Charpy tests have been used to indicate the impact toughness of isotropic materials, and they have yielded useful information on notch effects and brittle–tough transition temperatures. However, because of the complexity of the failure processes, these tests are only of limited value with composite materials and they give no indication of residual properties after impact. Dropweight tests (simulating dropped tools) and ballistic tests prove to be more useful.

12.2 IMPACT TESTING

Increasing use is being made of *instrumented impact testing*. This is usually done on dropweight machines or pendulum impact machines where the striker or supports are instrumented to measure the applied load. A representative specimen is then subjected to a typical impact that might be experienced during service. Thus a low velocity impact (a few m/s) of a few Joules would represent a dropped tool, whereas a high velocity impact of a tiny mass travelling at, say, 80 m/s would be typical of road debris hitting a racing car. The available impacting energy is usually much greater than that needed to break or to penetrate the specimen, and displacement can usually be related to time. However, some machines have independent means of measuring displacement. Thus traces of load–time can be obtained and these can be converted to give energy–time traces, so that features such as peak load and absorbed energy can be related to fracture processes occurring in the material.

Some typical load–time traces for specimens supported at their ends are illustrated in Figure 12.1. The area under a trace is the impulse between projectile and specimen and is equal to the change in momentum of the projectile. When there was no failure (Figure 12.1(a)), peak load occurs when the projectile is brought to rest and the second half of the curve corresponds to the projectile rebounding. When a flexural failure occurs (Figure 12.1(b)) the projectile is brought to rest and does not rebound, its energy causing damage to the composite. The shear failure (Figure 12.1(c)), results in a change of stiffness of the laminate, which can still deflect and the projectile still rebound, but with reduced momentum. The results for a specimen on circular supports (Figure 12.2) show that the failure process can be changed by the geometry of the specimen and the support conditions. Further examples are shown for a (0/90) laminate (Figure 12.3) and for a honeycomb sandwich panel (Figure 12.4) in which load peaks associated with damage to the two skin panels can be clearly distinguished.

Also of interest in impact testing are effects of strain rate on elastic properties, and these effects should be included in an analysis if appropriate. For example, projectiles travelling at different velocities can cause the failure mode to change because of the dynamic response of the specimen. Tensile tests on CFRP over a wide range of strain rates show little effect on failure stresses whereas GFRP does show a stain rate effect (Figure 12.5). Similar effects are seen when comparing strain rates used in static and fatigue tests: composites using carbon fibres or aramid fibres show no effect of strain rate whereas GFRP does.

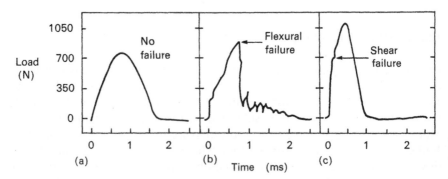

Figure 12.1 Load–time traces during dropweight impact using end supported specimens. (Courtesy DRA, Farnborough.)

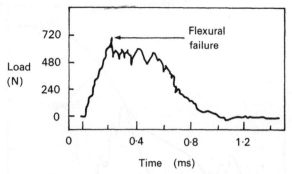

Figure 12.2 Load–time traces during dropweight impact using circular supports. (Courtesy DRA, Farnborough.)

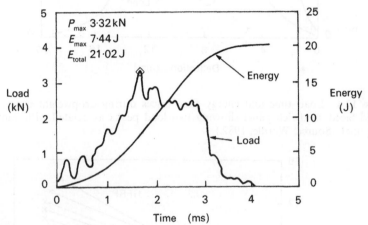

Figure 12.3 Load–time and energy–time traces from an instrumented dropweight machine testing a CFRP cross-ply laminate. (Source: Adams, 1985.)

12.3 IMPACT DAMAGE

The various forms of impact damage that can be observed are shown schematically in Figure 12.6 together with expressions for the energy needed to cause them, typical values for 2 mm thick CFRP and the area of fracture surface caused. It can be seen, by taking the ratio of delamination energy to flexural energy, that delaminations are more likely with short spans, thick laminates or with lower interlaminar shear strengths, whereas flexural failures are more likely with large spans or thin skins. Penetration is most likely for small projectiles moving at such a high velocity that the laminate cannot respond quickly enough in flexure and high stresses are generated close to the point of impact.

One of the main energy absorbing mechanisms in polymer matrix composites is the strain energy to failure of the fibres, i.e. the area under the

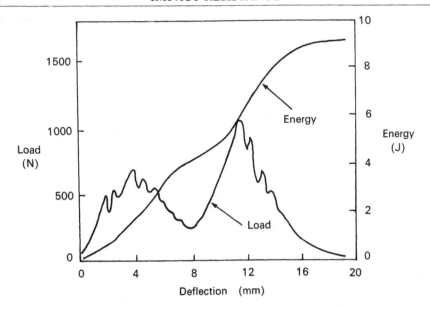

Figure 12.4 Load–time and energy–time traces during dropweight impact of an aramid faced sandwich panel showing two load peaks associated with damage to each panel. (Source: Wardle, 1982.)

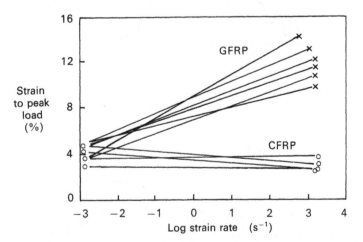

Figure 12.5 Effect of strain rate on the failure strain of several GFRP and CFRP composites. (Source: Kawata, 1981.)

stress–strain curve of the fibres. Lower modulus, higher strain fibres such as glass fibres and aramid fibres have a larger energy absorbing capacity. It is therefore possible to make hybrid laminates incorporating glass or aramid fibres to improve the impact resistance of CFRP. It is preferable to put the glass or aramid fibres on the surface because the strains are

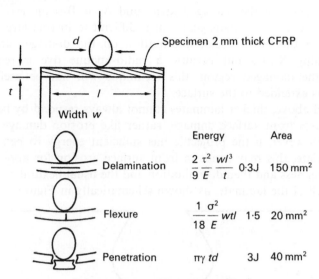

Figure 12.6 Geometry of impact test, primary failure modes, associated threshold energies of damage and damage areas. (Courtesy DRA, Farnborough.)

greater here in flexure and these surface plies can prevent fibre breakout on the back face.

If the impact results in visible damage this can be clearly seen on inspection and the appropriate remedial action can be taken. Delaminations however are not visible even though there may be extensive internal damage to the laminate (Figure 12.7), and consequently non-destructive evaluation is required. The damage is often worse towards the back (inside surface) of the laminate, which makes detection more difficult. Delaminations can be detected by ultrasonic techniques (see Chapter 15); C-scans

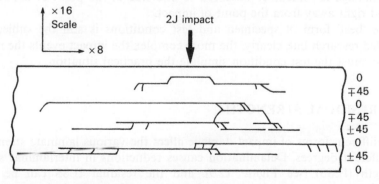

Figure 12.7 Multiple delamination in CFRP laminate due to dropweight impact. (Courtesy DRA, Farnborough.)

give a silhouette of the damaged areas and A or B-scans give information on the depth of the damage, but it is difficult to inspect large complex structures in this way. Useful information on the damage can also be obtained uing X-rays but because a radio-opaque dye is required to penetrate the damaged region, this technique will not be useful unless damage has extended to the surface.

As stated above, thicker laminates cannot always respond by flexing and impact causes front surface damage, rather like erosion damage by small particles. However, if the projectile has sufficient energy to penetrate the thick laminate, this may result in front surface erosion for approximately half the thickness and then delamination and the usual flexural response for the rear half of the laminate, as shown schematically in Figure 12.8.

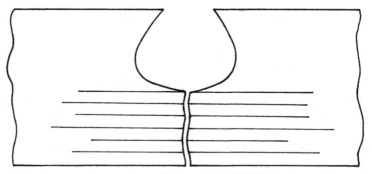

Figure 12.8 Impact damage in a 64-ply thick CFRP laminate due to ballistic impact of sufficient energy to penetrate the specimen. (Source: Cantwell, 1985.)

Similar work has been done on composite structures incorporating features such as *stiffeners*. It is found that impact midway between stiffeners requires 30 to 50 J of energy to cause *barely visible impact damage* (BVID), whereas near a stiffener, where the structure is stiffer, only a few Joules of energy are needed. Also, the stiffeners alter the local response of the panel so that damage is sometimes observed to one side of the point of impact or indeed right away from the point of impact.

The 'best' form of specimen and test conditions is still the subject of detailed research but, clearly, the more complex the impact events the more closely must the test condition simulate the practical situation.

12.4 RESIDUAL STRENGTH

The different forms of impact damage affect the various laminate strengths to different degrees. Delamination causes reductions in interlaminar shear strength (ILSS) (see Figure 12.9), and the residual ILSS can be predicted from the area of delamination using a fracture mechanics model and an interlaminar fracture toughness. The damage observed for both

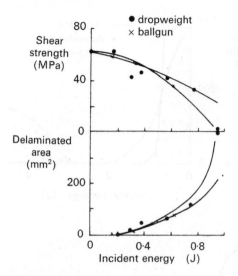

Figure 12.9 Effect of incident energy on residual interlaminar shear strength of CFRP $[(0/90)_3]_s$ laminate. (Courtesy DRA, Farnborough.)

drop-weight impact and *ballgun impact*, i.e. ballistic impact by a ball fired from a gas gun, is found to depend mainly on the energy of the projectile, even though the velocity and momentum can be very different. Residual strength reductions are also observed for tensile (Figure 12.10) and flexural residual strength (Figure 12.11), up to the point of penetration by the 6 mm ball from a gas gun. We see from Figure 12.11(b) that CMCs can show a similar form of behaviour to PMCs.

The maximum amount of damage is done for incident energies just sufficient to cause penetration and, since there is more damage on the back of the specimen, there is a greater reduction in flexural strength than in tensile strength. At higher velocities the ball penetrates the laminate relatively cleanly, losing little energy, causing little cracking away from

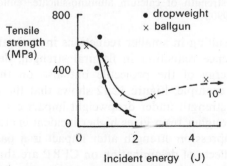

Figure 12.10 Effect of incident energy on residual tensile strength of CFRP $[(0/90)_3]_s$ laminate. (Courtesy DRA, Farnborough.)

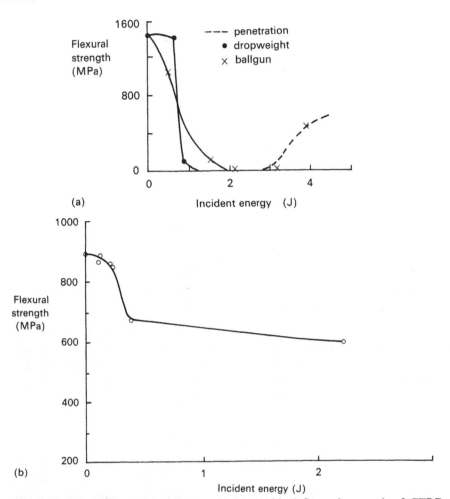

Figure 12.11(a) Effect of incident energy on residual flexural strength of CFRP $[(0/90)_3]_s$ laminate. (Courtesy DRA, Farnborough.); (b) Effect of incident energy on residual flexural strength of calcium alumino-silicate reinforced with SiC fibres (Source: Wong, 1990.)

the hole and resulting in smaller reductions in strength. The extent of the damage, and hence reduction in flexural strength, depends not only on the incident energy of the projectile but also on the properties of the composite. For example, Figure 12.12 shows that the threshold for reduction of flexural strength under dropweight impact can be affected by resin toughness; the tougher resin gives higher incident energy thresholds.

Residual compression strength after impact is a particularly important property. The effects of delamination on CFRP are shown in Figure 12.13; there is a marked reduction in compression strength for delamination damage produced at low energies (1 and 2 J) that have little effect on tensile

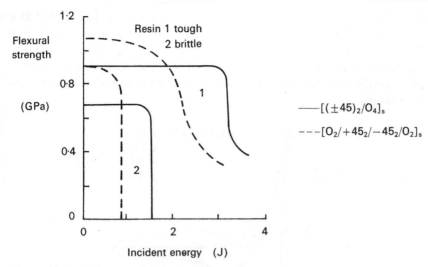

Figure 12.12 Effect of matrix toughness on residual flexural strength after drop-weight impact for two CFRP lay-ups. (Courtesy DRA, Farnborough.)

strength. Lower values of fibre surface treatment give tougher laminates in tension, but because of the larger areas of delamination, smaller residual compression strengths.

A factor that might affect the damage, and hence the residual strength, is the static load that a component is carrying when it is impacted. It might be expected that a component, when under load, will be less capable of further

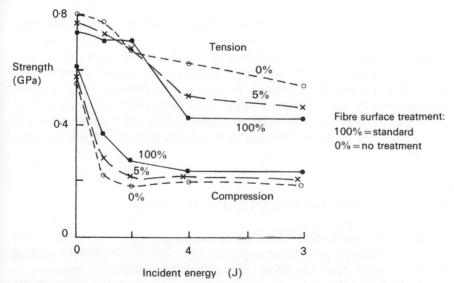

Figure 12.13 Effect of fibre surface treatment on residual strength of a $[(0_2/+45)_2]_s$ laminate after dropweight impact. (Courtesy DRA, Farnborough.)

elastic deformation under the additional impact loading and hence fail at a lower incident impact energy. Figure 12.14 shows that in tension there is a slight reduction in residual strength with tensile pre-load, but the main effect is a complete failure on impact if the tensile pre-load exceeds the expected residual tensile strength. Similar results have been obtained with compression pre-loads.

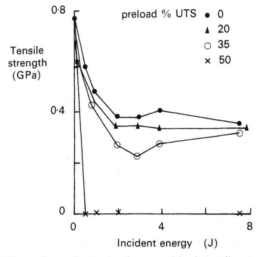

Figure 12.14 Effect of tensile preload on residual tensile strength of a CFRP laminate after ballgun impact. (Courtesy DRA, Farnborough.)

One concern has been the effect of fatigue loading on laminates containing small areas of delamination. Although damage significantly reduces the static strength it has been shown that at 10^6 cycles the fatigue strengths of damaged and undamaged specimens are similar. It seems that loads that are not great enough to cause static failures will not cause damage growth under fatigue loading. This issue needs to be more fully investigated before it can be taken as a general rule, but if design limits are set to take care of impact damage then it seems that fatigue will not be a problem.

12.5 EPOXY AND THERMOPLASTIC MATRICES

Most of the applications for high performance FRP use epoxy resin matrices. Although many thermoplastics offer certain advantages such as toughness and short moulding times, their environmental resistance is relatively poor. However, semicrystalline thermoplastics such as PEEK are little affected by solvents and have high melting points (Chapter 5).

On impact at low energies carbon fibre–PEEK shows an indentation on the front face with associated compression failures; sections through these compression lines show them to be characteristic compression failures with associated shear bands. C-scans of impacted laminates show less delamination but some 0° splitting not seen with epoxy matrices. It seems that the fibre–matrix bond could still fail in transverse tension, but that shear cracks did not grow so readily in the PEEK matrix. It has been shown that the delamination could extend a considerable distance in carbon fibre–epoxy but that in carbon fibre–PEEK the damage was more confined to the actual impact site. Because of the restriction of damage and the splitting which tends to blunt any cracks produced by impact, both the residual tensile and compressive strengths are significantly greater for carbon fibre–PEEK than for carbon fibre–epoxy.

12.6 SUMMARY

The present generation of fibre-reinforced polymer matrix composites, and particularly CFRP, are severely limited by poor impact resistance and poor hot/wet compression strengths. This makes design limits low compared with the materials' capability although significant mass savings can still be made. Improved understanding of the failure processes that occur during impact is expected to improve the ratio of design stress to material strength, but greater advances are likely to come from improvements in fibres (higher failure strains) and matrices (greater toughness).

FURTHER READING

Specific

Adams, D. F. (1985) *Composites*, **16**, 268–278.
Cantwell, W. J. (1985) *Ph.D. Thesis*, Imperial College, University of London.
Kawata, K. (1981) *Proc. Japan–USA Conf. on Composite Materials*, Tokyo.
Wardle, M. (1982) *Proc. ICCM-IV*, Oct. 1982, Japan Soc. Composite Materials.
Wong, W. C. (1990) *M.Sc. Dissertation*, Centre for Composite Materials, Imperial College, London.

PROBLEMS

12.1 Discuss the different methods of impact testing. Describe typical specimen responses and the associated damage.

SELF-ASSESSMENT QUESTIONS

Indicate whether statements 1 to 6 are true or false.

1. Izod and Charpy test methods provide only limited information when used for impact testing of composites.

   (A) true
   (B) false

2. The effects of strain rate on elastic properties need not be included when analysing impact data.

   (A) true
   (B) false

3. Delamination failures are more likely in thick laminates.

   (A) true
   (B) false

4. Lower modulus fibres with high failure strain will absorb more energy.

   (A) true
   (B) false

5. Delamination damage causes a greater reduction in residual tensile strength than in residual compressive strength.

   (A) true
   (B) false

6. Failure under impact is unaffected by static pre-load.

   (A) true
   (B) false

For each of the statements of question 7, one or more of the completions given are correct. Mark the correct completions.

7. Impact damage in CFRP laminates will be typically

   (A) delamination for long spans,
   (B) delamination for thick laminates,
   (C) flexural for short spans,
   (D) penetration for large masses at low velocity,
   (E) front surface erosion for thin laminates.

Fill in the missing words in the sentences. Each dash represents a letter.

8. − − − − − − − − − − − causes reductions in interlaminar shear strength.

9. Rear surface damage causes greater reduction in − − − − − − − − strength than in − − − − − − strength.

10. $------$ areas of delamination result in $-------$ residual compression strengths.

11. Complete failure occurs under impact if the tensile $-------$ exceeds the $--------$ tensile strength.

Each of the sentences in questions 12 to 17 consists of an assertion followed by a reason. Answer:

   (A) if both assertion and reason are true statements and the reason is a correct explanation of the assertion,

   (B) if both assertion and reason are true statements but the reason is not a true explanation of the assertion,

   (C) if the assertion is true but the reason is a false statement,

   (D) if the assertion is false but the reason is a true statement,

   (E) if both the assertion and reason are false statements.

12. CFRP is very susceptible to accidental impact damage *because* design strain limits have been set to one third the failure strain of the fibres.

13. Projectiles travelling at different velocities can cause a change of failure mode *because* the specimen alters its dynamic response.

14. Delamination damage is more likely in thick laminates *because* the energy needed to cause failure is directly proportional to thickness.

15. Penetration is more likely for small, high velocity projectiles *because* the laminate cannot deform quickly enough in flexure.

16. At high incident energies residual strength shows greater reductions *because* more damage is caused.

17. Residual strengths of carbon–PEEK laminates are greater than those for carbon–epoxy *because* the matrix is a thermoplastic.

ANSWERS

Self-assessment

1. A;   2. B;   3. A;   4. A;   5. B;   6. B;   7. B;   8. delamination;
9. flexural, tensile;   10. larger, smaller;   11. preload, residual;   12. B;
13. A;   14. C;   15. A;   16. E;   17. B.

13	# Fatigue and environmental effects

13.1 INTRODUCTION

Fatigue failure, that is failure under cyclic or fluctuating loading, has been discussed in previous chapters (sections 3.4.2, 4.4.2 and 5.4.2). Under cyclic loading a crack is initiated which then slowly grows until it reaches a critical size, at which point failure occurs. The reader will recall that these failures occur at stresses below those needed to cause static failures, as encountered in standard tensile or flexural testing, for example.

In this chapter we shall discuss the fatigue of fibre-reinforced polymers in some detail. The behaviour of these materials is more complicated than that of more homogeneous materials, such as the well-documented metals. The main reasons for this are the different types of damage that can occur (fibre fracture, matrix cracking, fibre–matrix interface failure, delamination), their interactions and their different growth rates. Among the parameters that influence fatigue performance are fibre type, matrix type, strain to failure and strength (of fibre and matrix), laminate configuration and cycling frequency.

At the end of the chapter a few sections are included on the dependence of mechanical performance on environmental conditions.

13.2 TEST METHODS

Before discussing the fatigue behaviour of composite materials, it is perhaps first worth considering the means of obtaining fatigue data on composite materials. The main requirement for a material's fatigue test coupon is that it should fail in a manner similar to the material in the comparable structural component. Ideally this should be combined with ease of use and cheapness in preparation. Test coupon profiles have been varied, in attempts to ensure failure away from the stress concentrations at the gripped portions of the specimen, ranging from those incorporating waists and cut-outs to simple, parallel-sided coupons. Waisting usually ensures failure away from the grips for static loading, but not necessarily in fatigue. Indeed, the plain parallel sided specimen frequently yields the longest fatigue lives and the best all-round behaviour, although failures do occasionally occur at the grips.

Waisting is usually restricted to across the width of coupons. Waisting in the thickness direction will disturb the lay-up of the laminate, except for unidirectional material, which then frequently leads to shear failures at the waists. These grow back during fatigue loading to the grips and trigger failure within the grips at short lifetimes.

Compression or tension–compression loading is more complex than tensile loading, with the added problem of stabilizing the coupon during compression. As for static loading this requires that either short, stable coupons be used, or an anti-buckling guide to support the coupon. Short, stable coupons, which may be parallel-sided or waisted, suffer from the disadvantage that the stress distribution in the short free length may be affected by the restraint at the grips. Reducing the specimen width to allow for this renders edge stresses more critical. Typically such specimens will be about 10 mm wide with a 10 mm free length and a minimum thickness of 1.5 mm.

Long coupons are to be preferred but when a compressive excursion is to be included in the fatigue cycle, it is necessary to provide supports to prevent buckling. No standard antibuckling guide is currently recommended, every laboratory generally having developed their own devices. The main factor to consider when designing guides is that the free unsupported area of the specimen should be a maximum, consistent with the requirement of preventing buckling, so as not to restrict any anticipated failure process. In addition, friction between the supports and the specimen must be minimized, perhaps by the use of PTFE tape on the contact surfaces.

Many laboratories use flexural fatigue testing as an alternative to axial loading, since it is easier to perform, requiring no supporting guides and, generally, lower capacity testing machines. Flexural test methods used for static loading are usually suitable for fatigue, but care should be taken to minimize friction at the loading rollers, and it is sometimes necessary to introduce backing rollers on the reverse side of the coupons.

Fatigue testing by shear loading is less common, but perhaps should be considered more than it is at present. The most widely used shear test is probably the short beam or interlaminar shear test. This can easily be modified for fatigue use by the introduction of backing rollers opposite the main rollers, particularly if the deflection is to be reversed.

Alternative shear fatigue test methods that are used are also based on modifications of methods used for static testing. First the tensile test on

±45° laminates, which induces shear along the fibres, has been used extensively as a static test for shear strength and could also be used for the generation of shear fatigue data. Second, the rail shear test, also widely used for static shear strength measurements, has also been used in fatigue. The rail shear specimen requires some modification for it to be suitable for fatigue testing and recent work has shown that the fatigue lives obtainable are very dependent upon the surface quality of the exposed edge of the coupon.

13.3 UNIDIRECTIONAL COMPOSITES

Although application of purely unidirectional composites is limited, because of the poorer mechanical properties in directions other than along the fibres, use has been made of composites with large fractions of fibres in one direction, such as in helicopter rotor blades and vehicle leaf springs. The study of such systems is, therefore, important especially as knowledge of their behaviour may help to predict the performance of multi-directional laminates. Typical plots of peak tensile stress versus log cycles to failure, the traditional *S–N* presentation of data, are shown in Figure 13.1. Data for three unidirectional materials are displayed, carbon fibre, glass fibre and aramid fibre-reinforced epoxy resin. The ratio of fatigue stress at high lifetimes to static strength is smallest for the glass fibre-reinforced material and greatest for the carbon fibre-reinforced material, i.e. the latter has a superior fatigue performance.

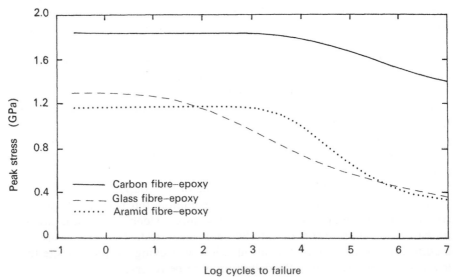

Figure 13.1 S–N diagrams for representative unidirectional PMC composite materials. (Source: Curtis, 1987.)

Since for unidirectional composite materials under tensile loading the fibres carry virtually all the load, the tensile fatigue behaviour might be expected to depend solely on the fibres, and since the fibres are not usually sensitive to fatigue loading, good fatigue behaviour should result. However, experimental evidence has shown that the slopes of the curves are determined principally by the strain in the matrix. Consequently plots of mean strain rather than stress versus log cycles to failure are frequently more meaningful for composite materials. All non-metallic fibres, as already mentioned, have a statistical distribution of strength, determined by flaws, thus a few of the weakest fibres will fail during fatigue loading. This gives rise to local high stresses in the matrix and at the fibre–matrix interface, which lead to the development of fatigue damage with increasing numbers of cycles.

Damage may also develop at local microdefects, such as misaligned fibres, resin rich regions or voids. Resin cracks frequently develop between the fibres, isolating them from adjacent material and rendering them ineffective load carriers, causing fibres to become locally overloaded and further static fibre failures to occur. Close to final failure, the matrix may show extensive longitudinal splitting parallel to fibres caused by resin and interfacial damage, leading to the brush-like failure characteristic of most unidirectional materials. The rate of this degradation process in the matrix and at the interface is a function of the bulk strain in the resin as well as the nature of the matrix. The use of very stiff fibres results in relatively low strains and thus shallow S–N curves, whereas less stiff fibres lead to greater matrix strains and steeper S–N curves. This effect of fibre stiffness on the form of the S–N curve is well illustrated by the data for carbon fibre and glass fibre-reinforced epoxy given in Figure 13.1. Aramid fibres have a stiffness in between that of glass and carbon fibres, thus intermediate fatigue behaviour would be expected. However, the fatigue damage mechanism is complicated in this material since aramid fibres are themselves fatigue sensitive and can defibrillate during fatigue loading. This causes the S–N curve to adopt the shape depicted in Figure 13.1, becoming much steeper at intermediate to long lifetimes.

In recent years manufacturers have sought to improve the mechanical properties of composite materials based on standard high strength carbon fibres in resin matrices. Carbon fibres with a breaking strain of 1.8% are now widely available, all the major manufacturers offering products in this class, and fibres with a breaking strain of over 2% have been produced on a laboratory scale and will be marketed shortly. Using higher performance carbon fibres in the same standard epoxy resin matrices generally results in little improvement in fatigue behaviour, as shown in Figure 13.2 which shows plots of peak strain versus log cycles to failure for several composites, all based on the same resin matrix but with differing carbon fibre reinforcements. Only small changes in the fatigue behaviour are apparent, usually just a slight shift along the ordinate in keeping with changes in the static

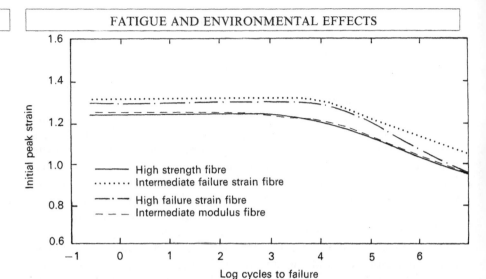

Figure 13.2 S–N fatigue data for unidirectional CFRP showing the limited effect of different fibres with the same epoxy resin. (Source: Curtis, 1986.)

strength differences measured. This is undoubtedly a consequence of the fatigue behaviour being dependent upon the matrix and interfacial characteristics, rather than the fibre strength.

Concurrent with fibre developments, resin manufacturers have sought to improve the performance of matrix materials. The principal goal has been increased toughness without detriment to the high temperature properties, particularly after moisture absorption. This has led to improvements in static strength, but usually poorer fatigue behaviour with steeper S–N curves, as shown in Figure 13.3. Thus fatigue behaviour in the next

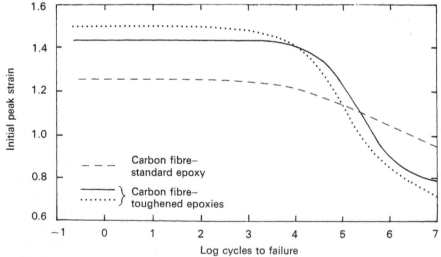

Figure 13.3 S–N data for unidirectional CFRP with the same fibre type in different matrices. (Source: Curtis, 1986.)

generation of composite materials may be somewhat more important than has been the case in the past. The poorer fatigue behaviour of these new composites may be due not only to the poorer performance of the tougher matrices used, but also to different failure processes being triggered, such as excessive fibre–matrix debonding or splitting.

13.4 MULTIDIRECTIONAL LAMINATED COMPOSITES

On increasing the percentage of non-axial fibres in a laminate, the longitudinal static tensile strength and stiffness are reduced since fewer fibres are available to support the applied loads. The slope of the tensile S–N curve increases in relation to the static strength (Figure 13.4) because the layers with off-axis fibres, whose mechanical properties are resin dependent, are more easily damaged in fatigue.

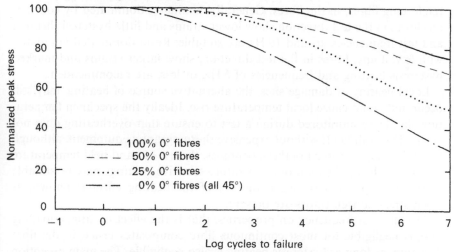

Figure 13.4 Normalized S–N curves for CFRP $(0/\pm45°)$ laminates, with varying percentages of 0° fibres, showing poorer fatigue performance as the proportion of non-axial fibres increases. (Source: Curtis, 1987.)

Transverse layers, with fibres at 90° to the test direction, develop cracks parallel to the fibres either during the first tensile load cycle or with increasing numbers of cycles but, since they support little axial load, this has little effect on the axial strength or stiffness of the material. Angle-ply layers, with fibres typically at $\pm45°$, can also develop intraply damage which will cause small reductions in strength and stiffness. The stress concentration at the ends of intraply cracks can lead to the initiation of delamination between layers, usually resulting in the decoupling of the 0° principal load bearing layers, which may lead to a general loss of integrity with potential for environmental attack and certainly to a reduction in compressive strength. Alternatively the cracks may occasionally propagate into adjacent primary load bearing layers and seriously weaken the material. Ultimate tensile fatigue failure of composite laminates is still determined by the unidirectional layers, thus the tensile $S-N$ curves for multidirectional laminated composite materials are relatively shallow, although steeper than for wholly unidirectional material.

13.5 FREQUENCY EFFECTS

As a general consideration in the fatigue testing of PMC materials, the test frequency should be chosen so as to minimize heating of the material. The source of this heating effect is hysteresis in the resin, also, perhaps, at the fibre–matrix interface and in a few cases, such as composites using polymer reinforcing fibres, in the fibres. Generally, laminates dominated by continuous fibres in the test direction show lower strains and little hysteresis heating and test frequencies around 10 Hz are suitable. Resin dominated laminates, with few, if any, fibres in the test direction, show larger strains and marked hysteresis heating and frequencies of 5 Hz, or less, are recommended.

Local heating at damage sites, the alternative source of heating, may still occur and could cause local temperature rise. Ideally the specimen temperature should be monitored during a test to ensure that overheating does not occur. This is difficult without expensive thermography equipment, although the strategic positioning of thermocouples, the use of hand held temperature sensors or the application of temperature sensitive coatings may be suitable alternatives, particularly when the site of the heating effect is known, as when stress concentrators are present.

The effect of frequency on properties, that is the effect of fatigue loading rate, is negligible for most continuous fibre composites tested in the fibre direction, as long as hysteresis heating is also negligible. The main exception is GFRP in which there is a significant rate effect (see Figure 12.5); the greater the rate of testing the greater the strength. A rate sensitivity for strength of over 100 MPa per decade rate has been reported. The reason for this is not entirely clear, but is believed to be due to the environmental sensitivity of the glass fibres rather than any visco-elastic effect. Certainly the effect has been found to change when the environment surrounding the glass fibres changes. Testing composites with no fibres in the test direction, where the resin matrix has visco-elastic behaviour, will often also result in a significant rate effect.

When collecting fatigue data on composite materials, therefore, the best policy is to carry out all fatigue tests at a constant rate of stressing. Thus low load tests are performed at relatively high frequencies and high load tests at lower frequencies.

13.6 EDGE EFFECTS AND STRESS CONCENTRATORS

Edge induced stresses can be a problem in many types of testing, but especially so in fatigue. Some tests, such as those investigating interlaminar effects, may aim to maximize edge effects, but in fatigue tests the policy is usually to try to minimize edge induced stresses and the damage that inevitably develops as a result. Both shear and normal stresses can develop at the coupon edges, these arising from the mismatch of properties between the layers as explained in a previous chapter (section 9.4.2). The magnitude of these stresses will change both with temperature, because layers have differing expansion coefficients, and with moisture content, as layers expand to different extents on absorbing external moisture. Edge induced damage, apparent in static loading, usually grows with increasing numbers of fatigue cycles. In the worst cases the layers can become completely delaminated, leading to potential environmental attack and certainly serious losses in compressive strength.

Stress concentrators such as notches, holes, fasteners, impact damage and other imperfections have less effect on tensile fatigue strength than on static strength. Depending upon the laminate configuration, these stress concentrations can reduce static tensile strength by up to 50%. In fatigue, however, damage zones develop at stress concentrations which can serve to reduce their magnitude. These zones usually consist of cracks along the fibres within layers, and interlaminar cracking between layers which, if they do not damage fibres, can lead to increased strength. Further cycling results in some loss in strength, but typically fatigue strength, calculated on a net stress basis, approaches that of the plain unnotched material after long lifetimes, resulting in fairly flat S–N curves. This is shown in Figure 13.5 in

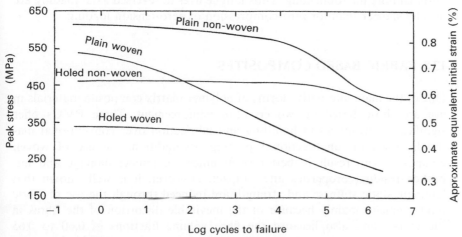

Figure 13.5 Effect of a stress concentrator (a hole) on S–N curves for woven and non-woven (0/90°) laminates. (Source: Curtis and Moore, 1985.)

which four S–N curves are plotted for two sets of composite materials, one made from non-woven material and the other from woven carbon fibre fabric. The static strengths of the notched coupons, both woven and non-woven, are significantly lower than those of the plain coupons, but after large numbers of fatigue cycles their fatigue strengths approach those of the plain coupons.

13.7 COMPRESSIVE LOADING

The Achilles heel of PMC materials undoubtedly lies with their compressive behaviour. As indicated in Chapter 8, the mechanism of compressive failure in such composites is not well understood. Generally much less information is available on the compressive fatigue of PMCs, mainly because the compressive testing of these materials presents many problems, not the least of which is the need to support the specimens against global macrobuckling, combined with limitations imposed on specimen geometry by the anisotropic nature of the materials.

Ultimately, the worst fatigue loading condition for composite materials is fully reversed axial fatigue, or tension–compression loading. The poorer behaviour of composite materials in reversed axial loading compared with tensile loading is because many of the laminate's plies, without fibres in the test direction, develop intraply damage and this causes local layer delamination at relatively short lifetimes. In tensile loading this is not serious, since the layers containing fibres aligned along the test direction continue to support the majority of the applied load. In compression, however, tensile induced damage of this type can lead to local layer instability and layer buckling, perhaps before resin and interfacial damage within the layers has initiated fibre microbuckling. Thus fatigue lives in reversed axial loading are usually shorter than for zero-compression or zero-tension loading.

13.8 FABRIC BASED COMPOSITES

One of the main alternative forms of polymer matrix composite materials in use are those based on woven fibre reinforcement. These PMCs offer significant advantages in handling and fabrication over conventional materials based on unidirectional prepreg. In addition, the use of woven composites can result in better containment of impact damage and improved residual properties after impact. However, it is well known that laminate static stiffness and strength are reduced through the use of fabric reinforcement, mainly because of the inevitable distortion of the fibres in the weave, and also because the fibre volume fractions of 0.60 to 0.65 found in non-woven continuous fibre material are not achievable in woven composites, a figure of 0.50 to 0.55 being more typical. This effect is also observed in fatigue, with strength at short lifetimes being reduced in the woven material. However, additional degradation processes lead to further reductions in strength (Figure 13.5) which result in steeper S–N curves than for the non-woven material. This is due to the stress concentrating effect of fibre crossover points in the fabric, which become sites for fatigue induced damage in the resin and fibre–resin interface. In the worst case, cross-plied CFRP, sustainable strains after one million cycles of reversed axial fatigue loading can be as low as 0.3%. Thus although woven composites have many processing advantages, their mechanical properties, particularly in fatigue loading, are generally poorer than the equivalent non-woven material.

13.9 HYBRID COMPOSITES

Another alternative form of PMC material in general use is the hybrid composite. This term is usually reserved for composites fabricated with two different fibres in the same resin matrix, e.g. glass and carbon fibres either separately in individual plies or intermingled in each ply. The main reasons for hybridization are improved toughness or stiffness matching. Some fibres, such as the polymer based types, are particularly tough and when hybridized with fibres such as carbon or glass can yield tough laminates without some of the disadvantages associated with polymer fibres alone. Alternatively some components, such as helicopter rotor blades, may require exact stiffness matching and this can more readily be achieved with hybrid composites, usually with mixtures of glass and carbon fibres.

The static mechanical performance of hybrid composites is dominated by the behaviour of the lower strain phase, often carbon fibres. If this is present in only small proportions then failure of this phase may be tolerable with their load being transferred to the higher strain phase. Usually, however, failure of the low strain phase triggers failure of the whole composite. In fatigue, as has been shown earlier, different fibres result in very different composite fatigue behaviour. Since fatigue performance is primarily strain dependent, in hybrid composites the behaviour is similarly dominated by the lower strain phase. This is because this phase is usually the stiffer and as such carries the greater part of the load. The other phase is usually subjected to strains below normal and shows a smaller fatigue degradation than when tested in isolation.

13.10 MOISTURE ABSORPTION AND ITS EFFECTS ON PROPERTIES

As well as having to withstand loading extremes, composite components have also to survive in a range of different environments of moisture and temperature. Relative humidities can vary from 0 to 100%. Temperatures for most uses range from $-40\,°C$ to $70\,°C$, although military aircraft have to withstand temperatures up to $130\,°C$ in flight and even greater around engines. The effect of these temperatures on composite properties needs to be known and, ideally, predicted.

As we saw in Chapter 5, most polymer matrix composites absorb moisture from humid atmospheres. This is usually confined to the resin matrix, but some fibres also absorb moisture. Epoxy resins absorb between 1 and 10% by weight of moisture, so a typical composite with 60% by volume of fibres will absorb around 0.3 to 3% moisture, but the systems in widespread use tend to absorb in the range 1 to 2% by weight.

The absorption of moisture is diffusion controlled and in thin sheets is a one-dimensional process which can be described by Fick's second law of diffusion

$$\frac{dc}{dt} = D\frac{d^2c}{dx^2},$$

where c is the moisture concentration, t is the time, D is the diffusion coefficient and x is distance into the panel. For a flat plate absorbing moisture through both surfaces the initial moisture uptake is proportional to the square root of time, as given below.

$$\frac{M}{M_m} = (4/h)(Dt/\pi)^{1/2}$$

where M is the moisture uptake at time t, M_m the maximum moisture content and h the plate thickness. As the moisture content increases the rate of absorption decreases and approaches a maximum for a particular environment. Once the diffusion coefficient and the maximum moisture content are known, the moisture content and indeed distribution can be determined for any composite for which Fickian diffusion applies. Many composite materials follow Fickian diffusion kinetics, at least at low temperatures and humidities. There is usually a maximum temperature/humidity above which the moisture kinetics become non-Fickian. Some materials exhibit non-Fickian behaviour in typical exposure environments, and means of modelling these have been proposed.

Absorption of moisture from humid environments at room temperature, although causing swelling, is usually a reversible process and with most advanced composites there is no degradation in room temperature properties. Because moist resins soften, resin dominated composite properties are reduced at elevated temperature, particularly as the glass transition temperature of the resin is approached. The reduction in the glass transition temperature of epoxy resins with moisture content is illustrated in Figure 5.19. Microcracking of the resin or fibre–matrix interface, blistering or delamination may also occur.

For multidirectional laminates tested in tension, temperature and moisture have little effect on the failure of the dominant $0°$ load bearing layers. In any resin dominated situation, such as applied tension on laminates with little or no $0°$ fibres, or more usually in compression where the resin is required to support the fibres against buckling, there can be significant effects due to temperature and moisture. For example, a $\pm 45°$ CFRP laminate, tested in tension at $130°C$ and 1.5% moisture, shows a 40% drop in strength and almost doubling of failure strain, compared with the room temperature dry state.

13.11 OTHER ENVIRONMENTS

In addition to temperature change and water, composites are often exposed to other environmental influences such as ultra-violet light, solvents, vapours, acids and alkalis. The resistance of the matrix is clearly crucial in protecting the fibre–matrix interface and the fibres themselves. In general epoxy resins are more resistant than polyesters whilst high performance thermoplastics, such as PEEK, are virtually insensitive to environmental attack.

Both epoxy and polyester resins more readily absorb organic solvents than aqueous solutions. Traditional glass fibre-reinforced polyester (GRP) can show a catastrophic reduction in strength when exposed to acidic solutions, especially if the composite is under load (a phenomenon known as stress corrosion). Principally this is due to the E-glass fibres cracking when attacked by the acid.

13.12 COMBINED FATIGUE AND ENVIRONMENTAL EFFECTS

The first issue that must be tackled in this specialist area of composites fatigue work is how to perform the test. Fatigue tests are inevitably medium- to long-term tests, sometimes lasting days or weeks, thus the possibility of test coupons drying out during the period of the test must be considered. Tests performed at room temperature generally lead to little change in moisture content, since the times involved for most materials to absorb or desorb significant amounts of moisture are long. The problem of coupon drying is particularly acute, however, when fatigue testing takes place at elevated temperatures. Precautions must then be taken to preserve the moisture content of the specimen. One possibility is to test in a chamber in which the temperature and humidity are controlled. Although this method is used, it involves expensive equipment perhaps beyond the budget of many laboratories. Alternative approaches that have been used include enclosing the specimen in a polythene bag in which a salt solution is used to maintain the required humidity, or sealing the specimen totally by encapsulation.

The effect of environmental exposure on fatigue behaviour depends on the sensitivity of the laminate to matrix properties. Thus carbon fibre laminates, having a strong fibre–matrix interface, show little sensitivity to moisture content when tested at room temperature. In materials in which the fibre–matrix bond strength is less strong and susceptible to environmental attack, the laminate lay-up is also important, laminates with few fibres in the test direction being more susceptible than those with many fibres in the test direction. Even then, if tested at room temperature, the effect is small. Data on materials tested in elevated temperature fatigue are scarce, but do suggest that steeper S–N curves are obtained, even for CFRP, when tested wet and hot in fatigue.

13.13 SUMMARY

Different fibres and resins behave differently in fatigue, the resin being particularly important. In general, if fibres are aligned with the main loads then the S–N curves are very flat. Stress concentrations can produce marked reductions in static strength but result in even flatter S–N curves, the fatigue strength not being greatly affected.

Because of relatively low static design allowables, fatigue strength is not usually a serious design criterion for PMC structures. However, composite materials and structures are still developing relatively quickly and if static design allowables are increased, fatigue behaviour could become more important in future.

The moisture content and distribution can usually be determined for PMC composites by the application of Fickian diffusion theory. The effects of environmental exposure vary, but generally laminates with large percentages of fibres in the test direction are little affected, whereas those with few fibres in the test direction may show a significant reduction in properties. In addition, testing composites in modes such as compression or shear, which impose significant stresses in the matrix, can also cause reductions in properties when combined with environmental exposure, regardless of the laminate configuration.

FURTHER READING

General

Curtis, P. T. (1985) *CRAG test methods for the measurements of the engineering properties of fibre reinforced plastics*, RAE TR85099. RAE, Farnborough (now DRA Aerospace).

Harris, B. (1977) Fatigue and accumulation of damage in reinforced plastics, *Composites*, **8**(4) pp. 214.

Jamison, R. D., Schulte, K., Reifsnider, K. L. and Stinchcomb, W. W. (1983) *Characterisation and analysis of damage mechanisms in the fatigue of graphite-epoxy laminates*, ASTM STP836.

Talreja, R. (1981) Fatigue of composite materials: damage mechanisms and fatigue life diagrams, *Proc. Roy. Soc. Lond.*, **A378**, 461.

Specific

Curtis, P. T. and Moore, B. B. (1985) *A comparison of the fatigue performance of woven and non-woven CFRP laminates*, RAE TR85059. RAE, Farnborough (now DRA Aerospace).

Curtis, P. T. (1986) *An investigation of the mechanical properties of improved carbon fibre composite materials*, RAE TR86021. RAE, Farnborough (now DRA Aerospace).

Curtis, P. T. (1987) *A review of the fatigue of composite materials*, RAE TR87031. RAE, Farnborough (now DRA Aerospace).

PROBLEMS

13.1 Describe the types of damage that occur during fatigue of unidirectional and multidirectional fibre-reinforced plastics. Discuss the influence of fibre, matrix and interface mechanical properties on damage development.

13.2 Describe the effects of moisture, and other fluids, on polymer matrices. How do the changes that occur affect the performance of FRP when tested in fatigue in hot, moist and hot/moist environments?

SELF-ASSESSMENT QUESTIONS

Indicate whether statements 1 to 7 are true or false.

1. Test coupons subjected to fatigue loading never fail in the same manner as under static loading.

    (A) true
    (B) false

2. The tensile fatigue behaviour of unidirectional composites is determined solely by the fibres.

    (A) true
    (B) false

3. Increasing the toughness of the resin matrix improves the fatigue performance of unidirectional FRP.

    (A) true
    (B) false

4. The tensile fatigue strength of GFRP increases with loading rate.

(A) true
(B) false

5. Stress concentrators have less effect on tensile fatigue strength than on tensile static strength.

(A) true
(B) false

6. In hybrid composites the fatigue performance is dominated by the fibre type with the lowest strain to failure.

(A) true
(B) false

7. Moisture absorption in FRP always obeys Fick's second law of diffusion.

(A) true
(B) false

For each of the statements of Questions 8 and 9, one or more of the completions are correct. Mark the correct completions.

8. Fatigue test coupons will give reliable results if

(A) they contain notches,
(B) they are waisted across the width,
(C) they are long,
(D) they are cycled at high speed,
(E) they fail in the same manner as a comparable structural component.

9. Unidirectional composites are tested because

(A) they are frequently used in practice,
(B) they are easy to make,
(C) their performance is dominated by the fibre,
(D) they give fundamental understanding,
(E) the results can be used to predict the performance of multidirectional laminates.

Each of the sentences in Questions 10 to 18 consists of an assertion followed by a reason. Answer:

(A) if both assertion and reason are true statements and the reason is a correct explanation of the assertion,
(B) if both assertion and reason are true statements but the reason is not a true explanation of the assertion,
(C) if the assertion is true but the reason is a false statement,

(D) if the assertion is false but the reason is a true statement,

(E) if both the assertion and reason are false statements.

10. Waisting of a test coupon is usually done across the width *because* the lay-up of a laminate is not disturbed.

11. Long specimens are preferred for tension–compression loading *because* anti-buckling guides are not required.

12. The tensile fatigue performance of unidirectional composites improves as fibre static strength is increased *because* the strain in the matrix is thereby reduced.

13. Unidirectional composites with tougher matrices have improved fatigue performance *because* the static tensile strength is increased.

14. The slope of the $S-N$ curve for multidirectional laminates is steeper than that for unidirectional composites *because* the off-axis layers are more easily damaged in fatigue.

15. Composites should be cycled at low frequencies *because* this minimizes heating of the specimen.

16. The high cycle fatigue strength of notched laminates is significantly lower than that of unnotched material *because* the notch causes extensive damage.

17. Woven fabric composites have better fatigue performance than those based on unidirectional prepreg *because* they are better at containing impact damage.

18. Fatigue performance of CFRP improves in a hot/moist environment *because* the resin matrix is more flexible.

ANSWERS

Self-assessment

1. B; 2. B; 3. B; 4. A; 5. A; 6. A; 7. B; 8. C, E; 9. B, D, E; 10. A; 11. C; 12. D; 13. E; 14. A; 15. A; 16. D; 17. D; 18. D.

14 | Joining

14.1 INTRODUCTION

Ideally, designers would like to make monolithic structures, i.e. structures without joints. For many reasons this ideal can never be realized, of course. There may, for example, be size limitations imposed by the materials or the manufacturing process; the structure may have to be disassembled for transportation; access may be required for inspection and repair, etc. The designer has two basic types of load-carrying joint at his disposal: *mechanically fastened* or *bonded*, both of which we shall discuss here. Most current experience relates to PMCs and although general trends may be relevant to MMCs they are unlikely to apply to CMCs. The data we shall present in this chapter relate only to PMCs.

Reaching a decision about the type to be used requires careful consideration of several parameters together with a knowledge of the service that the joint is expected to provide. Thus a requirement for disassembly would, almost certainly, preclude bonding; lack of adequate preparation facilities would also preclude bonding; a requirement to join thin sheet might rule out the use of mechanical fastening. Machining operations, which damage both fibres and matrix, will introduce sites at which failure may be initiated and are, therefore, to be kept to a minimum.

We can list several points both for and against each type of joint, as indicated below:

(a) *Mechanically fastened joints*

Points for:

  (i) no surface preparation of component is required,
 (ii) disassembly is possible without component damage,
(iii) there are no abnormal inspection problems.

Points against:

(i) holes cause unavoidable stress concentrations,
(ii) they tend to incur a large weight penalty.

(b) *Bonded joints*

Points for:

(i) stress concentrations can be minimized,
(ii) they incur a small weight penalty.

Points against:

(i) disassembly is impossible without component damage,
(ii) they can be severely weakened by environmental effects,
(iii) they require surface preparation,
(iv) joint integrity is difficult to confirm by inspection.

Load carrying joints often have an overlap configuration; see Figure 14.1(a). Such joint forms would be appropriate for joining flat laminates or tubular members, which can be either bonded (glued) together or mechanically fastened (Figure 14.1(b)). Other joint forms, such as flanges, are not considered here.

Because the strength of a joint is sensitive to a large number of parameters it is vital that consideration is given to the joints at the outset of any design. Failure to do so may result in the component being, say, of adequate stiffness but being impossible to join.

14.2 MECHANICALLY FASTENED JOINTS

In addition to material and configurational parameters the behaviour of mechanically fastened joints is also influenced by *fastener parameters* such as fastener type (screw, rivet, bolt), fastener size, clamping force, washer size, hole size and tolerance. These definitions are, of course, quite arbitrary and there will often be overlap between some of the parameters from each group. Of these fastener parameters the clamping force, i.e. the force exerted in the through-thickness direction by the closing of the fastener, is of critical importance. It is clear from this that a full theoretical description of the stresses in such a joint must include their three-dimensional nature, a fact which has limited the analytical treatment given to such joints.

14.2.1 Failure modes

In additional to fastener failure – in shear and/or bending – there are essentially four modes of failure, as illustrated in Figure 14.2. In all cases we

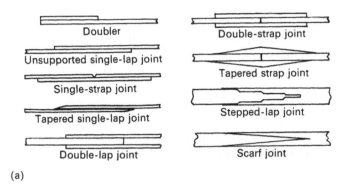

(a)

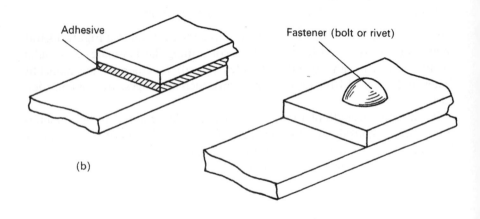

(b)

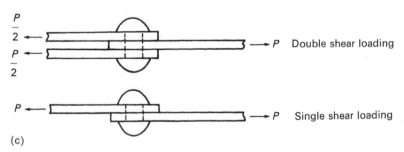

(c)

Figure 14.1 (a) Typical joint configurations. (Source: Hart-Smith, 1987a.); (b) Bonded and mechanically fastened joints; (c) Modes of loading; double shear and single shear. The applied force P is transmitted via the fastener from one side of the joint to the other.

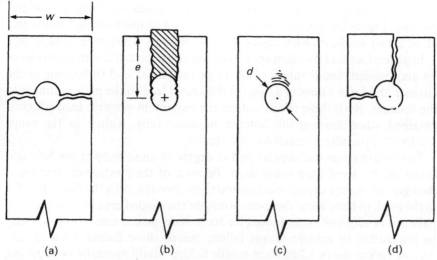

Figure 14.2 Modes of failure of mechanically fastened joints and definition of dimensions: (a) tension; (b) shear out; (c) bearing; (d) cleavage.

usually express the value of the fastener load (P) at failure in terms of a stress which is presumed constant over the appropriate area (P is defined in Figure 14.1(c)).

Tensile (or *tearing*) *failure* (Figure 14.2(a)) is related to the net area (($w-d)t$) through the fastener hole, corresponding to the member width in a single hole joint, or the pitch in a row of fasteners; thus

$$P_t = \hat{\sigma}_t(w-d)t, \qquad (14.1)$$

t being the minimum plate thickness, P_t the failure load, w and d are defined in Figure 14.2, and $\hat{\sigma}_t$ is known as the net tensile strength of the joint. Clearly the narrower the laminate the more likely it is we shall get tensile failure.

Shear out failure (Figure 14.2(b)) is related to the shear areas emanating from the hole edge parallel to the load and determined by the end (or edge) distance, e, which is typically about $4d$. The smaller the value of e the more likely we are to get shear failure. The shear out strength $\hat{\sigma}_S$ of the joint is given by

$$P_S = \hat{\sigma}_S \, 2et, \qquad (14.2)$$

where P_S is the failure load.

Bearing failure is based on the projected area of the hole (Figure 14.2(c)) and is determined by the diameter, the relationship of failure load P_b to bearing strength $\hat{\sigma}_b$ being

$$P_b = \hat{\sigma}_b \, dt. \qquad (14.3)$$

The bearing strength is normally greater than the compressive strength of the composite. At loads below failure the bearing stress $\sigma_b = P/dt$.

Cleavage (or *bursting*) *failure*, which is a mixed mode involving tension and bending (Figure 14.2(d)), cannot be expressed by a simple formula. Irrespective of the failure mode, we often express joint strength by the bearing strength ($\hat{\sigma}_b$).

In general we find that fastener failure can be prevented if extreme values of d/t are avoided. Small values of d/t (large thickness) lead to bending of the fastener and large values of d/t (small thickness) lead to the plate cutting into the fastener. Both these effects reduce the maximum possible load which is obtained when loading the fastener in shear only. Values in the range $1 < d/t < 3$ generally give satisfactory results.

The failure stress will depend on the degree of anisotropy at the hole and hence on the local fibre orientation. Because of their extremely low shear strength, we would expect unidirectional composites, with the fibres parallel to the load, to have a low shear-out strength; the shaded area in Figure 14.2(b) pulls out at very low loads. Indeed for some fibre–matrix combinations it may be impossible to achieve tensile failure before shear failure, even in very narrow specimens ($w < 2d$), when tensile failure would normally be expected, at ridiculously high end distances ($e > 10d$). Laminates containing a significant proportion of $\pm 45°$ fibres have high shear strength and low stress concentrations at the hole, hence they are relatively insensitive to end distance.

Bearing strength is determined by the compressive strength of the material and the through-thickness constraint afforded by the fastener. Without this constraint, as occurs in a pin-loaded joint, failure in a brush-like manner due to through-thickness expansion will take place at very low loads. Support for the load carrying 0° fibres is essential and laminates containing $\pm 45°$ and/or 90° fibres have a good bearing performance. It follows that GRP (i.e. glass fibre–polyester) laminates in the form of chopped strand mat (CSM), which are effectively isotropic, are better than the more directional woven roving (WR) materials.

'Homogeneous' lay-ups with a large number of ply interfaces give a higher bearing strength than 'blocked' laminates (i.e. those in which large numbers of plies of the same orientation are grouped together), and the latter may have only 50% the strength of the former.

14.2.2. Fasteners

All the conventional forms of fastener can be used with FRP. The simplest method, the *self-tapping screw*, is not however recommended as the low through-thickness strengths of the laminate can easily lead to thread stripping. If a demountable joint is required 'heli-coil' inserts, or something similar, must be used.

Rivets can be used quite successfully on laminates up to about 3 mm thick. The choice lies between solid and hollow types and whichever are chosen, care must be taken to minimize damage to the laminate during hole drilling and closing of the rivet. If countersunk rivets are used there will obviously be a limitation on the minimum laminate thickness. It is also beneficial to choose the largest possible countersink angle (say 120° rather than the more usual 90°) to reduce the possibilities of rivet pull-through.

Bolts provide the most efficient form of mechanical fastening because the through-thickness constraint prevents early failure in bearing. Even a bolt with 'finger-tight' washers raises the pin bearing strength by a large margin, further improvement being produced as the bolt is tightened. Hence, even if a nut loosens in service considerable load-carrying capacity is retained. For CFRP the optimum lay-up contains about 55% of ±45° fibres while for GFRP the optimum proportion of ±45° fibres appears to be rather higher. Most fastener manufacturers now produce bolts and rivets that are specially configured for use with composites.

14.2.3. Typical bearing strengths

Provided certain geometric requirements are met joint failure will be in bearing. The 'ideal' values given in Figures 14.3, 14.4 and 14.5 refer to a

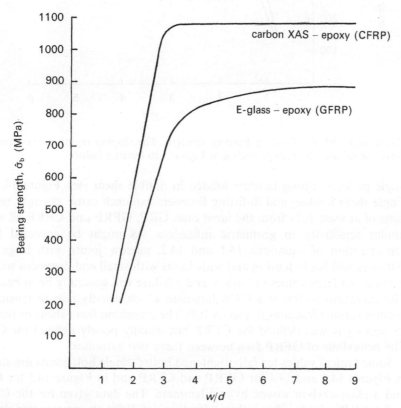

Figure 14.3 Variation of bearing strength with w/d. The plateau region corresponds to bearing failure, the sloping region to tensile failure.

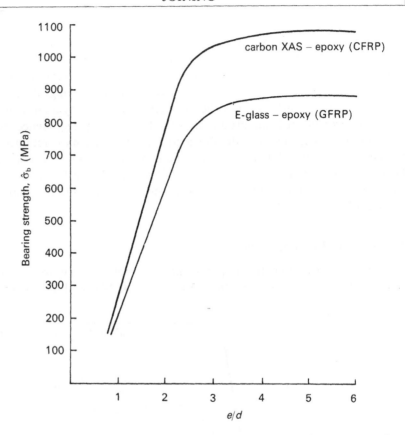

Figure 14.4 Effect of e/d on bearing strength. The sloping region corresponds to shear out failure, the plateau region at high e/d to bearing failure.

single perfectly fitting fastener loaded in double shear (see Figure 14.1(c)). Single shear loading and ill-fitting fasteners can each cause strength reductions of at least 10% from the ideal case. GRP, GFRP and CFRP all show similar sensitivity to geometric influences. As might be expected from consideration of equations 14.1 and 14.2, narrow joints with large end distances will fail in tension and wide joints with small end distances will fail in shear. At large values of both w and e failure will generally be in bearing. The exception to this is a CSM laminate which usually fails in tension at normal volume fractions (typically 0.2). The transition from shear or tension to bearing is well defined for CFRP but usually poorly defined for GRP. The behaviour of GFRP falls between these two extremes.

Some typical values for fully tightened bolted single hole joints are shown in Figures 14.3 and 14.4 for GFRP and CFRP and in Figure 14.5 for GRP and a glass–carbon woven hybrid laminate. The data given for the CFRP and GFRP refer to fibre volume fractions of 60%; the two materials can therefore be directly compared. The GRP and hybrid materials are of a

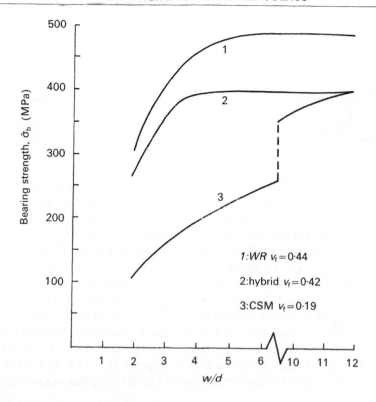

Figure 14.5 Variation of bearing strength with w/d for various glass fibre–polyester (GRP) laminates.

much lower fibre content, as quoted on the figure. However, when corrected for the difference in volume fraction, it is found that the performance of the GRP and GFRP are similar. Furthermore for the GRP the CSM is seen to be superior to the WR laminates, indicating the benefits of an 'isotropic' lay-up.

14.2.4. Comparison of metals and composites

Using typical values for bearing strength and density, *specific bearing strengths* (strength/density) may be obtained for steel and aluminium. Taking the 'plateau' values from Figures 14.3–14.5, similar data may be obtained for the composites (all adjusted to 60% fibre volume fraction). These specific bearing strengths are compared in Table 14.1 and it is seen that all the composites listed there show an advantage over the metals.

14.2.5. Efficiency and multi-hole joints

In most practical applications joints will consist of several fasteners. If the fasteners are arranged in a 'line', i.e. at right angles to the load, the data in

Table 14.1 Specific bearing strengths determined for single hole joints

	Density (Mg/m^3)	$\hat{\sigma}_b(MPa)$	Specific strength $[(Mg/m^3)/(MPa)]$
Steel S96	7.85	973	124
Al. alloy L72	2.70	432	160
CFRP	1.54	1070	695
GFRP $\}\ v_f=0.6$	2.10	900	428
GRP (WR)	2.10	682	324

the figures and in Table 14.1 for single hole joints can be used to determine the joint's performance, provided the spacing between holes (the 'pitch') is large enough for there to be minimal interaction between adjacent fasteners. For CFRP and GFRP this spacing should be $>4d$ whilst for GRP, particularly in the form of CSM, spacings of $6d$, or more, will be required.

When fasteners are arranged in a 'row', i.e. parallel to the load, each fastener will react only part of the load in bearing, the remainder being distributed around each hole as a 'by-pass' load. In these circumstances it becomes difficult to design a joint that performs significantly better than a single hole joint, as illustrated in Figure 14.6. In this figure the lower curve would have been obtained by plotting bearing stress, for a given lay-up, against d/w, rather than against w/d as in Figures 14.3–14.5. In Figure 14.6 gross section structural efficiency is obtained by dividing bearing strength by

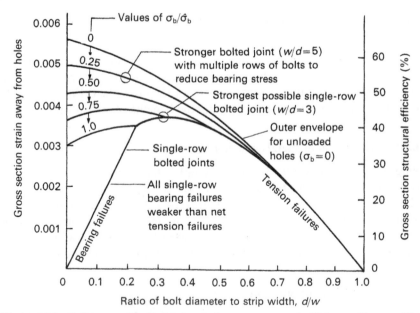

Figure 14.6 Influence of bolted joint design on structural efficiency. (Source: Hart-Smith, 1987a).

the tensile strength of the plain laminate, i.e. without a hole. The peak in this curve indicates the most efficient single hole joint, for that lay-up.

We can see from a comparison of the upper curves in Figure 14.6, which are for multi-hole joints, with the lower curve for a single-hole joint, that it is difficult to improve upon the optimum single-hole (or row) joint. Indeed, two fasteners will only be about 10% stronger than the single fastener case because, although the bearing stress is halved ($\sigma_b/\hat{\sigma}_b = 0.5$), the stress concentration arising from the by-pass load is nearly as high as that caused by the bearing load. Significant improvements in efficiency can only be achieved with complicated configurations having several fasteners, probably of different diameter, in tapering thickness laminates. For example, the point X in Figure 14.6 represents a situation where the bearing stress is only $0.25\hat{\sigma}_b$.

14.2.6. Theoretical stress analysis

Unlike bonded joints (see below) bolted joints have, until recently, received little theoretical treatment.

Two approaches are possible; either classical elasticity methods using complex variables or finite element methods. To give realistic results the friction between the bolt and the hole must be included in the analysis together with the length of the contact surface, a parameter which varies nonlinearly with load. The through-thickness stresses also influence the stress distributions as do the fact that stress–strain behaviour in shear is almost certainly nonlinear.

Clearly to include all the above effects results in an intractable problem. Data obtained from simplified analyses (two-dimensional) support the strength trends shown by experiment. The prediction of failure loads is, for the moment at least, only semi-empirical at best and any improvement will depend on the development of failure criteria that are more generally applicable, together with an easy-to-use three-dimensional stress analysis.

14.3 BONDED JOINTS

Bonded joints can be made by *glueing* together pre-cured laminates with a suitable adhesive or by forming joints during the manufacturing process, in which case the joint and the laminate are cured at the same time (*co-cured*). With co-cured joints the 'adhesive' is the matrix resin of the composite and we might expect that this monolithic type of construction will behave differently from joints with a separate adhesive. Whilst it is to the latter type of joint that this section is directed, many of the remarks, particularly those relating to geometry and failure modes, may also apply to co-cured joints.

14.3.1 Failure modes

A number of failure modes occur in bonded composite joints because of their anisotropic nature. In the adherends failure can be tensile, interlaminar or transverse, in the last two cases either in the resin or at the fibre–resin interface. These modes are shown in Figure 14.7, together with cohesive failure which can occur in the adhesive.

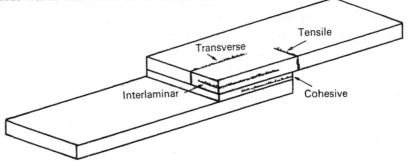

Figure 14.7   Failure modes in bonded joints.

There is, additionally, the possibility of failure at the adhesive–adherend interface. However such failures, which usually occur at low loads, should not take place in properly prepared joints.

14.3.2 Surface pretreatments prior to bonding

Polymer matrix composites are usually based upon epoxy or polyester resins which are highly polar and hence very receptive to adhesive bonding. Also they do not, of course, form oxides or corrode in moist environments. This combination of properties results in surface pretreatments being required which simply remove contaminants such as oils, mould lubricants or general dirt. There are two main techniques used to achieve this: (i) the *peel-ply method*, and (ii) *abrasion and solvent cleaning*, often conducted after a peel-ply surface has been exposed.

In the peel-ply method one ply of fabric, such as 'Dacron' or an equivalent, should be installed at the bonding surfaces, and removed just prior to bonding. In principle a clean bondable surface is exposed. However, many workers have commented that it is extremely difficult to ensure that the peel-ply has not left behind sufficient contamination, in the form of release agents used in the manufacture of peel-ply, to reduce both the strength of the adhesively bonded composite and greatly increase the coefficient of variation. If joints are prepared from the 'as-moulded' composite surface then the joint strength is dependent upon the extent of surface contamination; i.e. it is a function of the nature of the surface in contact with the composite during moulding.

To overcome the above problems it is necessary either to use a peel-ply which does not leave contaminants behind, which often means not using release agents and so makes the peel-ply difficult to remove, or to eliminate the residual contamination. This latter route has been the one successfully followed in industry and an *abrasion treatment,* followed by a *solvent wipe* to remove the abrasion products, has proved to be most effective. A light grit-blast with alumina particles has been found to be very efficient in removing the contamination, but methodical hand-abrasion, especially using commercial abrasive pads, may be equally effective.

14.3.3. Theoretical stress analysis

Early work in this field relating to bonded joints between metal adherends has been adapted for use with composite adherends. The application of classical theories to composites is complicated by their relatively low values of through-thickness shear and extensional moduli. However, it appears more important to account for the latter than the former. It is also essential to include the effect of residual thermal strains in the laminate arising from curing, and thermal mis-match when bonding to metals. Both linear and nonlinear analyses can be performed, the nonlinear approach being vital for calculating ultimate loads.

Although *linear analyses* cannot be used to calculate the ultimate strength of a joint they are nevertheless extremely useful as a basis for parametric studies. The results of a number of such studies show that maximum adhesive stresses in a joint can be reduced, and hence the strength increased, if we take the following steps: (a) use identical adherends or, if this is not possible, then equalize the in-plane and bending stiffnesses; (b) use as high an in-plane adherend stiffness as possible; (c) use as large an overlap as possible; (d) use an adhesive with the lowest possible tensile and shear elastic moduli; (e) use as homogeneous a ply lay-up (i.e. as many ply interfaces) as possible, with the fibres in the outer layer (next to the adhesive) oriented along the length of the joint.

The *nonlinear analysis* of *double lap joints* shows that the adhesive shear strain energy is the only significant factor determining joint strength. Thus a weaker ductile adhesive with a large area under the shear stress–strain curve will produce a joint of higher strength than a stronger brittle adhesive having a smaller area under the stress–strain curve. Also, for the same adherends, adhesives having the same failure stress, failure strain and strain energy should produce joints of equal strength.

We frequently assume the adhesive to behave in an elastic–plastic fashion which occurs once the maximum adhesive shear strain, at the joint ends, reaches its yield value γ_e, as defined in Figure 14.8(a). The stress will remain constant as a consequence of the assumed form of the stress–strain curve. If the joint is sufficiently long an 'elastic trough', of length s, will form as shown

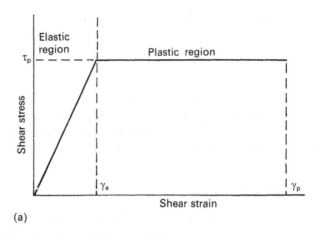

(a)

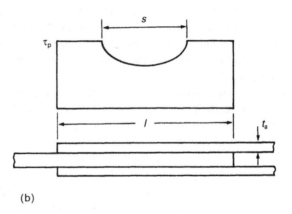

(b)

Figure 14.8 (a) Idealized adhesive shear stress–strain curve; (b) Consequent adhesive shear stress distribution for a double lap joint. (Source: Hart-Smith, 1987b.)

in Figure 14.8(b). The length of the plastic zone ($\frac{1}{2}(l-s)$) depends only on material parameters and is independent of total lap length, l.

The limiting value of joint strength, based on a failure criterion of total adhesive shear strain, occurs when the joint is just long enough for the elastic trough to form, i.e. $s=0$. Typically for CFRP the limiting overlap is about $30t_a$, where t_a is defined in Figure 14.8(b). There is no point in increasing the overlap beyond this critical value since no significant enhancement in strength will result. As adherend thickness increases, peel (through-thickness tensile) stresses become increasingly important and could become sufficiently large to limit joint strength and cause failure in an adherend with a low through-thickness tensile strength.

In contrast to double lap joints, the shear characteristics of the adhesive have little influence on the strength of *single lap joints*, which are determined mainly by adherend properties, peel stresses and overlap length. The length of the plastic zone in the adhesive is only one half that in an equivalent double lap joint and the elastic trough carries a significant proportion of the total load. Peel stresses are an order of magnitude greater than in a double lap joint and, in contrast to the latter situation, their effect can be minimized by increasing the lap length. For acceptable efficiency the overlap should be at least $80t_a$.

Scarf joints (Figure 14.1) between identical adherends will have a uniform distribution of adhesive shear stress and hence will, in contrast to other joint types, show a higher strength when used with a stronger brittle adhesive as opposed to a weaker ductile adhesive. Parametric studies show that joint strength increases indefinitely with lap length and hence it is always possible to choose a small enough scarf angle to transfer the full adherend strength. With non-identical adherends the main load transfer takes place at the end from which the more flexible adherend extends. It is thus possible for a plastic zone to exist in the adhesive only at this end of the joint.

Analysis of *stepped joints* shows that only over the end few steps will the adhesive strains exceed the elastic limit. As with double lap joints once an elastic trough has developed and a step reached its maximum load carrying capacity, further increase in step length is pointless. Once the end steps have reached their critical state, joint strength can only be increased by having more steps of smaller thickness. Analysis of both stepped and scarf joints between dissimilar adherends (e.g. composite and metal) must include the influence of the different adherend stiffnesses and thermal expansion coefficients.

14.3.4 Test results

In contrast to mechanically fastened joints, experimental data giving the strength of bonded joints in composites are limited and for polyester–glass composites in particular, virtually non-existent. Unfortunately, owing to the number of variables involved and their complex interactions it is difficult to quote 'typical' joints strengths. However, in single or double lap joints the failure load can be as high as 80 or 90% of the failure load of the basic laminate.

The need to include, in any analysis, the effects of adhesive nonlinearity is apparent when static strength predictions are compared with experimental results. For single lap joints the effect of joint rotation must also be included. Failure to account for these effects can result in experimental results being twice the predicted value.

When loaded in fatigue, the strength at 10^6 cycles is likely to be only 30% of the static value. Although carrying a higher uniform shear stress, the scarf joint performs better in fatigue than all other types, principally because shear and peel stress concentrations are minimal.

14.3.5 Adhesive properties

Clearly, to use the analytical methods described above, we need an accurate knowledge of the stress–strain properties of the adhesive. Two main methods are used for measuring the properties in shear, namely the *single lap joint with thick adherends* and the *napkin-ring torsion* test (see Figure 14.9). The former method has the merit of being relatively simple to perform with specimens that can be easily and cheaply made. However, the shear stress distribution along the joint is not completely uniform in spite of the joint geometry being chosen with this requirement in mind. The napkin-ring method, although more difficult and expensive to perform, produces essentially pure shear in the adhesive.

Unfortunately, elastic properties have probably not been determined for most of the large number of available adhesives. A typical epoxy adhesive will have a shear modulus around 1 GPa with a shear strength of

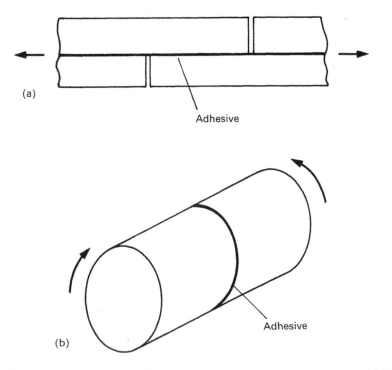

Figure 14.9 Adhesive test methods: (a) thick adherend; (b) napkin-ring. (Source: Adams and Wake, 1984.)

perhaps 50 MPa. A ductile adhesive might have a shear failure strain γ_p as high as 2.0, with $\gamma_p/\gamma_e = 20$ (γ_p and γ_e are defined in Figure 14.8(a)).

14.3.6 General design considerations

As indicated, the efficiency of a single lap joint is limited by the *peel stresses* caused by rotation of the joint. If suitably supported against this rotation the efficiency will be raised to that of a double lap joint. The limiting factor then becomes, as for double lap joints, the adherend thickness, with interlaminar tensile failure of the adherends being the predominant mode. This effect can be alleviated by tapering the adherends in the joint region. Simultaneous resistance to rotation and peeling can be achieved in a single lap joint by a combination of thickening and tapering, as shown in Figure 14.10.

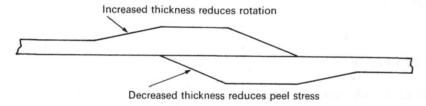

Increased thickness reduces rotation

Decreased thickness reduces peel stress

Figure 14.10 Profiling of adherend thickness to improve performance of single lap joint (Hart-Smith, 1987b.)

In view of the thickness limitation, scarf or stepped joints must be used for highly loaded adherends. Such joints are, of course, more difficult to make than single or double lap configurations. It is particularly difficult to maintain a uniform adhesive thickness in a scarf joint. Alignment and fit are not, however, so critical in a stepped joint. In both cases a ductile adhesive is better than a brittle one for overcoming manufacturing non-uniformities. The recently developed toughened acrylic adhesives appear to offer improved performance where peel is likely to be significant.

Surface preparation of the metal can affect the durability of composite-to-metal joints. The durability of composite-to-composite joints is adversely affected by absorbed moisture, both in the adhesive and the matrix of the composite itself; strength loss can be 50% or more compared to a dry joint.

To ensure acceptable reliability it is advisable to choose an adhesive whose shear strength is about 50% above that of the adherends. Additionally, the allowable value of adhesive maximum shear strength used for calculations should be reduced by some 20% from measured values to account for incomplete wetting of actual, as opposed to laboratory-made, joints.

Even in a fatigue environment, when adhesive strains are probably limited to the elastic region, ductile adhesives are preferable since they possess better fatigue properties and a higher static strength margin than do brittle adhesives.

Care must be taken when using results from tests on small (laboratory) specimens to design large joints. Because all dimensions, other than adhesive thickness, will scale-up, stress concentrations will increase with a consequent change in failure mode and a lower strength than expected.

Another problem that may arise upon bonding laminates is that *absorbed moisture* in the composite may be evolved during the adhesive bonding cycle and lead to poor adhesion and voids in the adhesive layer. The obvious method of overcoming this problem is to dry the composite prior to bonding, and this is the usual route followed, although it can be difficult to achieve in practice when undertaking repair work. Also, when undertaking adhesive bonded repairs, it may be necessary to remove absorbed hydraulic oils, etc., from the composite before satisfactory adhesion can be achieved.

14.4 REPAIR

14.4.1 Introduction

The approach to repair of composites to be described in this section is based on experience in the aircraft sphere (air lines, manufacturers, military), where most work has been done. The philosophy is quite general, however, and could obviously be applied in other fields. A repair is essentially a joint and it is clear that the requirements for repairability should be considered at the outset of a design, to ensure that the component is repairable (i.e. has adequate joint strength).

As already seen, many types of damage are possible in composite structures. Defects may be introduced during composite manufacture owing to errors in manufacturing procedure. Damage may be initiated during the life of the component as a result of exposure to the service environment, by accumulation of minor damage sustained in the normal use of the composite, from occasional exposure to an abnormal environment, or from impact or a local mechanical overload caused by misuse. One of the most important aspects is to establish the size and location of the area which is defective. As the reader will learn in the next chapter, ultrasonics and X-radiography are the most commonly utilized inspection techniques.

The objective of a repair action is to restore structural integrity to a damaged component. The particular procedure used will depend on the type of component, amount of joint efficiency required and considerations of surface smoothness. The repair environment will also affect the technique used in repair operations. At a depot facility, major repairs are possible, but at a field position the limits on repairability are much more restricted.

14.4.2 Bonded repair techniques

Bonded techniques can be used in situations which range from cosmetic to primary structural repairs. *Cosmetic repairs* refer to damage which is not structurally significant (e.g. dents, missing surface plies). The repair is made to restore surface smoothness. In these repairs, a potting compound or a liquid adhesive is spread into the damaged area and formed to the component's contour. *Injection repair* is another type of procedure which is used for minor disbonds or delaminations. In this procedure a number of holes are drilled to the depth of the damage. Filler resin is heated to decrease its viscosity and injected under pressure until the excess flows out of adjacent holes. Pressure can be applied to the repaired area to ensure mating of adjacent regions. If serious damage is encountered, then more rigorous repair procedures must be employed. In this case there are two types of bonded patches which we can use to repair structural damage, namely, flush scarf patches and external patches (Figure 14.11).

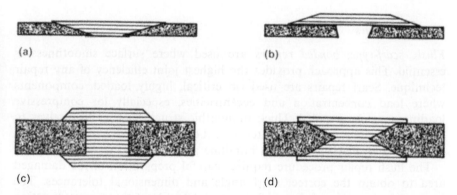

(a) (b)

(c) (d)

Figure 14.11 Typical repair schemes: (a) flush scarf patch in thin section; (b) external patch in thin section; (c) butt jointed insert with doublers in thick section; (d) double scarfed flush patch in thick section. (Source: Robson, 1992.)

After assessing the damage and before deciding upon a repair, the question of the parent laminate moisture condition becomes important. Moisture absorbed in the laminate, and/or entrapped moisture in honeycomb, can be very detrimental to the integrity of bonded repairs. Prebond drying (a minimum of 48 h at 76–93 °C), slow heat up rates, reduced cure temperatures and selection of adhesives less sensitive to moisture can minimize or eliminate the problems. The 120 °C curing adhesives, as a group, are more sensitive to prebond moisture at higher temperatures (above 70 °C) than 175 °C curing adhesives. Drying the parent laminate to an average moisture content of less than 0.5% is recommended. This can be very time consuming, taking over 24 h for a 16-ply laminate.

(a) External patches

As the name suggests, the *external patch* method consists simply of applying a piece of composite to the exterior surface of the component to be repaired; no attempt is made to retain a flush finish. The technique is simpler to apply and less critical in nature than a scarf approach. External patch repairs require less preparation than scarf repairs. Limited back-side access or substructure interference in a damaged area would also favour the use of an external patch repair.

In this approach the load is taken over and around the damaged area. The bending strains due to the eccentric load path must be considered in the patch design. The patch must also be capable of withstanding the high peel and shear stresses which develop at the edge of the damage. In order to minimize this effect, and to diffuse loads to the substrate, the patch plies are stepped in diameter. The low through-thickness tensile strengths of laminates imposes a thickness limitation which means that the technique is restricted to thinner laminates. Peeling can be minimized by small fasteners pitched at 25 mm spacing around the hole.

(b) Scarf repairs

Flush, scarf-type, bonded repairs are used where surface smoothness is essential. This approach provides the highest joint efficiency of any repair technique. Scarf repairs are used on critical, highly loaded, components where load concentration and eccentricities, especially for compressive loading, must be avoided. Thick monolithic structures lend themselves to such repairs since an external patch would cause excessive out-of-mouldline thickness and unacceptably high bondline peel and shear stresses.

The flush repair procedure requires careful preparation of the damaged area to obtain the correct scarf angle and dimensional tolerances. The laminate orientation of the patch must match that of the damaged section which has been removed. Scarf patches can be co-cured (cured on the damaged laminate) or precured (cured and then secondarily bonded on to the laminate) and they can either be single or double scarfs. The double scarf permits a shorter joint length and requires less material to be removed from the substrate. Taper ratios of 20:1 are typical. Rather than placing unidirectional material as the outermost ply, an outside cover of woven material or $\pm 45°$ layers should be used, making surface defects such as cuts, scratches and abrasions less strength-critical.

14.4.3 Bolted repair techniques

Bolted patch composite repair is an alternative approach to the bonded repair concept. Bolted repairs can be used in cases where bonded patch repair of a thick laminate may result in shear stresses beyond the limit of the adhesive strength. This method is also appropriate when a bonded scarf approach would be too complex in terms of preparation and material removal.

Bolted patches can be applied from one side, or from two sides with a backing plate (Figure 14.12(a)). If the plates are thick, and bolt tolerances are tight, the backing plate can also carry load. The patch must be thick enough in some cases to accept flush head fasteners. The patch may have bevelled edges to improve surface conformability.

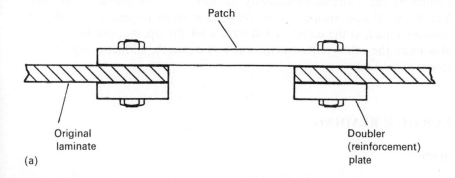

(a)

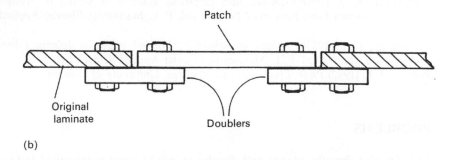

(b)

Figure 14.12 Bolted patch repairs: (a) external pach; (b) flush patch.

Another approach to bolted repair is a flush-type patch. In this case, the damage is removed and a section is inserted which is now flush with the surrounding undamaged area. Fasteners are applied through the patch to a doubler, as shown in Figure 14.12(b). One of the problems associated with the flush-type patch is the potential difficulty, with limited back-side access, of installing the large doublers that may be needed to accommodate the large number of fasteners required with such a repair.

Bolted repairs are not without limitations. The drilling operation can be time consuming and improper drilling may introduce additional damage.

Also, bolted repairs cannot be used on honeycomb structures. A further limitation may be the parent laminate; certain lay-ups may give too low a bearing strength to enable an effective repair to be made.

14.5 SUMMARY

We have discussed in this chapter the important issues that must be considered when designing joints in composite materials. Strengths can be limited by the material, particularly its anisotropic properties. It is, therefore, essential that adequate consideration is given to joints, including the need for repair, at the outset of a design, when lay-up, etc., can be chosen to also meet the other requirements such as strength, stiffness, impact resistance, and so on.

FURTHER READING

Specific

Adams, R. D. and Wake, W. C. (1984) *Structural Adhesive Joints in Engineering*, Elsevier Applied Science.

Hart-Smith, L. J. (1987a) Design and empirical analysis of bolted or riveted joints, in *Joining Fibre-Reinforced Plastics*, (ed. F. L. Matthews), Elsevier Applied Science.

Hart-Smith, L. J. (1987b) Design of adhesively bonded joints, in *Joining Fibre-Reinforced Plastics*, (ed. F. L. Matthews), Elsevier Applied Science.

Robson, J. E. (1992) *Ph.D. Thesis*, Imperial College, University of London.

PROBLEMS

14.1 Outline the advantages and disadvantages of using mechanical fastening to join composite laminates. Describe the typical failure modes that occur in such joints and how the mode and associated failure load is influenced by: fastener type, laminate lay-up and stacking sequence, joint geometry.

14.2 Discuss the relative performance of bonded single lap, double lap, stepped and scarf joints between composite adherends with particular reference to stress distributions in the joints, and to adhesive and adherend mechanical properties. Pay special attention to any factors that may limit, or enhance, joint strength and the issues that should be considered when predicting strength.

SELF-ASSESSMENT QUESTIONS

Indicate whether statements 1 to 6 are true or false.

1. Mechanically fastened joints require extensive surface preparation.

   (A) true
   (B) false

2. Bonded joints can be severely weakened by environmental effects.

   (A) true
   (B) false

3. Mechanically fastened joints in strongly anisotropic laminates will have very low strength.

   (A) true
   (B) false

4. Bolts provide the most efficient form of mechanical fastener.

   (A) true
   (B) false

5. The objective of pre-treatment prior to bonding is to produce a smooth surface.

   (A) true
   (B) false

6. The shear strain energy of the adhesive is the major factor determining the strength of bonded double lap joints.

   (A) true
   (B) false

For each of the statements of questions 7 and 8, one or more of the completions given are correct. Mark the correct completions.

7. The strength of a mechanically fastened joint is

   (A) dependent on the type of fastener,
   (B) dependent on the laminates' stacking sequence,
   (C) higher for rivets than bolts,
   (D) dependent on the fibres' failure strain,
   (E) reduced if the fasteners are a poor fit in the holes.

8. The strength of a bonded joint is

   (A) reduced if the adherends are not identical,
   (B) independent of the in-plane stiffness of the adherends,

(C) dependent on the overlap length,

(D) dependent on the laminates' stacking sequence,

(E) independent of the adhesive's elastic properties.

Each of the sentences of questions 9 to 16 consist of an assertion followed by a reason. Answer:

(A) if both assertion and reason are true statements and the reason is a correct explanation of the assertion,

(B) if both assertion and reason are true statements but the reason is not a true explanation of the assertion,

(C) if the assertion is true but the reason is a false statement,

(D) if the assertion is false but the reason is a true statement,

(E) if both the assertion and reason are false statements.

9. A bonded joint can be used when disassembly is required *because* the adhesive will be weakened by environmental exposure.

10. A bolted joint is stronger than a pin joint *because* no through-thickness clamping is provided by the pin.

11. Mechanically fastened joints in unidirectional composites have a high strength *because* the stress concentration factor is high.

12. Near-isotropic lay-ups have a high bearing strength *because* they are a 'homogeneous' lay-up.

13. A double lap bonded joint is more efficient than a single lap joint *because* the bonded area is larger.

14. Scarf joints have a higher strength when bonded with a strong brittle adhesive as opposed to a weaker ductile adhesive *because* the adhesive shear stress distribution is uniform.

15. When making a bonded repair pre-bond drying is essential *because* absorbed moisture can be detrimental to repair integrity.

16. Flush repairs cannot be made by bolting *because* countersunk fasteners give a lower strength than protruding head fasteners.

ANSWERS

Self-assessment

1. B;   2. A;   3. A;   4. A;   5. B;   6. A;   7. A,B,D,E;   8. A,C,D;
9. D;   10. A;   11. D;   12. C;   13. B;   14. A;   15. A;   16. D.

Non-destructive testing | 15

15.1 INTRODUCTION

It is accepted that all engineering components contain defects which may have been introduced from the raw materials, during manufacture of the component, during assembly of the component or by degradation during service. Not all defects are immediately harmful. Whether a defect makes a component unfit for service or not depends on many factors including the type, size and position of the defect and the service conditions. It is therefore important that we are able to detect and characterize defects in order that a decision may be made as to their severity. *Non-destructive testing* (NDT) is the name given to the various techniques which allow inspection of a material or component without impairing serviceability.

NDT techniques have been developed mainly for inspecting metal components and are used in quality control, in-service surveillance and in research. As far as composites are concerned, NDT techniques have been applied to ceramic, metal and polymer–matrix composites in research but only polymer matrix composites commonly undergo *non-destructive evaluation* (NDE) in industry. For this reason nearly all the examples given in this chapter will be for polymer–matrix composites (PMCs).

Most of the techniques normally used for the non-destructive testing of PMCs were originally developed for use with metals. Apart from visual inspection, five methods – ultrasonics, X-radiography, eddy currents, magnetic particle and dye penetrants – dominate the NDT of metal structures. Of these, magnetic particle inspection cannot be used with PMCs and eddy current testing is limited to those composites with conducting fibres, where it is sometimes used to determine fibre volume fraction and fibre orientation. Dye penetrants are used, though less than with metals as it is often desirable to repair a PMC component and it can be difficult to remove the penetrant before carrying out the repair. Therefore, the most frequently used NDT techniques for PMCs are ultrasonics and X-radiography.

We often employ PMC materials for the skins in honeycomb constructions and use adhesive bonding to join sections of composite structure. Therefore, NDT techniques for these constructions are required. The NDT of adhesive joints is particularly difficult since there are three types of defect which can occur even if we ignore failure of the adherends. These are

(a) a complete *disbond* (void),
(b) poor adhesion, i.e. a weak interface between the adhesive layer and the adherend, which leads to an *adhesive failure*, and
(c) *cohesive failure* due to poor cohesion, i.e. a weak adhesive layer.

There is no satisfactory test for adhesion strength but it is found that provided surface preparation is adequate, adhesive failure is not normally a problem. Cohesive strength is also difficult to assess though specialist ultrasonic techniques are used in this field. Disbonds are also a problem in honeycomb constructions and vibrations and thermal techniques can have significant advantages over ultrasonics and radiography in this area. These methods are also finding some uses in the inspection of solid composites.

The techniques discussed here are applicable both to quality control at the production stage and to in-service inspection, though some methods are difficult to apply in the field. Here again, the vibration techniques have the advantage of using portable equipment.

When composites are used in structures such as pressure vessels which are proof tested, acoustic emission monitoring during the proof test can give valuable information about whether and where damage is occurring. Since acoustic emission monitoring involves 'listening' to the propagation of damage, usually due to an applied stress, it is not strictly a non-destructive test. However, it performs a very useful role in quality assurance and is therefore included. Acousto-ultrasonics, which is based on a combination of the principles of both acoustic emission and ultrasonic testing, is also covered in this chapter.

Finally, it should be pointed out that since the failure mechanisms of composites are much less well understood than those of metals, it is often difficult to assess the significance of a defect once it has been detected by NDT. This tends to lead to very conservative design philosophies and there is currently much research activity aimed at determining what constitutes an acceptable defect.

15.2 ULTRASONIC INSPECTION

Ultrasonic inspection is the interrogation of materials or structures using *ultrasonic waves*, which are also known as *stress waves*. Some knowledge of these waves is necessary if we are to understand the principles of ultrasonics, and indeed acoustic emission and acousto-ultrasonics which will be discussed later in this chapter.

Stress waves are mechanical waves or vibrations and the usual relationship between frequency f and the wavelength λ applies:

$$\lambda = c/f, \tag{15.1}$$

where c is the velocity of ultrasound. The velocity is a function of the elastic properties of the material so it can be seen from equation 15.1 that for a given frequency of ultrasound the wavelength will vary with the material under investigation. For composites, a frequency in the range 1 to 10 MHz is normally used, though frequencies down to 500 kHz and up to 25 MHz are sometimes employed. These frequencies cannot be heard as only frequencies up to about 20 kHz are audible. In general high frequency (hence small wavelength) waves are more sensitive to defects as the theoretical minimum detectable defect size is of the order of the wavelength, whereas lower frequency waves have the advantage that they may penetrate a material to greater depths. The choice of frequency is therefore always a compromise between sensitivity and penetration.

There are several modes of propagation of waves in solids, the most important being

(a) *compressional* or *longitudinal waves* where particle motion is parallel to the direction of propagation;
(b) *shear* or *transverse waves*, where particle motion is at right angles to the direction of propagation; and
(c) *surface waves* of which there are a number of different forms, the most common being *Rayleigh waves* where particle motion is elliptical and restricted to a depth of about one wavelength from the surface.

As the stress waves travel through a material they are modified by the material itself, by the presence of defects and by encountering boundaries including the boundary between the component and the surrounding environment. At a boundary between two different materials a proportion of the waves is reflected and the rest transmitted. The proportion reflected depends on the angle of incidence and the *acoustic impedence* Z of the two materials. Z is given by the product of density ρ and wave velocity c:

$$Z = \rho c. \tag{15.2}$$

It can be shown that at an interface between two materials of impedances, Z_1 and Z_2, the percentage intensity, that is flow of energy per unit area, transmitted, I_T, is

$$I_T = \left[\frac{(4Z_1 Z_2)}{(Z_1 + Z_2)^2} \right] (100), \tag{15.3}$$

and the percentage intensity reflected, I_R, is

$$I_R = \left[\frac{(Z_1 - Z_2)}{(Z_1 + Z_2)}\right]^2 (100). \tag{15.4}$$

Later we will use these expressions to assess the efficiency of coupling media.

In ultrasonic testing we must inject stress waves into the material or component to be examined and then monitor the transmitted or reflected beam. Piezoelectric transducers or probes are used to both produce and receive the stress waves. These are capable of converting electrical pulses into vibrations when in the sending mode and of converting mechanical vibrations (stress waves) into electrical signals for analysis when receiving. It is essential that the stress waves propagate efficiently between the transducers and the component under investigation, that is there must be good acoustic coupling and a *coupling medium* is required between the transducer and the component to ensure satisfactory coupling. Applying equations 15.3 and 15.4 allows us to assess the feasibility of employing air and water as coupling media. The results given in Table 15.1 show that air is a poor coupling medium compared with water. Water is commonly used as a coupling medium, as are various greases and gels.

Table 15.1 Assessment of the efficiency of air and water as coupling media for PMCs using equations 15.3 and 15.4. (CFRP and GFRP are carbon and glass fibre-reinforced polymers respectively)

	Percentage intensity	
	Reflected	Transmitted
Air/CFRP	99.97	0.03
Water/CFRP	27.7	72.3
Air/GFRP	99.89	0.11
Water/GFRP	37.0	63.0

Clearly for reliable results it is important to ensure a constant level of coupling between the probe(s) and the test structure. This is often most conveniently achieved by immersion in a *water bath*. However, for large structures which would require large and expensive water tanks, and for honeycomb constructions which would float in water, a *jet-probe* (Figure 15.1) is frequently employed. In the jet-probe the water jet, which must flow in a laminar manner, acts as a waveguide for the ultrasound. Hand held probes with grease couplant are also used, particularly for in-service inspection when immersion is impossible.

Since most PMCs are in sheet form, the easiest way of assessing material quality by ultrasonic inspection is to scan the sheet and to measure the attenuation of the ultrasonic beam transmitted through it. This can be done

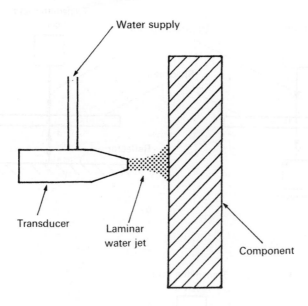

Figure 15.1 Jet-probe for ultrasonic testing.

by a single transmission of the beam and the use of a pair of probes held by means of a caliper on opposite faces of the sheet (Figure 15.2(a)); this method is called *through-transmission*. It is often convenient to use a single probe as both transmitter and receiver and to let the ultrasonic beam return through the specimen either by using a reflector plate (Figure 15.2(b)) or by examining the back-surface echo (Figure 15.2(c)). These techniques where the beam returns through the sample are termed *double through-transmission* or more usually *pulse-echo*. The *reflector plate* is glass as this has a large acoustic mismatch with water and therefore reflects most of the signal. In-service inspection, owing to constraints such as access only being readily available to one side of the component, may often be carried out using the *back-surface pulse-echo* method.

There are three ways of presenting ultrasonic results. These are known as A-, B- and C-scans. The *A-scan* is the simplest and is illustrated for back-surface echo testing of a sample with a defect in Figure 15.3. The amplitude of the signal reflected from the top surface, back surface and the defect are displayed on the y-axis of an oscilloscope while the x-axis corresponds to time, which can be converted to a depth scale. Thus the A-scan gives the depth of the defect and some information as to the size and nature of the defect may be obtained from the amplitude of the signal.

When a transducer is moved in a straight line over the surface of the sample under test a series of A-scans can be recorded as a function of position. If the information from the series of A-scans is stored by some

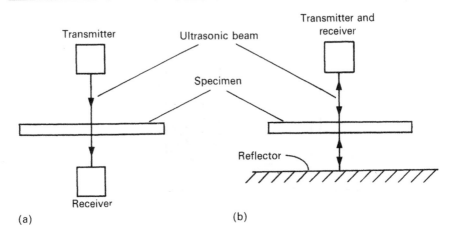

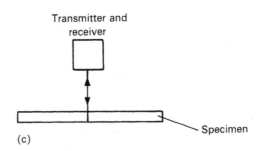

Figure 15.2 Transmitting and receiving transducer arrangements for ultrasonic testing: (a) through-transmission; (b) reflector plate in pulse-echo; (c) back surface pulse-echo.

means, such as a storage oscilloscope, we obtain a representation of the cross-section of the sample normal to the surface and on the line of the scan. Figure 15.4 shows that principle of this representation which is called a *B-scan*.

The test results from PMCs in sheet form are most conveniently presented in *C-scan* form. This involves mechanically or electrically scanning the transducer over a plane parallel to the sample in a rectilinear raster (Figure 15.5). The transducer position and the amplitude of the signal are sent to an $X - Y$ recording system. The amplitude is compared with various pre-set attenuation levels which are shown as changes in the trace density. Therefore, a 'map' of the attenuation level at different positions is produced as shown in Figures 15.5 and 15.6 which gives the spatial location in the plane of the sheet, but not the depth location, of any defects.

Attenuation increases can indicate the presence of delaminations, increased void content, the presence of foreign materials, moisture ingress and

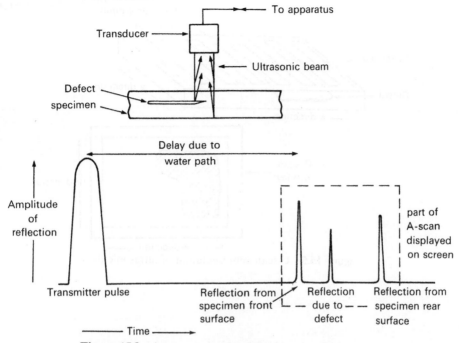

Figure 15.3 A-scan representation of ultrasonic data.

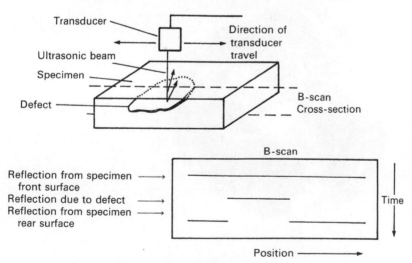

Figure 15.4 B-scan representation of ultrasonic data.

disbonds in adhesive joints. Translaminar cracks (normal to the plane of the sheet) can be revealed on a C-scan with high resolution probes but conventional pulse–echo testing using shear wave probes, or radiography, are probably more suited to this application.

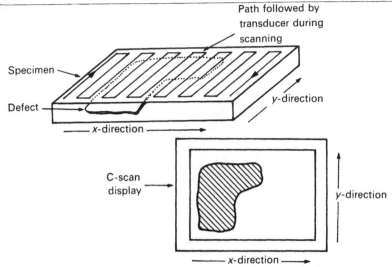

Figure 15.5   C-scan representation of ultrasonic data.

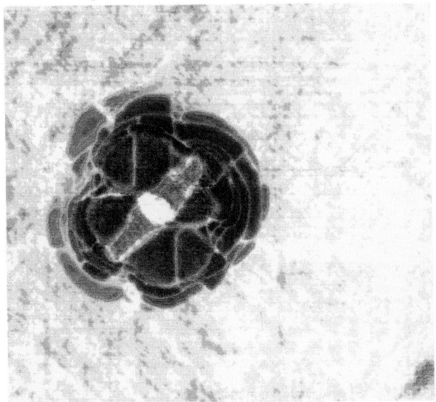

Figure 15.6   Ultrasonic C-scan of CFRP laminate showing an area of impact induced delamination; white indicates negligible damage and black bad damage. (Courtesy P. Cawley.).

Honeycomb constructions can present problems with ultrasonic testing since there is a severe impedance mismatch between the skins and the air in the cells. The pulse is therefore reflected back at these interfaces and so will only propagate down the cell walls. When there is a disbond between the top skin and the core, the level of the reflection from the bottom of the skin increases as no energy is transmitted to the core. Therefore, in this instance, a defect is indicated by an increase in signal strength. If access to both sides of the construction is possible, the through-transmission technique is sometimes used to inspect the bond between the upper and lower skins and the core in a single test. In this case, a fault is indicated by lack of transmission. Except when low frequencies of the order of 0.5 MHz are used, it is difficult to obtain satisfactory results from measurements of the echo from the lower face of a honeycomb construction. Through-transmission must therefore be used if the lower face is to be inspected.

There are a number of specialist techniques which employ ultrasound but the only one we will discuss is the *Fokker bond test*. The Fokker bond tester is an ultrasonic technique which uses shifts in the through-thickness resonant frequency of a bonded joint to detect disbonds or poor cohesion. The technique therefore works on a different principle from the attenuation measurements discussed earlier. However, since the frequencies used are well into the ultrasonic range, a coupling fluid is still required.

To summarize we note that the need for a fluid to couple the ultrasonic transducer to the structure means that the application of ultrasonic testing can be difficult, particularly in the field. However, designers have displayed considerable ingenuity in, for example, the design of water jet probes and the construction of special jigs for the inspection of complex shapes, and the technique is the most widely used for the inspection of composites.

15.3 RADIOGRAPHY

Radiography is based on the measurement of the attenuation of electromagnetic radiation after having travelled through the sample under test. The intensity of the transmitted radiation (I) depends on the intensity of the incident radiation (I_o), the thickness (x) and properties of the sample according to

$$I = I_o \exp(-\mu x), \tag{15.5}$$

where μ is the *linear absorption coefficient* of the material and is a function of the wavelength of the radiation and the density, ρ, of the material. The ratio μ/ρ is called the *mass absorption* or *attenuation coefficient* and is a material constant irrespective of the state of the material. Some typical values for μ and μ/ρ are given in Table 15.2 and the reader can see that the trend is for the absorption coefficients to increase with increasing atomic weight. Also μ

Table 15.2 Linear and mass absorption coefficients for some common elements. (Source: Bryant and McIntire, 1985)

Element	Energy of radiation (MeV)	Absorption coefficients	
		Mass μ/ρ (kg/m^2)	linear, $\mu(m^{-1})$
Hydrogen	0.05	3.35	28.1×10^{-4}
	0.15	2.65	22.2×10^{-4}
Carbon	0.05	1.86	41.3
	0.15	1.35	30.0
Oxygen	0.05	2.11	2.81×10^{-2}
	0.15	1.37	1.82×10^{-2}
Aluminium	0.05	3.57	96.4
	0.15	1.38	37.3
Silicon	0.05	4.27	100
	0.15	1.44	33.8
Iron	0.05	19.3	1520
	0.15	1.96	154
Lead	0.05	57.3	6500
	0.15	19.2	2180

and μ/ρ fall as the energy of the incident radiation is raised, from which we deduce from equation 15.5 that the penetrating power of the radiation is greater the higher its energy.

The mass absorption coefficient of a compound or a mixture is given by a simple law of mixtures:

$$(\mu/\rho)_{\text{material}} = (\mu/\rho)_A w_A + (\mu/\rho)_B w_B + \cdots \quad (15.6)$$

where w is the proportion by weight and the subscripts A, B, etc. refer to the constituents. The main constituents of polymers are hydrogen, carbon and other light elements with low μ/ρ values, and therefore the mass absorption coefficients for polymers are also low. This means that radiation of relatively low energy is sufficient to examine reasonable thicknesses of PMCs, whereas higher energies are needed for metals and metal matrix composites. For example 1 MeV X-rays will go through about 500 mm of polymer but only 150 mm of steel. If we compare the mass absorption coefficients of carbon and glass fibres we find that they are much higher for glass than for carbon owing to the presence in glass of elements of higher atomic weight such as silicon, calcium and aluminium (Table 15.2).

The monitoring of materials and components by radiography is based on the same principle as that for the detection of the fracture of bone, namely that the difference in attenuation as the radiation passes through the defect will be sufficient to reveal the defect by the changes in the blackening of a photographic film.

Radiographic interrogation of carbon fibre-reinforced polymers is made difficult by the low X-ray absorption of the material. This problem is overcome by using soft (low voltage, low energy) X-rays and high contrast recording techniques. Conventional low energy X-ray tubes containing beryllium windows and capable of operating between 5 and 50 kV are commonly employed. Radiography provides an excellent means of detecting transverse cracks and foreign inclusions such as swarf in solid composites. In honeycomb constructions the position of core inserts and shims, and damage to the core, can be detected. Radiography can also reveal porosity but it is not as sensitive as ultrasonic inspection for this purpose.

Carbon fibres and resin matrix materials have very similar mass absorption coefficients so X-radiography cannot be used to detect the fibres. However, carbon fibre prepreg is sometimes supplied with lead glass tracer filaments. These *radio-opaque tracers* enable radiography to be used to reveal lay-up directions, fibre buckling and the inclusion of any 'illegal' end butts between plies. A combination of oblique shots also allows the lay-up order, i.e. stacking sequence, to be determined.

We have concentrated on the problems in non-destructive testing of CFRP as a consequence of the low mass absorption coefficients of the constituents, however it is interesting to note radio-transparency also has some benefits as it makes CFRP ideal for constructing medical X-ray tables.

Radio-opaque penetrants are sometimes used to aid the detection of defects provided that ingress of the penetrant is possible. Thus the presence and extent of delaminations at the edge of composite components and of internal damage away from edges, but extending to the surface, may be monitored using penetrants. Unfortunately, some penetrants, such as tetrabromoethane (TBE), are highly toxic which has limited their application. Recently, however, less toxic penetrants, e.g. zinc iodide, have been developed and their use is expanding.

As mentioned earlier, glass fibres have larger values of mass absorption coefficient than carbon and polymers, so radiography is widely used with glass fibre composites for the determination of fibre distribution and fibre alignment. Voids and delaminations are also much easier to detect using radiography in glass-reinforced polymers than in carbon fibre composites because of the higher absorption of the glass giving greater contrast with the defect.

As a general rule radiographic techniques are better able to detect volume defects than narrow planar defects such as cracks. As far as planar defects are concerned the orientation of the defect to the incident beam is critical as illustrated in Figure 15.7; the best orientations for detection are low angles between the incident beam and the plane of the defect, and the worst situation is when the angle is 90°. Let us take as an example a 25 mm sheet; we can see from Figure 15.7 that at an angle of 5° a fine crack of only 0.05 mm width may be detected whereas at 45° the crack width would have

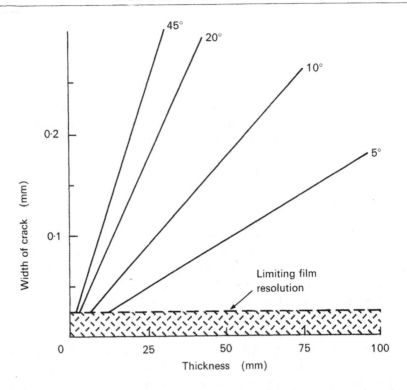

Figure 15.7 Detectable crack width as a function of thickness of sample under test for various angles between the incident radiation and the plane of the crack. (Source: Birchon, 1975.)

to be 0.26 mm before it may be detected. Radiography is used less with PMCs than with metal structures, partly because of the problems with obtaining high contrast images owing to low absorption coefficients, but also because composites are often used in sheet form and many of the defects of interest lie in the plane of the sheet. It follows that unless the beam is incident on the edge of the sheet, and in most instances this is an unrealistic option, this is the defect orientation for which radiography is least sensitive (Figure 15.8).

Two modifications to conventional radiography are real-time radiography and computed tomography. Developments in a number of areas including image intensifiers and digital processing for image enhancement have enabled *real-time radiographs* to be obtained. At present the sensitivity and resolution of the real-time images are not as good as those achieved with films which are also less expensive. Nevertheless, real-time radiography has considerable potential for rapid inspection of components as would be required, for example, on an assembly line.

In *computed tomography* a flat fan-shaped beam of X-rays penetrates a thin slice of the sample under test and the intensity of the transmitted beam is recorded as a function of position across the beam to give an absorption

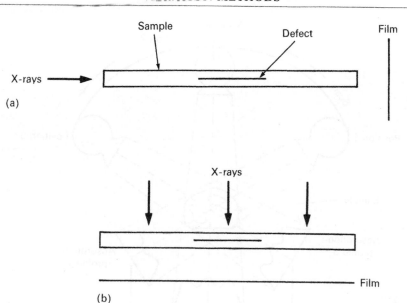

Figure 15.8 Radiographic examination of a planar defect lying in the plane of a PMC sheet: (a) desirable but unrealistic orientation of the beam and defect; (b) undesirable but experimentally viable orientation of the beam and defect.

profile of the transmitted beam. The X-ray source is moved around an axis normal to the plane of the beam and absorption profiles taken at regular intervals (Figure 15.9). Computer analysis of the absorption profiles enables a cross-sectional image of the sample to be constructed without any interference from the underlying or overlying material. This is the basis for the well-known body scanner which can typically achieve a spatial resolution of 1 mm and a relative density resolution of 0.02%. The densities of body tissues are similar to those of PMCs and medical scanners have been successfully employed for composites. Scanners are also produced specifically for non-medical applications; these are less expensive than the body-scanners but do not have the same relative density resolution.

Finally, radiography cannot be discussed without mentioning that as radiation is capable of penetrating the body there is a potential hazard to health. The need for safety precautions reduces the attractiveness of radiography, particularly for field use.

15.4 VIBRATION METHODS

We have seen how it is possible to use ultrasonic inspection, and to a lesser extent X-ray radiography, to detect many types of defect which are found in composites, honeycomb constructions and adhesive joints. However, a

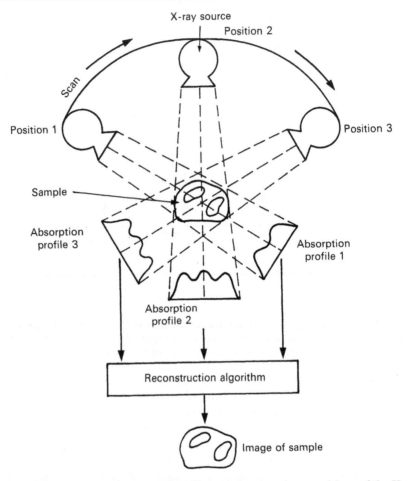

Figure 15.9 Computed tomography; diagram showing three positions of the X-ray source and the corresponding absorption profiles which are used to construct the image of the sample containing two defects.

coupling fluid is always required with ultrasonics which can be very inconvenient, particularly in the field, and honeycomb constructions can be difficult to inspect because the ultrasonic waves will only propagate through the cell walls, which are a small fraction of the plan area. There is therefore considerable interest in alternative methods such as *low frequency vibration methods*.

Low frequency vibration methods are classified as either global or local. *Global* methods involve the measurement of the natural frequencies and/or damping of the whole component and as such monitor the integrity of the whole component from a single measurement. In *local* tests the vibration properties of a region of the structure around the test point are measured. Monitoring a complete structure by a local test method is therefore slower than using a global method but is capable of locating and sizing a defect.

15.4.1. Global methods

Natural (resonant) frequencies and/or *damping* are properties of the whole structure and are independent of the position at which they are measured. Therefore a single test location can interrogate the whole component. This was the basis of the 'railway wheel-tap' test which simply involved a man striking the wheel and interpreting the characteristics of the resulting audible sound.

Modern signal processing equipment means that global tests can now be carried out objectively, quickly and reliably. A whole structure can be tested in less than five seconds so, if the detection of small, localized defects is not important, the method is very attractive.

Gross defects such as incorrect lay-up in structures fabricated from prepreg, errors in winding angle in filament wound components, incorrect fibre volume fraction and general environmental degradation produce readily measurable changes in the resonant frequencies (Figure 15.10). It should be borne in mind that resonant frequencies are sensitive to dimensions as well as to the presence of defects and consequently global methods can only be used for detection of defects in quality control if the components are produced to close dimensional tolerances.

The damping of PMCs is higher than that of metals and hence the pecentage change owing to the presence of a defect is smaller in the former. Nevertheless, for PMCs damping is still more sensitive to defects than are resonant frequencies and also less affected by small changes in dimensions. Unfortunately these advantages are more than outweighed by the difficulties in measuring damping accurately and in the foreseeable future damping measurements are unlikely to become commonplace.

15.4.2 Local methods

Probably the oldest test for the inspection of laminated constructions is the *'coin-tap'* test. This vibration method requires an operator to tap the area of the structure to be investigated with a coin and listen to the resulting sound. Defective areas sound 'dead' and an experienced operator can detect defects such as delaminations in composite materials, disbonds in adhesive joints and most types of defect found in honeycomb constructions. Until recently, however, the test has been purely subjective, which has severely limited its application.

The change in sound which the coin-tapper hears is caused by a local reduction in the stiffness of the structure in the region of the defect. This means that a 'tap' on the defective zone produces a 'soggy' response, i.e. the

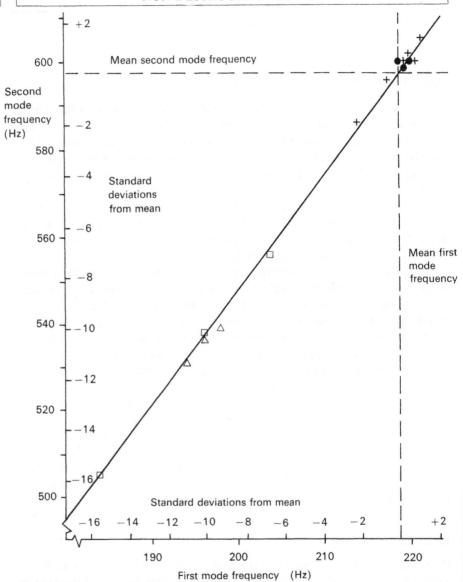

Figure 15.10 Plot showing the differences in the first and second mode flexural frequencies for good and faulty filament wound carbon fibre-reinforced polymer (CFRP): + good tube, ● tube with cut fibres, Δ tube with misaligned fibres, □ tube with low volume fraction of fibres. (Source: Cawley *et al.*, 1985.)

impulse has a low maximum force and a long time span, as shown in Figure 15.11(a). This impulse does not contain much energy at high frequencies and so does not excite the higher modes of the structure as strongly as an impact on a good area (Figure 15.11(b)). The structure therefore sounds 'dead' when tapped near a defect. An automated version of the coin-tap test, which will

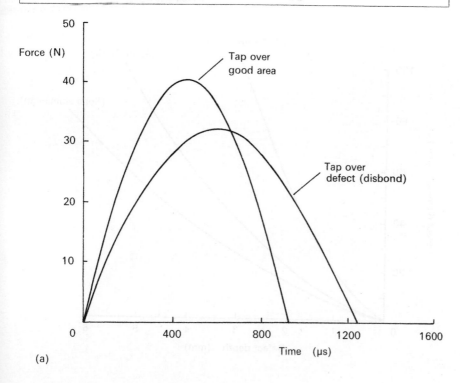

(a)

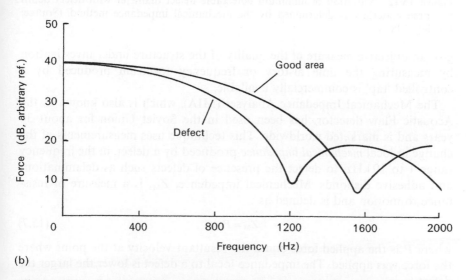

(b)

Figure 15.11 Local vibration method: (a) force–time records for impacts on defective and sound regions of an adhesively bonded structure showing that the maximum force is smaller and the time span greater in the defective area; (b) force-frequency record. (Source: Cawley and Adams, 1987.)

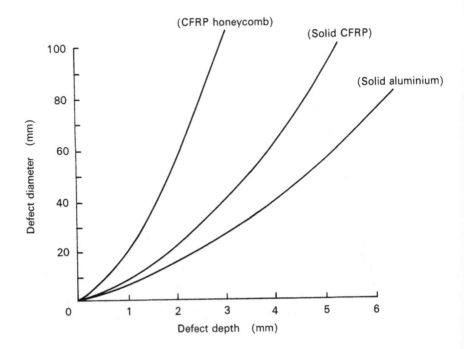

Figure 15.12 Variation of minimum detectable defect diameter with defect depth for three materials as determined by the mechanical impedance method. (Source: Cawley, 1987.)

give an objective measure of the quality of the structure under investigation by measuring the time history or frequency spectrum produced by a controlled 'tap' is commercially available.

The Mechanical Impedance Analyser (MIA), which is also known as the Acoustic Flaw detector, has been used in the Soviet Union for about 30 years and is marketed worldwide. This technique uses measurements of the change in local *mechanical impedance* produced by a defect, in the frequency range 1 to 10 kHz, to detect the presence of defects such as delaminations and adhesive disbonds. Mechanical impedence, Z_M, is a measure of resistance to motion and is defined as

$$Z_M = P/c, \qquad (15.7)$$

where P is the applied force and c the resultant velocity at the point where the force was applied. The impedance local to a defect is lower the larger the defect.

Several instruments work on the principle that when a layer of a structure is separated from the base layer(s) by a delamination or disbond, the separated layer can resonate and so, at certain excitation frequencies, it will vibrate at considerably higher than normal amplitudes. This increase in vibration level can readily be detected. Provided the resonant frequencies of the layers above defects of the size which must be detected lie within the operating range of the instrument, the technique can be a very rapid and attractive means of inspection.

All these vibration techniques work at frequencies below 30 kHz so there is no need to use a coupling fluid between the transducer and the structure. Also, the equipment required is usually portable so the methods are very attractive, particularly for field use. They are well suited to the inspection of honeycomb constructions and can detect, for example, barely visible impact damage (see Chapter 12) in solid laminates. However, they all suffer from the disadvantage that their sensitivity falls off with increasing defect depth, as indicated in Figure 15.12 which shows minimum detectable defect diameter versus depth curves for the mechanical impedence method. The coin-tap and membrane resonance methods have similar characteristics.

15.5 THERMAL METHODS

Most thermal methods involve the measurement of stationary temperature fields such as produced by variations in the insulation on a building. However the thermal method most used for the testing of composites, and hence the one that we will concentrate on, is based on measuring the response of the structure when subjected to a rapid pulse of heat. The rapid pulse produces transient effects and the recording of these is called *transient thermography*.

If a laminate is subjected to a heat pulse on one of its external surfaces the surface temperature will obviously rise. The temperature rise is governed by the amount of heat energy and the time of application, and by the thermal properties of the material. Heat diffuses through a defective material at various rates determined by the local thermal properties. Defects such as disbonds and delaminations result in lower heat diffusion and therefore produce local transient differences in temperature which are monitored using an infra-red camera.

Two experimental arrangements are possible which are termed single-sided and double-sided methods. In the *single-sided method* we place the infra-red camera so that it monitors the surface subjected to the heat pulse; with this arrangement the defects appear as a 'hot spot' as they restrict cooling by diffusion (Figure 15.13(a)). Transients can also be recorded by heating the back surface of the structure and measuring temperature changes at the front. This is the *double-sided method* and defective areas are cooler owing to the lower conduction through the defect (Figure 15.13(b)). *Cooling transients* can be used in a similar manner by applying an aerosol freezer spray to the surface to be tested. There are significant differences

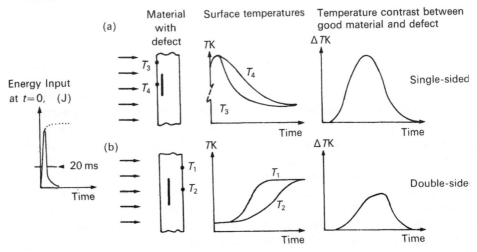

Figure 15.13 Schematic representation of transient thermography: (a) single-sided method; (b) double-sided method. (Courtesy C. Hobbs.)

between one- and two-sided examination: heating the back face and monitoring the temperature at the front enables deeper defects to be detected, heating and monitoring the temperature at the same surface, however, can produce better results with near surface defects.

It is important to note that *thermal transients* must be used because a defect would have a negligible effect on steady state heat transfer. In practice these thermal transients have to be recorded since a temperature difference sufficiently large for detection may only exist for a brief period, typically of 500 ms, and the use of video recording techniques has greatly simplified the recording process. The sensitivity of the method is reduced as conductivity increases and difficulties can also arise if the surface to be tested has areas of different emissivity, though this effect can be reduced by spraying the surface under examination to render it matt black.

Transient thermography has several attractive features:

(a) it is non-contacting and non-invasive although care must be taken not to overheat the surface;
(b) relatively large areas can be monitored with one heat pulse, and
(c) the data are readily stored and analysed and can be presented in pictorial form.

Another method which uses an infra-red camera is *vibro-thermography*. This involves vibrating the test structure at resonance and using the camera to detect hot spots. These correspond to areas of high energy dissipation (damping) produced by cracks or other defects which result in surfaces rubbing together during vibration. Unlike the heating method, this technique relies on a steady state temperature difference and is therefore not suitable for materials with high conductivity. Successful results have been reported with glass fibre composites but the conductivity of carbon fibre is too high for the method to be sufficiently sensitive.

These thermal techniques provide very rapid means of inspection and are useful when large areas of composite are to be tested and where the high cost of the camera and other capital items is justified.

15.6 ACOUSTIC EMISSION

When a dynamic process such as cracking occurs in a material, some of the released elastic strain energy can generate stress waves. These stress waves propagate through the material and eventually reach the surface, so producing small temporary surface displacements. In extreme cases, for example in the well known cracking of ice, the stress waves may be of high amplitude and low frequency and consequently easily audible. This is the origin of the term 'acoustic emission'. However, in most cases the stress waves are of low amplitude and high frequency and sensitive transducers are required to detect and amplify the very small surface displacements. The transducers which are generally used are of the same type as those employed in ultrasonic testing, namely piezoelectric crystals which convert a surface displacement into an electrical signal. The electrical signal is subsequently amplified and the signal resulting from a single surface displacement will be similar to that shown in Figure 15.14. Acoustic emission involves quantifying the numerous signals which may be detected during a test. A number of signal analysis techniques may be used but only those commonly employed will be described.

15.6.1 Signal analysis techniques

The simplest method of obtaining an indication of acoustic emission activity is to count the number of amplified pulses which exceed an arbitrary threshold voltage V_t. This is *ring-down counting* and the signal in Figure 15.14 would correspond to twelve ring-down counts. As the signal shown in Figure 15.14 was produced by a single surface displacement it is sometimes convenient to record a count of unity rather than the multiple count obtained by ring-down counting. This mode of analysis is known as *event counting.*

Counting techniques are extremely sensitive and are capable of detecting early stages of damage in composites under static and dynamic loading (Figure 15.15). They have been found to be particularly useful in the proof testing of PMC structures in conjunction with a parameter referred to as the *Felicity ratio.* If the load is removed during a proof test and then the structure reloaded, emissions may be detected at loads below that previously

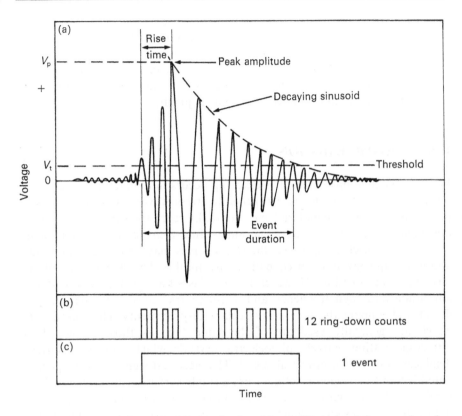

Figure 15.14 Acoustic emission monitoring: (a) amplified signal from a piezoelectric transducer resulting from a single surface displacement; (b) ring-down counting; (c) event counting.

attained. This is called the Felicity effect and the ratio of load at which the structure emits on reloading to the previous load is the Felicity ratio. The Felicity ratio is a measure of the damage to the composite: the lower the ratio the greater the damage (Figure 15.16).

Counting techniques are simple and sensitive to damage but, as a general rule, are not a good guide to the type and extent of damage. More detailed and comprehensive information on the emissions emitted over a period of time may be obtained from histograms of the number of events against peak voltage V_p (in practice, this is normally quoted in dB where: $dB = 20 \log (V_p/V_o)$ and V_o is the lowest detectable voltage), and number of events against event duration (also known as pulse width).

The peak voltage histogram, also called *amplitude distribution*, from a composite is usually complex, consisting of a number of peaks each attributable to a particular micro-damage mechanism. Figure 15.17 demonstrates how amplitude distributions are an aid to the determination of the extent of the different micro-damage mechanisms as load is increased. As a

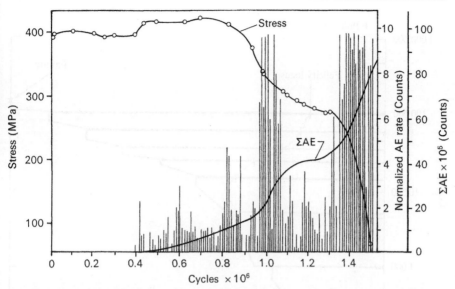

Figure 15.15 Acoustic emission counts and stress changes (due to change in the compliance of the specimen with crack length) as a function of the number of cycles for aluminium reinforced with 50% boron in a strain controlled fatigue test. (R. S. Williams and K. L. Reinfsnider, 1974.)

general rule, high amplitude events are indicative of high energy, deleterious damage.

When the *event duration*, or *pulse width*, histogram is obtained for a PMC we normally search for the presence of events of long duration. For example, in full-scale testing of PMC pressure vessels it is found that poor quality vessels, i.e. those with low burst pressures, produce long duration events at low pressures and a larger number of these events as the pressure increases (Figure 15.18).

15.6.2 Location

A stress wave may travel a considerable distance from the source before it is attenuated below the detection level of standard AE equipment. The attenuation depends on three factors, namely

(a) geometrical beam spreading, which is proportional to $(distance)^{-1/2}$;
(b) energy absorption by the material; and
(c) wave dispersion due to the frequency dependence of velocity.

The energy absorption by a polymer is much larger than that by a metal but, nevertheless, a transducer on a polymer matrix composite may still monitor a reasonable area of structure. Furthermore, by measuring the times of arrival of the elastic stress waves emanating from a given source at

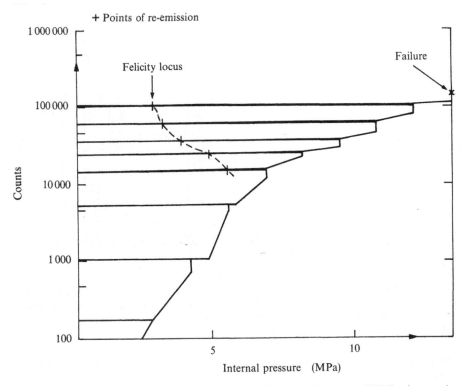

Figure 15.16 Acoustic emission monitoring of a proof test on GFRP pipe-work showing a decreasing Felicity ratio with increasing load. (Source: Fowler and Gray, 1979.)

several transducers, it is possible to determine the position of the source in an analogous manner to the determination of the epicentre of an earthquake in seismology. The number of transducers required depends on the size and complexity of the structure: one-dimensional location, on say a pipeline, would require two transducers in the area of interest (Figure 15.19(a)), two-dimensional location on a plane requires at least three transducers and the monitoring of a complex structure, such as a pressure vessel, may need many transducers as illustrated in Figure 15.19(b).

Location by acoustic emission can be carried out during proof or in-service testing and is a powerful technique as large areas of structure can be monitored. Although acoustic emission can locate an active defect, with current knowledge it is unlikely to be able to identify and size the defect.

15.6.3 Acousto-ultrasonics

Acousto-ultrasonics is a relatively recent method of non-destructive inspection which, as the name suggests, is based on a combination of the principles

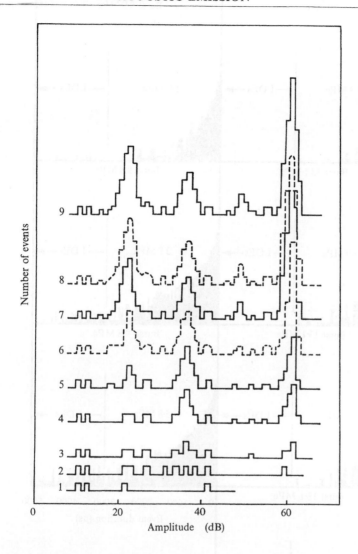

Figure 15.17 Amplitude distributions at loads increasing from 1 to 9 (arbitrary units) from a three-point bend test on epoxy resin reinforced with steel fibre. (Source: Berthelot and Billand, 1983.)

of both acoustic emission and ultrasonics. Acousto-ultrasonics utilizes pulsed ultra-sound stress wave stimulation in materials, as in ultra-sonics, combined with the sensing and measurement of the transmitted signal by acoustic emission techniques. Generally it is found that the detected acoustic emission signal, termed the *stress wave factor*, for given pulsing conditions, i.e., a fixed number of pulses of constant amplitude, decreases with defect content of a composite. This is demonstrated

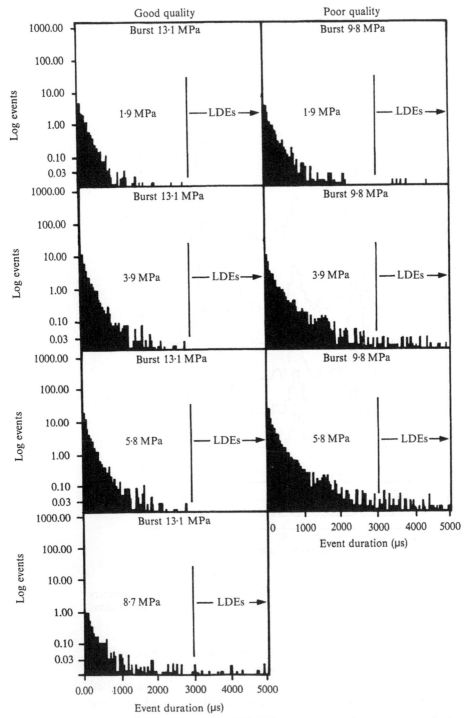

Figure 15.18 Event duration (pulse width) histograms at various pressures during the proof testing of good and poor quality filament wound carbon fibre–epoxy pressure vessels. (Source: Gorman and Rytting, 1983.)

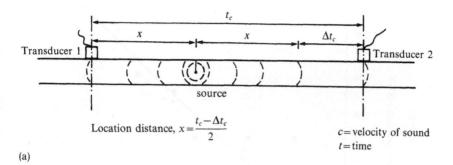

Location distance, $x = \dfrac{t_c - \Delta t_c}{2}$

c = velocity of sound
t = time

(a)

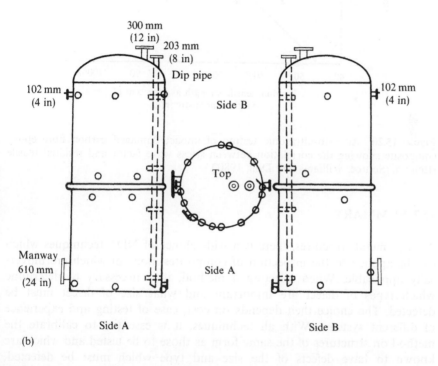

(b)

Figure 15.19 Location of defects by acoustic emission (a) linear location using two transducers; (b) sensor positioning guidelines (shown as open circles) for location on a vertical vessel. (CARP, 1982.)

by the data of Figure 15.20, which show that the stress wave factor decreased as the residual strength of a carbon fibre/epoxy composite fell because of impact damage.

It must be emphasized that acousto-ultrasonics gives an averaged measure of the condition of the material between the transmitting and receiving transducers and location of a defect can only be achieved by moving, usually by hand, the two transducers over a structure.

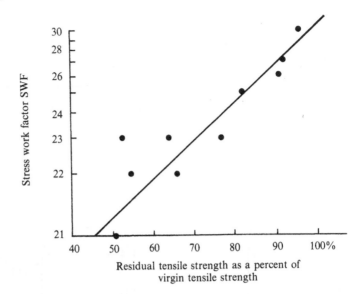

Figure 15.20 Acousto-ultrasonic testing of impact damaged carbon fibre–epoxy composite showing the correlation between stress wave factor and residual tensile strength. (Source: Williams and Doll, 1980.)

15.7 SUMMARY

As with metal structures, there is a wide choice of NDT techniques which can be applied to the inspection of composites, none of which are universally applicable. When selecting a method, it is necessary to determine which types of defect are important and what size of defect must be detected. The choice then depends on cost, ease of testing and experience of different systems. With all techniques, it is essential to calibrate the method on structures of the same form as those to be tested and which are known to have defects of the size and type which must be detected. Acoustic emission monitoring is most valuable when composite structures can be proof tested.

FURTHER READING

General

Boving, K. G. (ed.) (1989) *NDE Handbook – Non-destructive Examination Methods for Condition Monitoring*, Butterworths.

Halmshaw, R. (1987) *Non-destructive Testing*, Metallurgy and Materials Science Series, Arnold, London.

Miller, R. K. and McIntire, P. (eds) (1987) 'Nondestructive Testing Handbook, Vol. 5, Acoustic Emission Testing. American Society for Nondestructive Testing.

Mix, P. E. (1987) Introduction to Nondestructive Testing, Wiley.

Specific

Berthelot, J. M. and Billand, J. (1983) 1st. Int. Symp. on AE from Reinforced Composites, Soc. Plastics Industry, California. USA.

Birchon, D. (1975) Non-destructive Testing, Engineering Design Guides, Oxford University Press.

Bryant, L. E. and McIntire, P. (eds). (1985) Nondestructive Testing Handbook, Vol. 3, Radiography and Radiation Testing, American Society for Nondestructive Testing.

CARP (Committee on AE from Reinforced Plastics) (1982) Recommended Practice for AE Testing of GRP Tanks/Pressure Vessels, Soc. Plastics Industry.

Cawley, P. (1987) NDT Int., 20, 209.

Cawley, P. and Adams, R. D. (1987) The non-destructive testing of honeycomb structures by the coin-tap test, in Proc. ICCM, ECCM London, 1987. (eds F. L. Matthews, N. C. R. Buskell, J. M. Hodgkinson and J. Morton), Vol. 1, Elsevier, p. 1, p. 415.

Cawley, P., Woolfrey, A. M. and Adams, R. D. (1985) Composites, 16, 23.

Fowler, T. J. and Gray, E. (1979) Development of an acoustic emission test for FRP equipment, ASCE Convention and Exposition, Boston, USA.

Gorman, M. R. and Rytting, T. H. (1983) 1st. Int. Symp. on AE from Reinforced Composites, Soc. Plastics Industry, California, USA.

Williams, J. H. and Doll, B. (1980) Materials Evaluation, 38, 33.

Williams, R. S. and Reifsnider, K. L. (1974) J. Comp. Mat., 8, 340.

PROBLEMS

15.1. Ultrasonic equipment which is capable of working at 1 MHz and 10 MHz is available for testing a composite structure. Calculate the wavelength of the stress waves in the composite at the two available frequencies given that the velocity of the stress waves in the composite is $4000 \, m \, s^{-1}$.

Which frequency would you select in the following circumstances: (a) when detection of small defects is the most important criterion, (b) when penetration through thick sections is the most important criterion.

15.2 A 3 cm thick composite with a mass absorption coefficient of $2.5 \, kg/m^2$ is to be interrogated using 0.05 MeV X-rays. The composite has a spherical defect of 1 cm diameter which contains air. Calculate the ratio of the intensity of the radiation transmitted through the defective region of the composite to that transmitted through good regions of the composite. The following data will be required:

density of air $(kg \, m^{-3})$ 1.3
density of composite $(kg \, m^{-3})$ 1800

mass absorption coefficient of oxygen (kg m^{-2})   2.11
mass absorption coefficient of nitrogen (kg m^{-2}) 1.94
air is 76wt.%N_2, 24wt.%O_2.

SELF-ASSESSMENT QUESTIONS

Indicate whether statements 1 to 5 are true or false.

1.   Unfortunately NDT of a component usually renders it unfit for further use.
   (A) true
   (B) false

2.   Computed tomography is a specialized radiographic technique which enables a cross-sectional image of a sample to be constructed.
   (A) true
   (B) false

3.   An attractive feature of low frequency (< 30 kHz) vibration techniques is that there is no need for a coupling fluid.
   (A) true
   (B) false

4.   Stress waves play an important role in both acoustic emission and ultrasonic testing.
   (A) true
   (B) false

5.   Acousto-ultrasonics is based on a combination of the principles of both acoustic emission and ultrasonics.
   (A) true
   (B) false

For each of the statements of questions 6 to 10, one or more of the completions given are correct. Mark the correct completions.

6.   Stress waves

   (A) are electromagnetic waves,
   (B) are mechanical waves,
   (C) travel at the velocity of light,
   (D) travel at the velocity of sound,
   (E) may be surface waves,
   (F) may be longitudinal waves,
   (G) may be transverse waves.

7. The B-scan in ultrasonic testing

(A) may be considered as a series of A-scans,
(B) may be considered as a series of C-scans,
(C) involves moving a transducer in a rectilinear raster,
(D) involves rotating a transducer in ever decreasing circles,
(E) involves moving a transducer in a straight line,
(F) can give the depth of a defect in a PMC sheet.

8. Global low frequency vibration methods

(A) may involve the measurement of natural frequencies of the whole component,
(B) may involve the measurement of the damping of the whole component,
(C) may involve the measurement of the ring-down count from the whole component,
(D) involve the measurement of the vibration properties of a region local to the test point,
(E) are exemplified by the 'coin-tap' test,
(F) are exemplified by the 'railway wheel-tap' test.

9. Location of active defects by acoustic emission

(A) involves the measurement of the Felicity ratio,
(B) involves the measurement of times of arrival of stress waves,
(C) requires two transducers in the area of interest for one-dimensional location,
(D) requires at least three transducers in the area of interest for one-dimensional location,
(E) may be carried out during proof testing.

10. The figure below gives the results from a local low frequency vibration method; the two curves A and B correspond to tapping at two points on a carbon fibre-reinforced polymer skinned honeycomb structure. Indicate whether:

The x-axis is

(A) strain,
(B) stress,
(C) time,
(D) velocity;

curve A corresponds to

(E) a faulty tap,
(F) an acceptable tap,
(G) a tap over a good area of the structure,
(H) a tap over a defective area of the structure; and

curve B corresponds to

    (I)   the tap inducing resonance of the structure,
    (J)   the tap producing damage in the structure,
    (K) a tap over a good area of the structure,
    (L) a tap over a defective area of the structure.

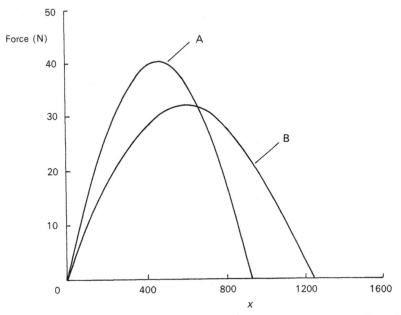

Each of the sentences in questions 11 to 15 consist of an assertion followed by a reason. Answer:

  (A) if both assertion and reason are true statements and the reason is a correct explanation of the assertion,
  (B) if both assertion and reason are true statements but the reason is not a correct explanation of the assertion,
  (C) if the assertion is true but the reason is a false statement,
  (D) if the assertion is false but the reason is a true statement,
  (E) if both the assertion and reason are false statements.

11.   Air is not used as the coupling agent in ultrasonic testing of carbon fibre-reinforced polymers *because* it has low linear and mass absorption coefficients.

12.   Water tanks are not often used for the ultrasonic testing of honeycomb constructions *because* honeycomb structures tend to float in water.

13.   Most thermal methods used on PMCs are based on measuring transient heating effects *because* defects significantly enhance heat flow and have a marked effect on steady state heat transfer.

14. Vibro-thermography, in spite of its name, is not classed as a thermal method but as a low frequency vibration method *because* it involves vibrating the test structure.

15. Acoustic emission monitoring of composites usually yields complex peak voltage histograms (amplitude distributions) *because* each micro-damage mechanism produces a characteristic amplitude distribution.

16. Select the correct word from each of the groups given in italics in the following passage.

Radiography is based on the measurement of the (*enhancement/attenuation/production*) of electromagnetic radiation by the component under test. The intensity of the transmitted radiation depends on the properties of the material being interrogated: the relevant property is the (*mass absorption coefficient/elastic modulus/mechanical impedence*). Because of the (*high/intermediate/low*) absorption of X-rays by (*glass/carbon/steel*) fibres, these fibres cannot be readily detected in a PMC. (*Radio-active/Radio-transparent/Radio-opaque*) penetrants are sometimes used to detect the presence and extent of (*swarf/delaminations/transverse cracks*) at the edge of a PMC component.

ANSWERS

Problems

15.1 $\lambda = 4$ mm at 1 MHz and 0.4 mm at 10 MHz; (a) 10 MHz; (b) 1 MHz.
15.2. 89.4.

Self-assessment

1. B; 2. A; 3. A; 4. A; 5. A; 6. B, D, E, F, G; 7. A, E, F; 8. A, B, F;
9. B, C, E; 10. C, F, G, L; 11. B; 12. A; 13. C; 14. D; 15. A; 16. attenuation, mass absorption coefficient, low, carbon, radio-opaque, delaminations.

Appendix

A.1 MATRICES AND DETERMINANTS

Sometimes it is convenient to set out numerical information in a rectangular array. In mathematics, if the array satisfies certain conditions, which we shall specify later, then it is called a matrix. It is normally written inside square brackets, e.g.

$$\begin{bmatrix} 3 & -2 & 5 \\ 1 & 0 & 8 \end{bmatrix}.$$

An $(m \times n)$ matrix is one with m rows and n columns, so the above matrix is a (2×3) matrix. A square matrix is one with the same number of rows as columns, i.e. $m = n$. A matrix with only one row or column is called a vector.

Example
$[4, 6, -1]$ is a (1×3) matrix. Since $m = 1$ it is called a row vector, or row matrix.
   Likewise

$$\begin{bmatrix} -1.2 \\ -2.5 \\ 0.8 \end{bmatrix}$$

is a column vector or column matrix as it has only one column.

$$\begin{bmatrix} 2 & 6 \\ 1 & -3 \end{bmatrix}$$

is a (2×2) matrix and is obviously square.

To save space when writing it is often convenient to put a column matrix into the form of a row matrix. In this case we often use a different type of bracket, e.g. $\{-1.2 \quad -2.5 \quad 0.8\}$ denotes the column matrix given above.

A.2 DETERMINANTS

From every square matrix we can calculate a single value called its determinant. The definitions for the calculation of a determinant and for the addition and multiplication of matrices have been chosen so that matrices and determinants are of some practical use, e.g. in the solution of simultaneous equations. Non-square matrices do not have an associated determinant.

A.2.1 The calculation of a determinant of a 2×2 matrix

The smallest square matrix to consider is the (2×2) matrix which in its most general form may be written

$$\begin{bmatrix} a & b \\ c & d \end{bmatrix}.$$

Its determinant, written

$$\begin{vmatrix} a & b \\ c & d \end{vmatrix},$$

is defined to be the value $(ad - bc)$. Thus

$$\begin{vmatrix} 4 & 1 \\ 11 & 2 \end{vmatrix} = 4 \times 2 - 11 \times 1 = -3.$$

We always replace the [ ] brackets of a square matrix by straight lines || when we want to denote its associated determinant. To calculate determinants of larger matrices, we use 'co-factors'. Each element of a square matrix has a co-factor (see below).

A.2.2 General method for finding a co-factor

Example
To find the co-factor of the element 6 in the first row second column of the (3×3) matrix

$$\begin{bmatrix} 1 & 6 & 4 \\ 2 & -2 & 1 \\ 4 & 5 & 3 \end{bmatrix},$$

(a) Remove the row and column of the given element. A smaller array is left.	Removing first row and second column we get $$\begin{bmatrix} 2 & 1 \\ 4 & 3 \end{bmatrix}.$$
(b) Calculate the determinant of this array.	$\begin{vmatrix} 2 & 1 \\ 4 & 3 \end{vmatrix} = 2 \times 3 - 4 \times 1 = 2.$
(c) The resulting value is the co-factor of the given element unless the position of that element corresponds to a negative sign in the diagram. $$\begin{vmatrix} + & - & + \\ - & + & - \\ + & - & + \end{vmatrix}$$ We must then change the sign of the resulting value to obtain the co-factor.	The sign for position of given element (first row, second column) is negative so we change the calculated value from 2 to -2. This is the co-factor of 6 in the given matrix.

The element in the ith row and jth column position of a matrix called $\mathbf{A}$ is written as a_{ij}. Its co-factor, if it exists, is often written as A_{ij}. The sign associated with the co-factor is $(-1)^{i+j}$.

A.2.3 General method for calculating the value of a (3×3) determinant

Example To evaluate	$\begin{vmatrix} 1 & 4 & 3 \\ 2 & -1 & 4 \\ 11 & 5 & 9 \end{vmatrix},$
(a) Take any row (or column).	We will use the first column.
(b) calculate the co-factors of each element in the chosen row (or column).	Co-factor of 1 is $+\begin{vmatrix} -1 & 4 \\ 5 & 9 \end{vmatrix} = -29.$
	Co-factor of 2 is $-\begin{vmatrix} 4 & 3 \\ 5 & 9 \end{vmatrix} = -21.$
	Co-factor of 11 is $+\begin{vmatrix} 4 & 3 \\ -1 & 4 \end{vmatrix} = 19.$

(c) Multiply each element of the chosen row (or column) by its co-factor and add the resulting terms together to obtain the value of the determinant.

$$(1 \times -29) + (2 \times -21) + (11 \times 19) =$$

$$-29 \quad - \quad 42 \quad + \quad 209 \quad = 138.$$

A.2.4 The Rule of Sarrus

This is a quick method of evaluating a 3rd order determinant, e.g.

$$\begin{vmatrix} a_1 & b_1 & c_1 \\ a_2 & b_2 & c_2 \\ a_3 & b_3 & c_3 \end{vmatrix}.$$

Write down the determinant and repeat the first two columns

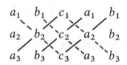

The result is then the sum of the products indicated by the dotted lines minus the sum of the products indicated by the full lines.

A.3 PROPERTIES OF MATRICES

A.3.1 Equality of matrices

Two matrices A and B with elements a_{ij} and b_{ij} respectively are only equal if they are of the same order and if all their corresponding elements are equal, i.e. $a_{ij} = b_{ij}$ for all defined values of i and j.

A.3.2 Addition of matrices

If two matrices A and B have the same number of rows and the same number of columns we can add them by adding corresponding elements. The new matrix C satisfies $C = A + B$ where $c_{ij} = a_{ij} + b_{ij}$ for all values of i and j which define A, B and C.

Example

If $A = \begin{bmatrix} 3 & 1 & 4 \\ 0 & 9 & -2 \end{bmatrix}$ and $B = \begin{bmatrix} 4 & 0 & -3 \\ -1 & 8 & 5 \end{bmatrix}$ then

$$A + B = \begin{bmatrix} 3+4 & 1+0 & 4+(-3) \\ 0+(-1) & 9+8 & -2+5 \end{bmatrix} = \begin{bmatrix} 7 & 1 & 1 \\ -1 & 17 & 3 \end{bmatrix}.$$

Similarly

$$A-B=\begin{bmatrix} -1 & 1 & 7 \\ 1 & 1 & -7 \end{bmatrix}.$$

A.3.3 Multiplication by a number

Multiplying a matrix A by a number (scalar) k gives a matrix B whose elements b_{ij} are k times the elements of A, *viz.* $b_{ij}=ka_{ij}$.

Example

$$2\times\begin{bmatrix} 3 & 1 & 4 \\ 3 & 0 & 2 \\ 1 & 2 & 4 \\ 5 & 2 & 7 \end{bmatrix}=\begin{bmatrix} 6 & 2 & 8 \\ 6 & 0 & 4 \\ 2 & 4 & 8 \\ 10 & 4 & 14 \end{bmatrix}$$

A.3.4 Multiplication by another matrix

The product of two matrices A and B is only defined when the number of columns in A is the same as the number of rows in B. If A is an $(m\times n)$ matrix and B is a $(p\times q)$ matrix, the product AB (not BA) is defined only if $n=p$. The resulting product C $(=AB)$ is an $(m\times q)$ matrix. To obtain c_{ij}, each term in the ith row of A is multiplied by the corresponding term in the jth column of B and all resulting products added together.

Examples

(a) $[3\ \ 1\ \ 2]\begin{bmatrix} 6 \\ 7 \\ -2 \end{bmatrix}=[3\times6+1\times7+2\times-2]=21$

(b) If $A=\begin{bmatrix} 3 & 1 & 2 \\ 4 & 0 & 5 \end{bmatrix}$ and $B=\begin{bmatrix} 6 & -1 \\ 7 & 11 \\ -2 & 8 \end{bmatrix}$ then

$$AB=\begin{bmatrix} 3\times6+1\times7+2\times-2 & 3\times-1+1\times11+2\times8 \\ 4\times6+0\times7+5\times-2 & 4\times-1+0\times11+5\times8 \end{bmatrix}=\begin{bmatrix} 21 & 24 \\ 14 & 36 \end{bmatrix}$$

$$BA=\begin{bmatrix} 6\times3+-1\times4 & 6\times1+-1\times0 & 6\times2+-1\times5 \\ 7\times3+11\times4 & 7\times1+11\times0 & 7\times2+11\times5 \\ -2\times3+8\times4 & -2\times1+8\times0 & -2\times2+8\times5 \end{bmatrix}$$

$$= \begin{bmatrix} 14 & 6 & 7 \\ 65 & 7 & 69 \\ 26 & -2 & 36 \end{bmatrix}$$

Note

(a) In the product **AB**, **B** is pre-multiplied by **A** or **A** is post multiplied by **B**. Matrix multiplication is non-commutative, $\mathbf{AB} \neq \mathbf{BA}$ in general.

(b) The distributive law of multiplication holds provided the order of products is retained and the products exist, i.e. $\mathbf{A(B+C)} = \mathbf{AB} + \mathbf{AC}$.

(c) The leading diagonal of a square matrix runs from the top left to the bottom right of the matrix. The matrix which has 1 on each element of the leading diagonal and 0 elsewhere is called the unit matrix and is denoted by **I**, e.g. the (3×3) unit matrix is

$$\begin{bmatrix} 1 & 0 & 0 \\ 0 & 1 & 0 \\ 0 & 0 & 1 \end{bmatrix}$$

Multiplication by **I** leaves the original matrix unchanged, i.e. $\mathbf{IA} = \mathbf{A}$ and $\mathbf{AI} = \mathbf{A}$.

A.3.5 Simultaneous equations in matrix form

Since

$$\begin{bmatrix} 1 & 4 & 3 \\ 2 & 1 & -3 \\ 6 & -5 & 2 \end{bmatrix} \begin{bmatrix} x \\ y \\ z \end{bmatrix} = \begin{bmatrix} x+4y+3z \\ 2x+y-3z \\ 6x-5y+2z \end{bmatrix},$$

the equations

$$x+4y+3z = 3,$$
$$2x+y-3z = -5,$$
$$6x-5y+2z = 17,$$

can be written

$$\begin{bmatrix} 1 & 4 & 3 \\ 2 & 1 & -3 \\ 6 & -5 & 2 \end{bmatrix} \begin{bmatrix} x \\ y \\ z \end{bmatrix} = \begin{bmatrix} 3 \\ -5 \\ 17 \end{bmatrix}.$$

In general simultaneous equations can always be written in the form $\mathbf{Ax} = \mathbf{b}$ where **A** is a square matrix and **x** and **b** are column matrices.

A.3.6 Derived matrices

Given any square matrix $\mathbf{A}$, a number of new matrices can be found from it. These include the following:

(a) *The Transpose of a Matrix*

If we interchange the rows and the columns of a matrix $\mathbf{A}$ we obtain its transpose $\mathbf{A}^t$, e.g. the transpose of

$$\mathbf{A} = \begin{bmatrix} 1 & 4 & 6 \\ 2 & 0 & 5 \\ 3 & 12 & 9 \end{bmatrix} \text{ is } \mathbf{A}^t = \begin{bmatrix} 1 & 2 & 3 \\ 4 & 0 & 12 \\ 6 & 5 & 9 \end{bmatrix}$$

If $\mathbf{A}^t = \mathbf{A}$ we have a square symmetric matrix, i.e. $a_{ij} = a_{ji}$. If $\mathbf{A}^t = -\mathbf{A}$ we have an anti-symmetric matrix.

(b) *The adjoint of a square matrix*

To obtain adj($\mathbf{A}$) we replace each term of the square matrix $\mathbf{A}$ by its co-factor and then transpose.

Example

To obtain the adjoint of $\mathbf{A} = \begin{bmatrix} -1 & 3 & 4 \\ 2 & 1 & 6 \\ 4 & 0 & 5 \end{bmatrix}$ we first replace each

element by its co-factor, e.g. we replace the -1 by $+\begin{vmatrix} 1 & 6 \\ 0 & 5 \end{vmatrix} = 5.$

This gives the matrix of co-factors $\begin{bmatrix} 5 & 14 & -4 \\ -15 & -21 & 12 \\ 14 & 14 & -7 \end{bmatrix}.$

Transposing we get adj($\mathbf{A}$) $= \begin{bmatrix} 5 & -15 & 14 \\ 14 & -21 & 14 \\ -4 & 12 & -7 \end{bmatrix}.$

Alternatively, transpose $\mathbf{A}$ and then obtain co-factors.

Not all matrices have adjoint matrices.

(c) *The inverse of a matrix*

This is the most important derived matrix. It is denoted by $\mathbf{A}^{-1}$ and can be obtained by dividing the adjoint matrix by the determinant of $\mathbf{A}$. Thus

$$\mathbf{A}^{-1} = \frac{\text{adj } \mathbf{A}}{|\mathbf{A}|}.$$

Clearly when $|\mathbf{A}| = 0$ the inverse is not defined. In this case $\mathbf{A}$ is called a singular matrix.

Example

To find the inverse of $\mathbf{A} = \begin{bmatrix} -1 & 3 & 4 \\ 2 & 1 & 6 \\ 4 & 0 & 5 \end{bmatrix}$.

We have in the previous example calculated the adjoint matrix. We can find the determinant $|\mathbf{A}| = 21$.

Then $\mathbf{A}^{-1} = \dfrac{\text{adj } (\mathbf{A})}{21} = \dfrac{1}{21} \begin{bmatrix} 5 & -15 & 14 \\ 14 & -21 & 14 \\ -4 & 21 & -7 \end{bmatrix}$.

Inverse matrices have one very important property. When we multiply a matrix by its inverse, we get a unit matrix, $\mathbf{I}$, i.e. $\mathbf{AA}^{-1} = \mathbf{A}^{-1}\mathbf{A} = \mathbf{I}$.
We see from A.3.5 above that

$$\begin{bmatrix} x \\ y \\ z \end{bmatrix} = \begin{bmatrix} 1 & 4 & 3 \\ 2 & 1 & -3 \\ 6 & -5 & 2 \end{bmatrix}^{-1} \begin{bmatrix} 3 \\ -5 \\ 17 \end{bmatrix}.$$

Evaluation of the inverse will allow us to find x, y and z, and to solve the set of simultaneous equations.

Index

Page numbers appearing in **bold** refer to figures and page numbers appearing in *italic* refer to tables.

Printed and bound by CPI Group (UK) Ltd, Croydon, CR0 4YY

03/10/2024

01040437-0015